Geophysics: Principles and Concepts

Geophysics: Principles and Concepts

Edited by **Karl Seibert**

SYRAWOOD
PUBLISHING HOUSE

New York

Published by Syrawood Publishing House,
750 Third Avenue, 9th Floor,
New York, NY 10017, USA
www.syrawoodpublishinghouse.com

Geophysics: Principles and Concepts
Edited by Karl Seibert

International Standard Book Number: 978-1-68286-051-9 (Hardback)

Printed in the United States of America.

Contents

Preface

Geophysics as a discipline, studies in detail the physical characteristics of earth and its environment. This field explores the internal structure of earth, surface dynamics and other phenomena of earth. The chapters provided in this book exhaustively discuss the theories of earth's structure and shape, seismology, fluid dynamics of oceans, and magnetic fields. The book presents researches and studies performed by experts across the globe. It aims to serve as a reference text for students and professionals alike and contribute to the growth of the discipline.

Significant researches are present in this book. Intensive efforts have been employed by authors to make this book an outstanding discourse. This book contains the enlightening chapters which have been written on the basis of significant researches done by the experts.

Finally, I would also like to thank all the members involved in this book for being a team and meeting all the deadlines for the submission of their respective works. I would also like to thank my friends and family for being supportive in my efforts.

Editor

Microscopic dynamics of a large-scale pedestrian evacuation model

Lim Eng Aik

Institut Matematik Kejuruteraan, Universiti Malaysia Perlis, Perlis, Malaysia.
E-mail: e.a.lim80@gmail.com.

Numerous models have been developed for estimating the time required to evacuate from a variety of places under various conditions. For high-traffic places, such as commercial and industrial buildings, it is vital to be able to accurately calculate the evacuation time required in order to ensure the safety of the occupants. To this end, various models of pedestrian dynamics have been proposed, either as a whole system or focusing only on the psychological interaction between pedestrians. However, most of these published studies do not take into account the pedestrian's ability to select the exit route in their models. To resolve these issues, we have developed a model to simulate evacuations from a hall using the social force model that incorporated with the degree of pedestrians' impatience along with the distance to exits and the density of the crowd, in determining pedestrians' selection of evacuation routes. For validation, the results obtained with the proposed model are compared with published data. Finally, the model is applied to predict a specific adjustment to the hallway that would improve the output of the system. Simulations show that reasonable improvement is achieved, with an additional 14.2% pedestrians being evacuated within a 12 min interval.

Key words: Social force model, exit route selection, crowd dynamic, density effect, intelligent agent.

INTRODUCTION

Pedestrian modeling is among the most interesting areas in transportation science. Understanding the dynamics of pedestrian flow is key to the design of public places that aims to reduce the loss of life and property in the event of disasters. However, pedestrian evacuation is a complex process complicated by human behaviors and emotions such as panic. It is difficult to capture scenes of pedestrian flows during evacuation for research purposes, and it is nearly impossible to simulate real-life evacuations.

Researchers, therefore, rely on simulation models to study pedestrian behavior during evacuation. Various useful models have been put forward, among which are models based on particle flow (Helbing, 1992; Helbing et al., 2003; Sakai et al., 2006; Raja and Pugazhenthi, 2011), social force (Helbing et al., 2000; Helbing and Molnar, 1995; Helbing et al., 2001; Okazaki and Matsushita, 1993; Teknomo et al., 2001) and cellular automata (CA) (Siamak et al., 2011; Alizadeh, 2011; Kirchner et al., 2004; Burstedde et al., 2001; Varas et al., 2007; Fang et al., 2010; Liu et al., 2009; Lim, 2011). Particle flow and social force models use physical models

to simulate movement. CA models divide the environment into cells between which pedestrians move. Each model has strengths and weaknesses. For example, the social force model (SFM) produces smoother movements than the CA model owing to its continuous nature.

Kosinski and Grabowski (2010) introduced an intelligent agent system using the Langevin equation with an additional SFM term to represent the level of panic during evacuation that takes the speed of movement in the exit area as the decisive factor. However, the model produces less realistic pedestrian movements as compared to the social force model. Parisi and Dorso (2005) devised a model to study the level of panic with different exit widths. However, their model is limited to crowds of about 200 pedestrians and focused on the effect of exit width on panic levels during evacuation. Durupinar et al. (2011) introduced an impatience rule for multi-agents systems for route selection. Similar to Parisi and Dorso (2005) work, the model is the only application for simulating limited number of pedestrian in the system due to it heavy computational algorithm. Frank and Dorso

(2011) investigation of the impact of human behavior during evacuation shows clogging and cluster forming during escape from a room with a single exit and a fixed obstacle. They found that the distance of the obstacle from the exit influenced the level of panic as well as the flow of evacuation. Following this work, Zheng et al. (2011) also suggested that psychological term such as impulsive or impatience behavior should be considered for improving evacuation model in order to make the model more accessible to reality. Ding et al. (2011) proposed a psychological force into SFM which is mainly applied for pedestrian avoidance in rail transit lane area, but not for improving pedestrian intelligent in selection less congested route for exiting the area.

In situations of panic, the crowd tends to jostle to try and escape in the shortest time. Congestion close to the exit may be avoided by placing an obstacle to disperse the crowd (Helbing et al., 2005; Kirchner et al., 2003). The size of the obstacle, as well as its distance to the exit, ought to be considered carefully to attain optimal evacuation time. Otherwise, the exit area may become overcrowded, or congestion may grow to a critical level outside this area. Properly done, an obstacle may improve pedestrian flow up to 30% or twice the flow that may occur without the obstacle (Helbing et al., 2005; Escobar and Rosa, 2003). The reduction in evacuation time is achieved when pedestrians are required to walk a longer path in order to avoid the obstacle. Detrimental results may occur if the obstacle is placed too near the exit. To achieve optimal evacuation time, the obstacle should be moved slightly away from the exit and placed in the center of the area next to the exit (Frank and Dorso, 2011; Yanagisawa et al., 2009). Helbing et al. (2000) proposed placing a column as obstacle near the exit area. Understanding evacuation dynamics would therefore allow safer designs of public facilities.

The objective of the present paper is to study the microscopic mechanism involved in the hallway flow during evacuation from a hall. In our study, we use a social force model to examine the psychological factor of impatience in pedestrians in the hallway area during mass evacuation from a hall. The social force model (Helbing and Molnar, 1995) produces realistic movements in simulating the evacuation process by taking into consideration discrete characteristics of pedestrian flow, thus allowing individuals' physical variables to be set, such as mass, shoulder width, desired speed and target destination. The forces of interaction that may cause high pressure capable of bringing down a brick wall or causing suffocation can thus be determined. These continuous flow characteristics cannot be simulated with CA models.

The efficiency of an obstacle in relieving congestion is affected by pedestrians' behavior during the evacuation process. Most studies assume a fixed route taken by pedestrians without considering the degree of impatience, which may arise in such situations (Kirchner et al., 2003; Escobar and Rosa, 2003; Yanagisawa et al., 2009). A more realistic scene would be that some

impatient pedestrians will alter their initial preferred route and head for the nearest exit available, pushing through other pedestrians in the process. Subsequently, we introduce a rule for exit route selection influenced by the degree of impatience. Next, the social force model of Helbing et al. (2000) is discussed, then we present and discuss the results of simulations performed with our model. Finally, we conclude the paper with suggestions for future research.

RULE FOR EXIT ROUTE SELECTION

During an evacuation, an impatient pedestrian will tend to show characteristics, such as walking faster than their normal speed, pushing nearby pedestrians and rushing toward the nearest exit available.

An impatient pedestrian's action can be expressed mainly in terms of changes in their speed. A pedestrian's degree of impatience, $n_i(t)$, at time t can be expressed as:

$$n_i(t) = \frac{v^{(i)}(t) - v^{(i)}(0)}{v_{max}^{(i)} - v^{(i)}(0)} \qquad (1)$$

where $v^{(i)}(t)$ is the speed of pedestrian i at time t, $v^{(i)}(0)$ is the initial speed of pedestrian i and $v_{max}^{(i)}$ is the maximum speed desired by pedestrian i.

Research has shown that pedestrians' emotions, such as impatience, affect their choice of escape route (Helbing et al., 2000; Okazaki and Matsushita, 1993; Fang et al., 2010). Impatient pedestrians would behave in such a way that causes an unstable flow, leading to delays or congestion during evacuation.

Exit selection is mainly based on distance. To select a route, a pedestrian would consider two factors: (1) their distance from an exit and (2) the presence of people flocking to that exit. Assuming there are $m(x = 1, 2, 3, ..., k, ..., m)$ evacuation exits, the probability of pedestrian i selecting exit k as the evacuation route can be defined as:

$$P_k^{(i)} = (1 - n_i)P_1 + n_i P_2 \qquad (2)$$

where P_1 is the probability of reaching the nearest exit, P_2 is the probability of people flocking to that exit and n_i is the degree of impatience of pedestrian i. When n_i approximates 0, pedestrian i is in normal mood. However, when n_i approximates 1, pedestrian i is in an extremely impatient mood, rushing to get out as fast as possible.

The probability of reaching the nearest exit, P_1, can be defined as:

$$P_1 = \frac{d_{(i)}}{d_{max}} \qquad (3)$$

where $d_{(i)}$ is the distance of pedestrian i to exit k, and d_{max} is the maximum distance measured from all pedestrians to exit k. Equation 3 indicates that the shorter the distance of pedestrian i to exit k, the higher the probability of pedestrian i selecting exit k as the evacuation route. Conversely, if the distance is longer, the probability of selecting exit k decreases.

The probability of the flocking phenomenon occurring in the exit area is defined as:

$$P_2 = 1 - \frac{N_{(k)}}{\sum_{\phi=1}^{m} N_{(\phi)}} \qquad (4)$$

where $N_{(k)}$ represents the number of pedestrians that select exit k as the evacuation route, while $\sum_{\phi=1}^{m} N_{(\phi)}$ is the total number of pedestrians that select exit ϕ as the evacuation route. The degree of impatience of pedestrian i in the system, n_i, can be expressed as:

$$n_i = \left| \frac{v^{(i)}(t) - v^{(i)}(0)}{v_{\max}^{(i)} - v^{(i)}(0)} \right| . \qquad (5)$$

According to Helbing (1992), Helbing et al. (2000) and Helbing and Molnar (1995), the maximum desired speed $v_{\max}^{(i)}$ that can be achieved by a pedestrian is 3 m/s. When an unexpected situation occurs, such as flocking, congestion or panic, pedestrian speed $v^{(i)}(t)$ drops to less than 1.5 m/s. The velocities of a pedestrian are uniformly distributed in the range of 0.5 to 1.5 m/s in simulations, while the initial speed $v^{(i)}(0)$ of a pedestrian is approximately 1.0 m/s. Hence, the decision of pedestrian i to select exit x as the evacuation route is determined by comparing probability $P_x^{(i)}$ (x = 1, 2, 3, ..., k, ... m) against the highest value of $P_x^{(i)}$ as the criterion of selection.

THE SOCIAL FORCE MODEL

According to the social force model, pedestrians' movement is determined by their desire to arrive at the destination, as well as the effects of the surroundings on them (Helbing et al., 2000; Helbing and Molnar, 1995). Recent social force models include the social force and granular force, while earlier models are based on one force known as the desire force.

Suppose that a pedestrian is moving at a desired speed of v_d in a given direction $\vec{e}_d$. In actual situations, pedestrians always walk a bit out of the actual desired path and they never walk exactly at the desired speed v_d. A pedestrian's actual speed $v(t)$ is influenced mainly by environmental factors (e.g., obstacles, exit size). Hence, pedestrians have to increase or decrease their speed with the intention of reaching the destination at the desired speed v_d. This acceleration or deceleration corresponds to the desire force, as it is dictated by their will and motivation. Therefore, the desire force of pedestrian i can be defined in mathematical terms as:

$$f_d^{(i)}(t) = \frac{v_d^{(i)}(t)\vec{e}_d^{(i)}(t) - v_i(t)}{\tau} \qquad (6)$$

where all parameters are assumed to be functions of time, and τ represents the relaxation time required to achieve the desired speed. The value of τ is determined by experiment.

The pedestrian's reactions to environmental stimuli are represented by social forces. Although, there exist stimuli such as family members or friends that generate attraction, they are not included in our model. However, the basic rule that pedestrians tend to preserve their private space between other pedestrians still applies (Helbing and Molnar, 1995). When people get closer to one another, the repulsive force will become stronger. In other words, the repulsive force is mainly dependent on the inter-pedestrian distance d, which can be modeled as an exponentially decaying function defined as:

$$f_s^{(ij)} = A_i n_{ij} e^{(r_{ij} - d_{ij})/B_i} \qquad (7)$$

where i and j correspond to any two pedestrians, d_{ij} is the distance between the centers of mass of the two pedestrians, $n_{ij} = \left(n_{ij}^{(1)}, n_{ij}^{(2)} \right)$ represents the unit vector in direction $\vec{ji}$ and r_{ij} = $r_i + r_j$ is the sum of pedestrian radii for pedestrians i and j. Parameters A_i and B_i are determined by experiment (Helbing et al., 2000).

Equation 7 is also applicable to environmental factors (e.g., obstacles). Pedestrians will tend to maintain a distance from obstacles to avoid being injured. Hence, r_{ij} and d_{ij} in Equation 7 are substituted by r_i and d_i, corresponding to the pedestrian's radius and their distance to the obstacle, respectively.

The final term in the social force model, which expresses the sliding friction that appears between pedestrians in contact with each other and with walls, is known as the granular force. By assuming the pedestrian's relative velocity as a linear function, the granular force can be expressed as:

$$f_g^{(ij)} = \kappa g(r_{ij} - d_{ij}) \Delta v_{ij} \cdot t_{ij} \qquad (8)$$

where $\Delta v_{ij} = v_j - v_i$ is the speed difference between pedestrian i and pedestrian j. If pedestrian i comes to a wall, then v_j becomes zero in Equation 8. The function $t_{ij} = \left(-n_{ij}^{(2)}, n_{ij}^{(1)} \right)$ is the unit tangential vector orthogonal to n_{ij}, and κ is an experimental parameter. The $g(.)$ function is set to zero when the argument value is negative (that is, $r_{ij} < d_{ij}$), and it is equal to the argument value for any other cases.

Body compression may occur in extremely crowded situations (Helbing et al., 2000). However, as reported by Parisi and Dorso (2005), body compression forces play no significant role during the evacuation process. Hence, it is not taken into consideration in the proposed model. A more detailed explanation of $f_s(t)$ and $f_g(t)$ can be found in the literature (Ding et al., 2011; Escobar and Rosa, 2003; Kosinski and Grabowski, 2010; Lim, 2011). Table 1 lists typical values for the experimental parameters in Equations 6 to 8. Consequently, both the desire and granular forces control pedestrians' dynamical characteristics by changing their speed. The movement of pedestrian i can be expressed as:

$$\frac{dv_i}{dt}(t) = f_d^{(i)}(t) + \frac{1}{m_i}\left[\sum_{i \neq j} f_s^{(ij)}(t) + \sum_{i \neq j} f_g^{(ij)}(t) \right] \qquad (9)$$

where m_i is the mass of pedestrian i. The subscript j represents all other pedestrians except pedestrian i and environmental factors.

The magnitude of the desired speed v_d in Equation 6 corresponds to the pedestrian's movement at free-flow speed. Additionally, the moving direction $\vec{e}_d$ sets the level of anxiety for the pedestrian, eager to reach a particular exit. An impatient pedestrian will tend to change their initial desired direction for the nearest exit available (Escobar and Rosa, 2003).

Table 1. Variables for evacuation simulations.

Parameter	Symbol	Value	Units
Force at $d_{ij} = r_{ij}$	A_i	2000	N
Characteristic length	B_i	0.08	m
Pedestrian mass	m_i	70	kg
Contact distance	r_{ij}	0.5 ± 0.2	m
Acceleration time	τ	0.5	s
Friction coefficient	κ	2.4×10^5	$kg\ m^{-1}\ s^{-1}$

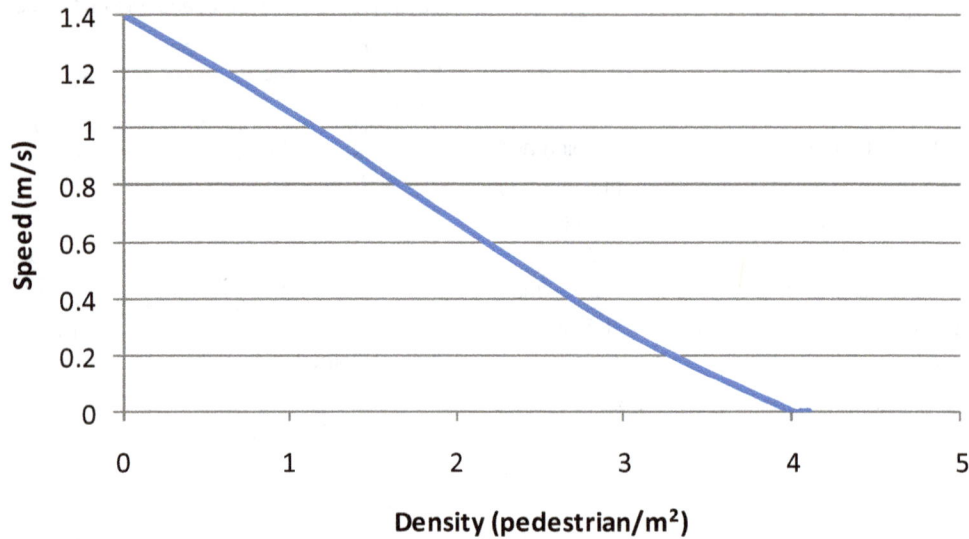

Figure 1. Relationship of pedestrian speed and density.

Effect of pedestrian density on speed

The higher the pedestrian density in an area, the slower pedestrians move as compared to their desired free-flow speed in order to avoid collision and injury. In the proposed model, pedestrians are moving at maximum speeds in the hallway area, but they slow down when there are obstacles or other pedestrians nearby. Fruin (1971), studying the effect of density on speed, concluded that, as pedestrian density increases, the speed of movement drops correspondingly. According to Fruin's study, pedestrian speed approximates zero when density approaches 4 pedestrians/m^2 (Figure 1).

Based on the findings of Fruin (1971), we introduce the density effect in Equation 10 as defined by Siamak et al. (2011), for the purpose of evaluation. The path is defined as the movable neighborhood of the desired route. The three regions of the path located near the exits are all fixed at equal sizes for evaluation purposes.

$$\text{Density Effect} = \begin{cases} \dfrac{\beta_{path}}{\beta_0}, & \beta_{path} < \beta_0 \\ 1, & \text{otherwise} \end{cases} \qquad (10)$$

In Equation 10, β_{path} is the path area density, while the margin area density represented by β_0 is equal to the size of the path area.

SIMULATION RESULTS

From the simulation results of the proposed model, the speed-density relation is determined and then compared to results reported by Schadschneider et al. (2008). In addition, simulation snapshots were compared with actual photographs taken by a surveillance camera installed at the corner of the scene. The relationship between pedestrian flow rates and evacuation time is then examined. Finally, the model is applied to predict whether placing an obstacle in the center of the hallway area would significantly enhance the overall system output.

Speed-density graph comparison

As suggested by Siamak et al. (2011), we use the speed-density graph as shown in Figure 1 earlier to compare our results with those of other studies. We use two graphs found in Schadschneider et al. (2008), from Fruin's and Predtechenskii-Milinskii's (PM) studies. Fruin's and PM's graphs produce the highest (1.4 m/s) and the lowest (1.0 m/s) desired speeds, respectively. Fruin's results indicate that the crowd will stop moving

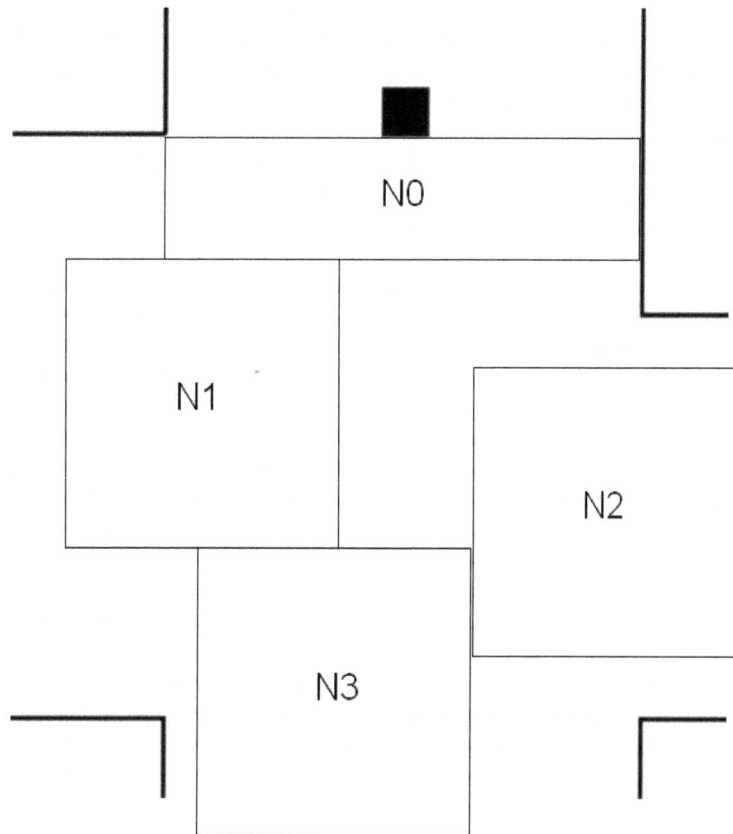

Figure 2. Sections of the hallway area for speed-density evaluation.

when density increases to 4 pedestrians/m^2, while PM's findings show that pedestrians will continue moving at a speed below 0.2 m/s even when density reaches 8 pedestrians/m^2. High densities exceeding 4 pedestrian/m^2 can sometimes be observed in the hallway area. Therefore, we infer that the speed-density relation for the hallway area is similar to that described in PM's graph.

According to Siamak et al. (2011), every evacuation area is made up of sections with different properties. To determine the speed-density relation for each section of the evacuation area in our study, we adopted the guidelines of Siamak et al. (2011). We divided the hallway area into sections as shown in Figure 2. Section N0 is the most crowded area, where pedestrians stream out of the hall's two adjacent doorways and move in different directions toward the exits in N1, N2 and N3 sections. Slowdown in N0 is caused partly by pedestrians in N1 heading for N2 or N3. Slower movement is also observed in spots where pedestrians are close to the walls in N2 or heading for N1. Pedestrians crossing over to different sections causes further delay. Since N1, N2 and N3 exhibit none of the extreme characteristics of N0, they are suitable for evaluating the efficiency of our model in simulating evacuation under different conditions (Figure 3).

Figure 4 compares the speed-density curves for the N1, N2 and N3 sections in our model against Fruin's and PM's graphs. We can see that the N1, N2 and N3 curves behave similarly to PM's curve, with an initial desired speed of around 1.2 m/s. We can thus infer that our model is comparable to PM's. The maximum density in our model is 4.25 pedestrians/m^2. Even with extremely high demand levels, densities higher than 5 pedestrians/m^2 cannot be achieved in the N1, N2 and N3 section.

Visual comparison

It was mentioned earlier that the hallway area has certain characteristics in different sections that lead to congestion in the N0 section. We performed evacuation simulations that were based mainly on videos of movements in the hallway area when participants leave a hall after an event. Most people would move toward one of the exits as fast as possible. The crowd in N0 creates resistance to pedestrian flow behind this area. This resistance leads to serious congestion in N0, whereas a less congested scene is observed in other sections.

Figure 5 shows the snapshots of simulations at medium, high and very high flow rates, along with actual

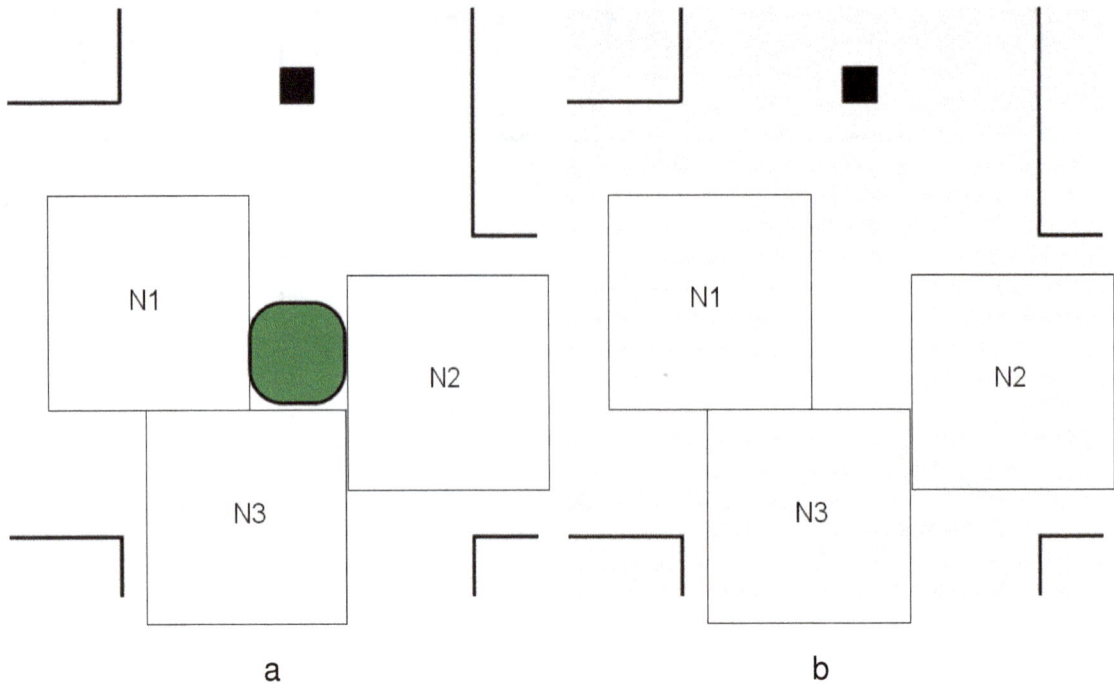

a

b

Figure 3. Speed-density evaluation of the hallway area (a) with and (b) without an obstacle.

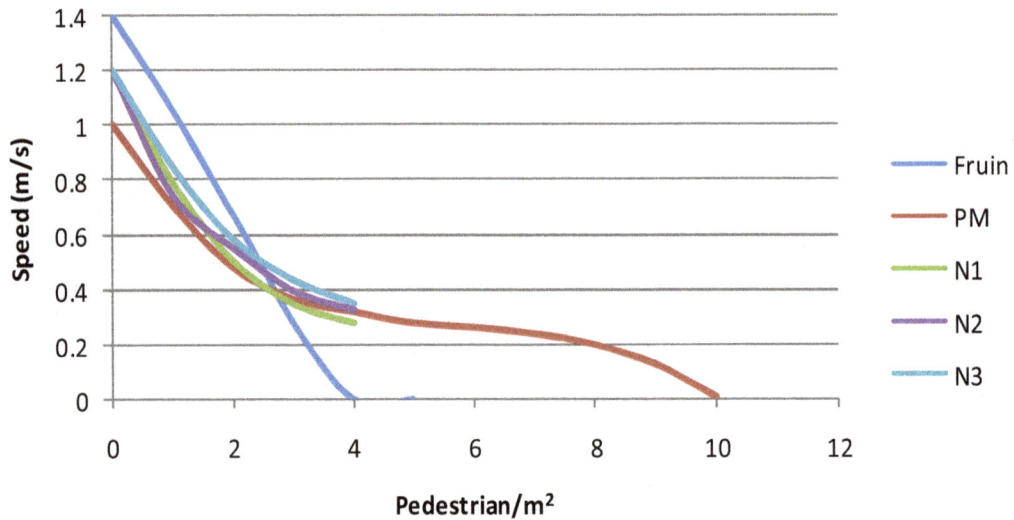

Figure 4. Speed-density relation in the N1, N2, and N3 sections.

scenes taken from surveillance photographs showing the flow of people leaving a hall. The area just outside the hall is most congested, corresponding to the situation in N0 in our simulation.

Evacuation time and flow rates

As the number of people in the hallway area builds up, movement slows down. To see the effect of this build-up on evaluation time, we ran the simulation for five different flow rates, which are 25, 75, 175, 225 and 275 pedestrians/min, corresponding to low, medium, high, very high and extremely high flow rates. Figure 6 shows the results of these simulations. As the flow rate increases, evacuation time decreases.

Additionally, as the flow rate increases up to 75 pedestrians/min, evacuation time drops significantly. The point before which a sharp drop in evacuation time occurs around 50 pedestrians/min. Simulations also show

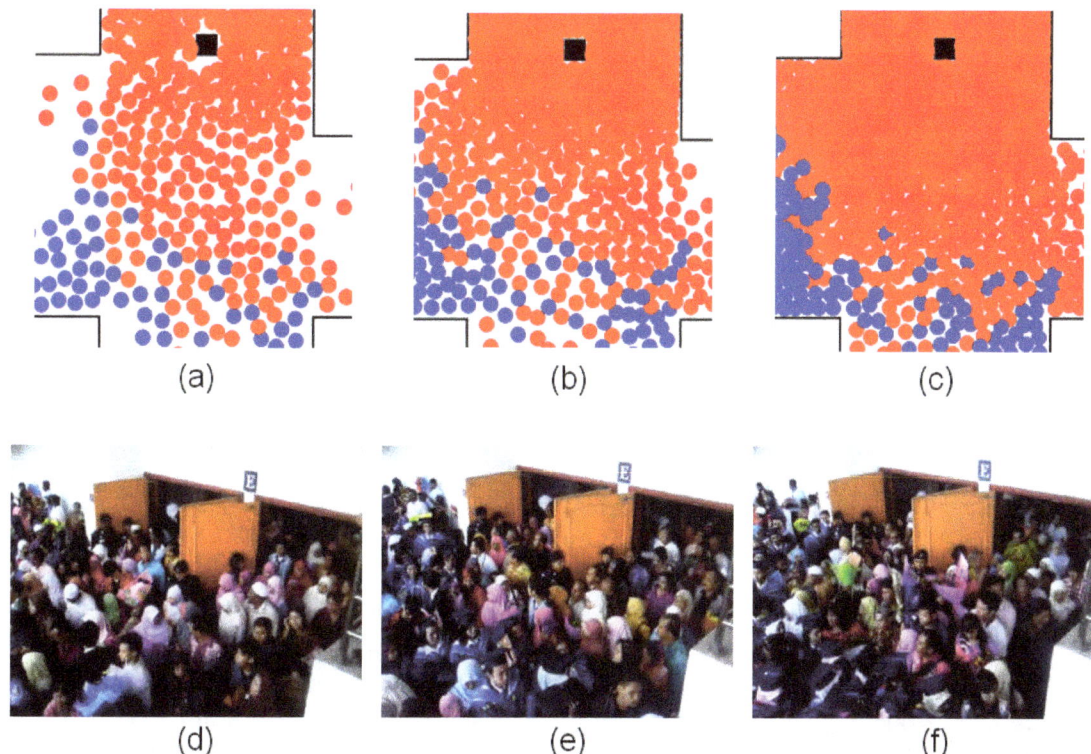

Figure 5. Simulation snapshots of (a) medium, (b) high and (c) very high pedestrian flow rates. Surveillance photographs of real scenes (d to f).

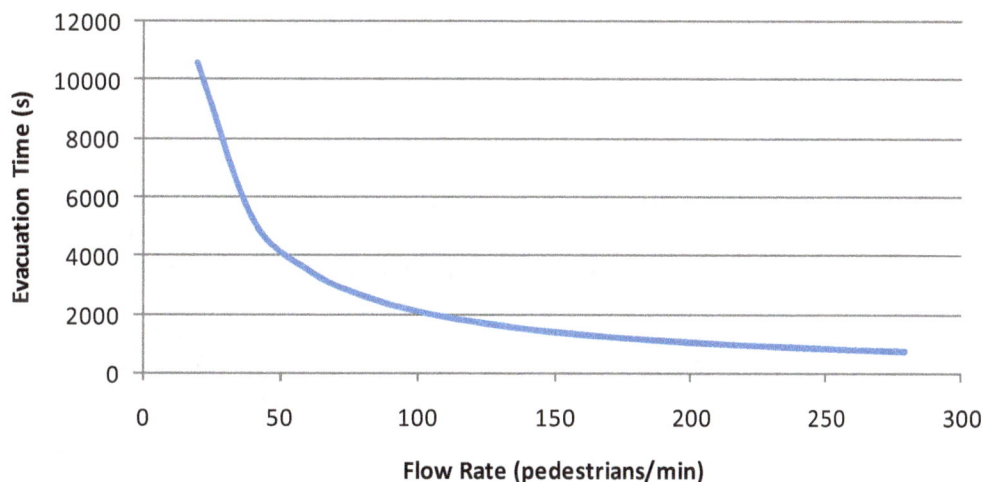

Figure 6. Evacuation time for different pedestrian flow rates.

that congestion quickly builds up in N0 once the flow rate reaches over 75 pedestrians/min. Congestion also begins to spread to N1 and N2, which delays evacuation. Such a level of congestion could develop into a hazardous situation as impatient or panicked pedestrians start to push one another.

The average pedestrian speed and evacuation time in the simulations match data obtained through the surveillance camera for the real situations. However, it is suggested that more data should be obtained to allow the parameters to be calibrated.

Application of the model

We then applied the model to evaluate the extent of

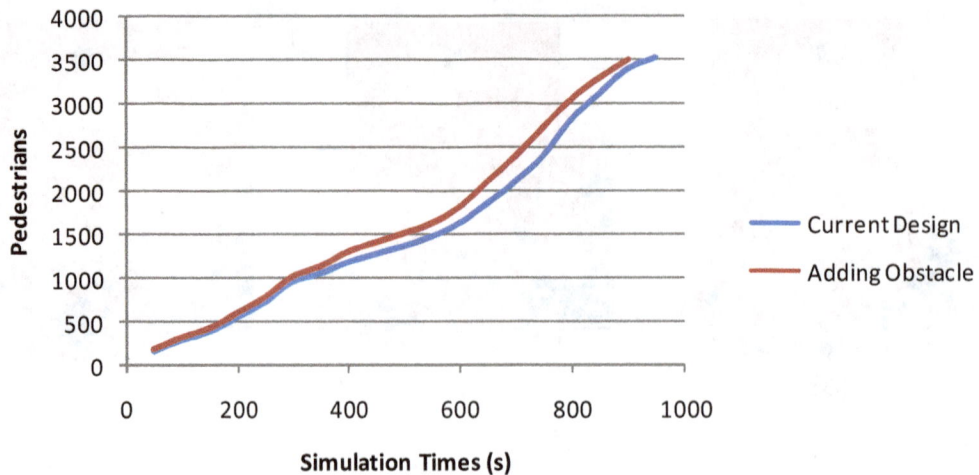

Figure 7. Comparison of simulations with the current design of the hallway and a what-if scenario where an obstacle is added.

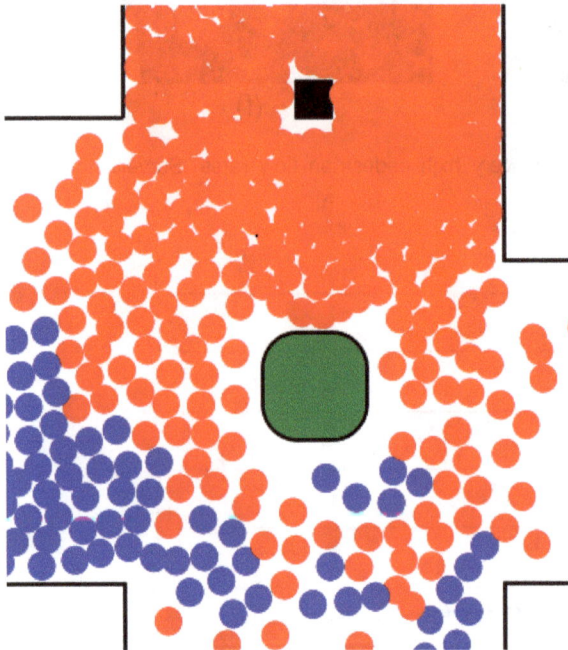

Figure 8. Simulation snapshot of the what-if scenario.

improvement in throughput with the implementation of changes designed to improve flow. In an experiment with a what-if scenario, an obstacle was placed in the center of the hallway area. Figure 7 shows that this improves the speed of evacuation. The reason is that the obstacle splits up pedestrian flow and reduces collisions between pedestrians from N0 and those crossing from N1 to N2. It also creates a smoother flow for pedestrians moving from N0 to N2 and to N1.

In the absence of the obstacle, pedestrians heading for N3 would walk past the central area in collision with others. With the obstacle in the center, the number of pedestrians crossing over to N1 and N2 is reduced. Most of the pedestrians going from N1 to N2 would take the route behind the obstacle to avoid the crowd in N0. The outcome is smoother flows and less congestion in the center of the hallway area. Overall, congestion in the hallway area and evacuation time are reduced. Figure 8 shows a simulation snapshot of the what-if scenario. In such a scenario, at a flow rate of 175 pedestrians/min, the simulation yields 318 pedestrians, or 14.2% additional output, within a 12 min interval.

Conclusion

The findings show that our social force model is suitable for simulating complex and highly crowded situations. We incorporated into the model the elements of impatience and route selection ability for more realistic simulations. We performed simulations of evacuation from a hall to the hallway area to investigate pedestrian behavior during the evacuation process. This model was applied to a what-if situation to predict whether changes to the hallway area would produce a better outcome as compared to the current design. For future research, pedestrian dynamics during evacuation that involve grouping will be explored, and additional factors such as age and other psychological characteristics of individuals will be included, with the aim of reproducing more realistic simulations for analysis and study.

REFERENCES

Alizadeh R (2011). A dynamic cellular automaton model for evacuation process with obstacles, Safety Sci., 49: 315-323.
Burstedde C, Klauck K, Schadschneider A, Zittartz J (2001). Simulation of pedestrian dynamics using a two-dimensional cellular automaton,

Phys. A. 295: 507-525.

Durupinar F, Pelechano N, Allbeck JM, Gudukbay U, Badler NI (2011). How the ocean personality model affects the perception of crowds, IEEE Computer Graphics Appl., 11: 22-31.

Ding QY, Wang XF, Shan QC, Zhang XY (2011). Modeling and simulation of rail transit pedestrian flow, J. Trans. Systems Eng. Inform. Technol., 11(5): 99-106.

Escobar R, Rosa ADL (2003). Architectural design for survival optimization of panicking fleeing victims, Springer-Verlag, Berlin, pp. 97-106.

Fang ZM, Song WG, Zhang J, Wu H (2010). Experiment and modelling of exit-selecting behavior during a building evacuation, Phys. A. 389: 815-824.

Frank GA, Dorso CO (2011). Room evacuation in the presence of an obstacle, Phys. A, 390: 2135-2145.

Fruin JJ (1971). Pedestrian planning and design, Metropolitan Association of Urban Designers and Environmental Planners, New York, pp. 39-46.

Helbing D (1992). A fluid-dynamic model for the movement of pedestrians, Complex Systems. 6: 391-415.

Helbing D, Isobe M, Nagatani T, Takimoto K (2003). Lattice gas simulation of experimentally studied evacuation dynamics, Physical Review. E 67: 067101.

Helbing D, Farakas I, Vicsek T (2000). Simulating dynamical features of escape panic, Nature. 407: 487-490.

Helbing D, Molnar P (1995). A social force model for pedestrian dynamics, Phys. Rev., E51: 4282-4286.

Helbing D, Molnar P, Farkas J (2001). Self-organizing pedestrian movement, Environment and Planning B: Planning Design, 28: 361-383.

Helbing D, Buzna L, Johansson A, Werner T (2005). Self-organized pedestrian crowd dynamics: experiments, simulations, and design solutions, Transportation Sci., 39: 1-24.

Kirchner A, Klupfel H, Nishinari K, Schadschneider A, Schreckenberg M (2004). Discretization effects and the influence of walking speed in cellular automata models for pedestrian dynamics, J. Statistical Mech., 10: 10011-10032.

Kirchner A, Nishinari K, Schadschneider A (2003). Friction effects and clogging in a cellular automaton model for pedestrian dynamics, Phys. Rev., E67: 056122.

Kosinski RA, Grabowski A (2010). Intelligent agents system for evacuation processes modeling, Proceeding of IEEE/WIC/ACM International Conference, Web Intell. Intell. Agent Technol., pp. 557-560.

Lim EA (2011). Exit-selection behaviors during a classroom evacuation, Int. J. Phys. Sci., 6(13): 3218-3231.

Liu SB, Yang LZ, Fang TY, Li J (2009). Evacuation from a classroom considering the occupant density around exits, Physical, A 388: 1921-1928.

Okazaki S, Matsushita S (1993). A study of simulation model for pedestrian movement with evacuation and queuing, Proceedings Int. Confer. Eng. Crowd Safety. pp. 271-280.

Parisi DR, Dorso CO (2005). Microscopic dynamics of pedestrian evacuation, Physical. A 354: 606-618.

Raja P, Pugazhenthi S (2011). Path planning for a mobile robot in dynamic environments, Int. J. Phys. Sci., 6(20): 4721-4731.

Sakai S, Nishinari K, Lida S (2006). A new Stochastic cellular automaton model on traffic flow and its jamming phase transition, J. Phys. Math. General, 39: 15327-15339.

Schadschneider A, Klingsch W, Klupfel H (2008). Modeling and applications, Meyer, B. (Editor), Encyclopedia Complexity System Sci. Springer, Berlin, p. 3142.

Siamak S, Fazilah H, Abdullah ZT (2011). A cellular automata model for circular movements of pedestrians during Tawaf, Simulation Modeling Practice Theory, 19: 969-985.

Teknomo K, Takeyama Y, Inamura H (2001). Microscopic pedestrian simulation model to evaluate lane-like segregation of pedestrian crossing, Proceeding of Infrastructure Planning Conference, Japan, pp. 208-218.

Varas A, Cornejo MD, Mainemer D, Toledo B, Rogan J, Munoz V, Valdivia JA (2007). Cellular automaton model for evacuation process with obstacles, Physica. A 382: 631-642.

Yanagisawa D, Kimura A, Tomoeda A, Nishi R, Suma Y, Ohtsuka K, Nishinari K (2009). Introduction of frictional and turning function for pedestrian outflow with obstacle, Physical Review. E 80: 036110.

Zheng Y, Jia B, Li XG, Zhu N (2011). Evacuation dynamics with fire spreading based on cellular automaton, Physical A. 390: 3147-3156.

A seismological view to Gökova region at southwestern Turkey

Doğan KALAFAT[1]* and Gündüz HORASAN[2]

[1]National Earthquake Monitoring Center, Kandilli Observatory and Earthquake Research Institute, Boğaziçi University, 34680 Çengelköy, Istanbul, Turkey.
[2]Department of Geophysical Engineering, Engineering Faculty, Sakarya University, 54040, Sakarya, Turkey.

Recent seismic activity in Gökova region can be characterized by earthquake swarms, which mostly occured during 2004 and 2005. This activity was continued for seven months and 1558 seismic events were recorded at this period. The b-value of Guthenberg-Richter relation is investigated for this earthquake swarm and a high b-value is found as 1.73±0.08 using the maximum likelihood method. In this study, seismogram and spectrum characteristics of the earthquake events in Gokova region are analyzed. Accordingly, low frequency waveform and spectrum are obtained for shallow events whereas deeper events are observed to be characterized by high frequency waveform and spectrum. Apart from direct P and S waves, we noticed the presence of strong reflection phase on the seismograms. These reflected phases come from ~17 to 18 km depth in Gökova region.

Key words: Gökova region, reflected phase, earthquake swarm.

INTRODUCTION

The Gökova province is located in the southeast Aegean Sea along the coast of southwest Anatolia which is a region including the major horst and graben systems such as Gediz, Büyük Menderes, Gökova, Burdur and Acıgöl grabens. The gulf of Gökova is surrounded by Bodrum Peninsula to the north, Datça Peninsula to the south and the island of Kos to the west (Figure 1). It has about 25 km maximum N–S width and 100 km E–W length .

During the Early–Middle Miocene period thick volcano sedimentary associations were formed within approximately NS trending fault-bounded continental basins under an E–W extensional regime (Yılmaz et al., 2000). After starting N–S extension, intracontinental plate alkaline volcanic province of western Anatolia was formed during Late Miocene to Quaternary time (Aldanmaz, 2002; Tonarini et al., 2005). Approximately E–W trending grabens and their basin-bounding active normal faults are the most prominent neotectonic features of Western Anatolia (Bozkurt, 2001).

In this region, east-west and northwest-southeast-trending rifts and related faults are the dominant neotectonic features (Şengör et al., 1984). Among these, the Gökova, Yatağan-Muğla and Milas-Ören rifts are most prominent (Figure 1; Görür et al., 1995).

The Milas-Ören and Yatağan-Muğla rifts are older than the Gökova rift. The east-west faults of the Gökova rift everywhere cut the northwest-southeast faults of the Milas-Ören and Yatağan-Muğla rifts. The structural relationships between the northwest-southeast and the East-west Gökova rifts, can be useful to explain the north-south extension of the Gökova region.

Recent studies based upon surface morphology, fault mechanism solutions, seismicity and marine seismic reflection data (Şaroğlu et al., 1995; Görür et al., 1995; Eyidoğan at al., 1996; Kurt et al., 1999; Uluğ et al., 2005) provide evidence of active normal faults in the area. A normal fault trending east-west is a prominent feature of the Gökova rift. The southern border of the Gökova graben is characterised by low-angle faults with listric type (Kurt et al., 1999).

Muğla, Bodrum, Yatağan and Gulf of Gökova are some of the most seismically active areas of the western Turkey. In the instrumental period seismic activity in the Gökova region includes the earthquakes of 23 April, 1933 (Ms = 6.4), 23 May, 1941 (Ms = 6.0), 13 December, 1941

---

*Corresponding author. E-mail: kalafato@boun.edu.tr.

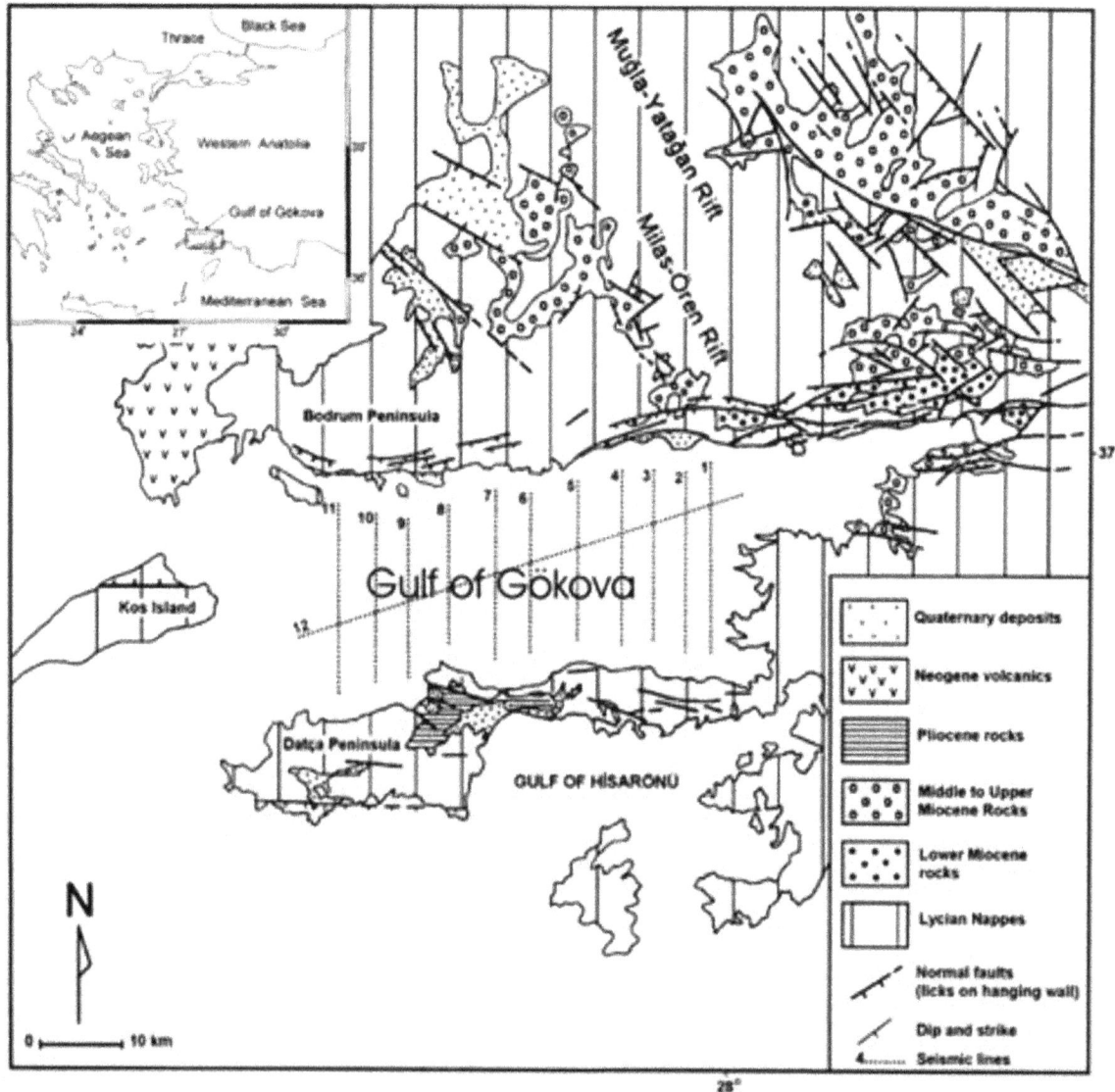

Figure 1. Land geology map of the Gökova province (modified from Görür et al., 1995; the original map also includes inferred submarine faults). Notice the E–W-oriented new graben system (the Gulf of Gökova and its margins) and NW–SE-oriented older graben systems (Muğla–Yatağan Rift and Milas–Ören Rift) (from Kurt et al., 1999).

(Ms = 6.5), 25 April, 1959 (Ms = 5.9) and 5 October, 1999 (Ms = 5.2) events. In August 2004 series of earthquakes (3 August M = 5.0; 4 August M = 5.4; 4 August M = 5.0) occured in Gökova Gulf. Another earthquake sequence continued from 20 December, 2004 (Mw = 5.3) to 10 January, 2005 (Mw = 5.4). In Figure 2 we can clearly see that the rate of seismicity increased in time after August 2004.

The released energy is predominantly in swarm type in this region (Figures 3 and 4). Earthquake swarms occur in many regions of the world like Long Valley caldera, USA; Yellowstone, USA; West Bohemian, Germany.

During the individual swarms, numerous events occurred consecutively as multiplets, that is event occur at nearly the same position and have vary similar source mechanisms (Horálek et al., 2000). The occurance of multiplets is a phenomenon obivously observed in geothermal or volcanic regions (Lees, 1998). It has been suggested that earthquake swarms occured because of stress perturbations associated with the migration of magmatic or hydrothermal fluids through new or previously formed crustal inhomogeneities including crustal fractures (Hill, 1977; Toda et al., 2002; Waite and Smith, 2002). There is a correlation between higher b values and the location of hydrothermal features in the western half of Yellowstone (Farrell, 2009). This would indicate that the high b-values may be due to both the highly fractured (heterogeneous) crust and the high temperatures as well as high pore pressures that allow hydrothermal fluid flow. Therefore, the high b-values

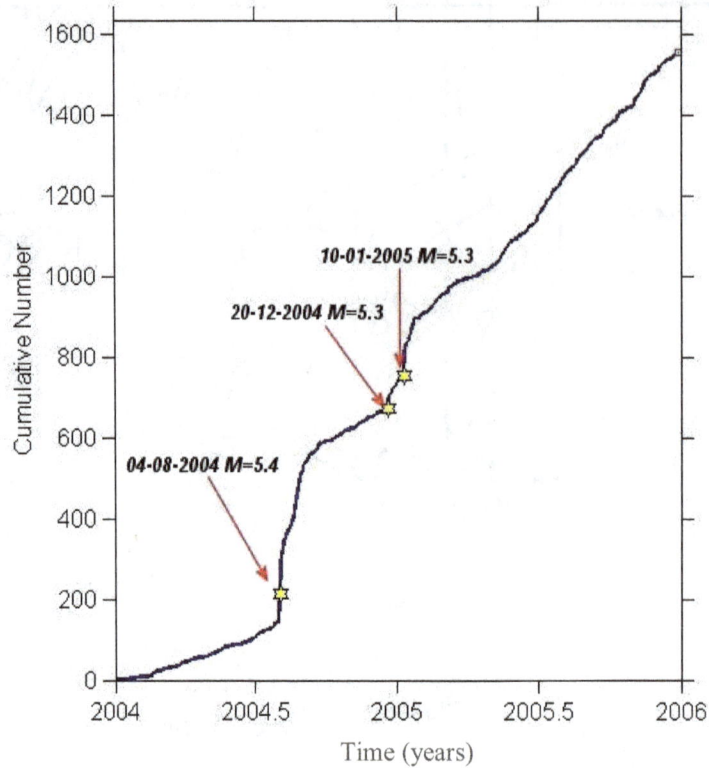

Figure 2. Cumulative number of earthquakes recorded at Gökova, as a function of time, January 2004 through December 2005. Stars represent the 4 August 2004, M = 5.4; the 20 December 2004, M = 5.3 and the 10 January 2005, M = 5.3 earthquakes.

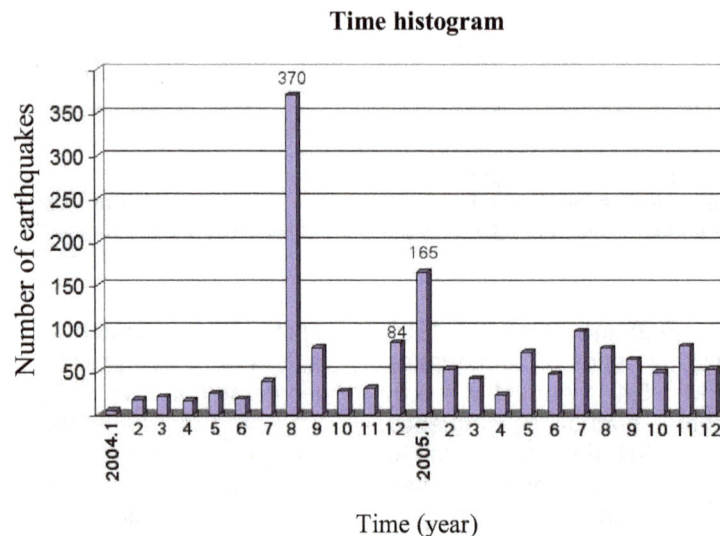

Figure 3. Plot of the number of events located for two years (2004-2006) in Gökova region.

could be an indication of the highly fractured crust that facilitates the movement of hot, hydrothermal fluids.

West of Datça (Cnidus) peninsula lies near the active volcanic centres of Nisyros and Yali (Figure 5), from which ash deposits crop out in patches around the Datça area. Major eruptive activity has occurred on Nisyros in

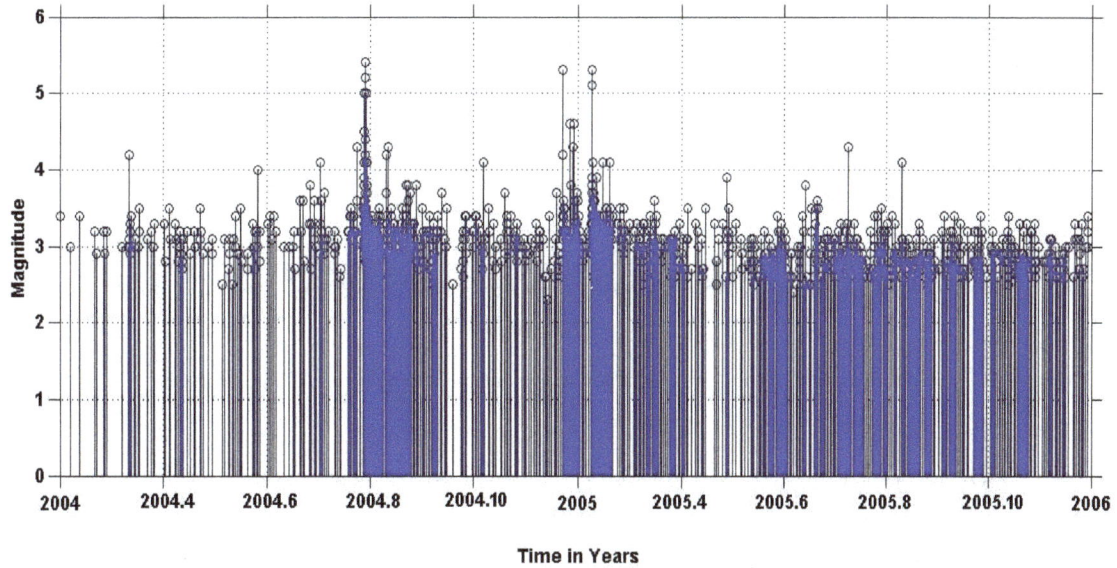

Figure 4. Plot of earthquake magnitudes as a function of time for an earthquake swarm that occured in August 2004, and from December 2004 to January 2005.

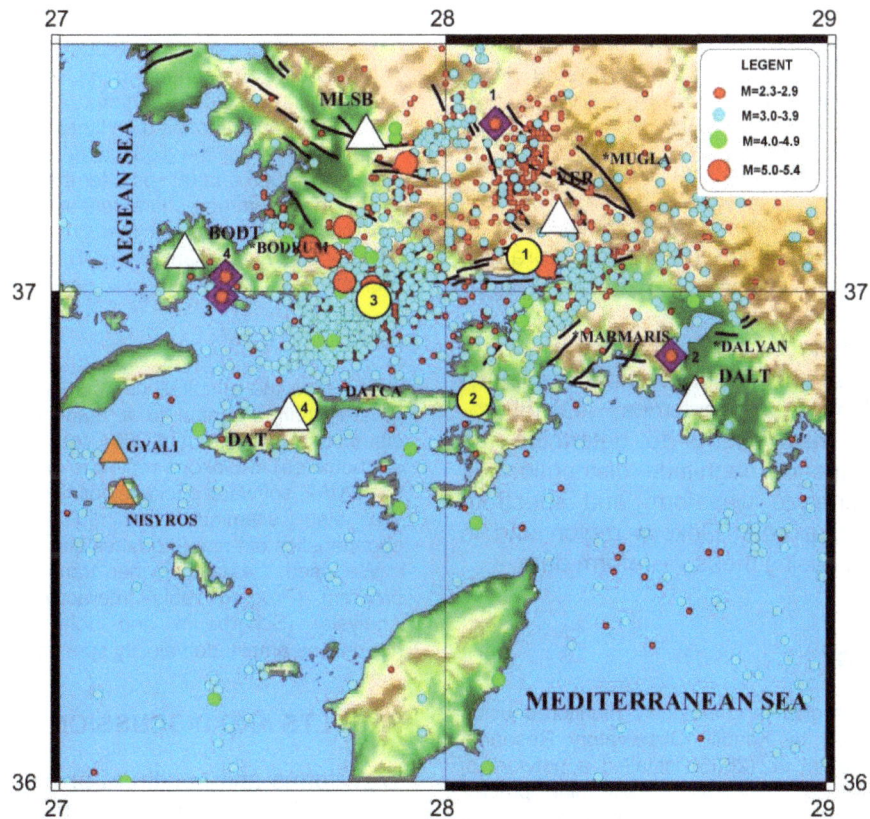

Figure 5. This map shows the epicentre distribution of earthquakes located by the seismic network of KOERI during a period from the begining of 2004 to the end of 2005 and location of seismic events investigated in the article (filled big yellow circles) and seismic stations (filled white triangles). Purple diamonds (1-4) indicate the following hot water springs in the area (Çağlar et al., 2000); 1. Bozhöyük (Yatağan), 2. Sultaniye (Köyceğiz), 3. Karaada (Bodrum), 4. Tavşanburnu (Bodrum). The orange triangles indicate Nysros and Yali volcanoes.

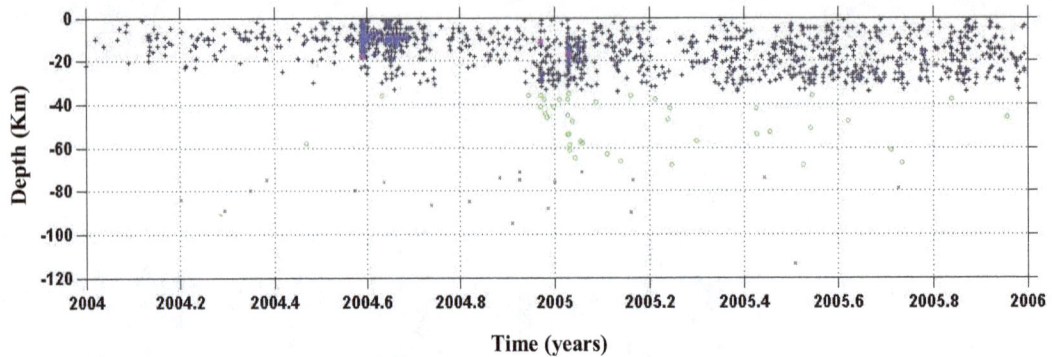

Figure 6. Time-depth plot of earthquakes recorded at Gökova region between 2004 and 2005.

recent times (AD, 1887, 1873 and possibly around 1422) and these violent volcanic events may have been associated with intense seismic activity (Stiros, 2000).

West of Bodrum peninsula is covered by volcanic rock outcrops (Ercan et al., 1984; Robert et al., 1992; Kurt and Arslan, 2001; Çubukçu, 2002; Genç, 2001; Güleç and Hilton, 2006). The volcanic rocks of the Bodrum Peninsula, in SW Turkey and NE of the Hellenic Arc, outcrop over an area of 138 km^2 (Ulusoy et al., 2004). Ulusoy et al. (2004) indicated that a monzonitic intrusion is exposed in the western part of Bodrum peninsula. They investigated the structure of the Bodrum caldera using "Satellite Pour Observer la Terre" (SPOT) image, digital elevation model (DEM), aerial photographs as well as field data.

There are numerous hot water springs in the SW of Turkey (Figure 5) such as Sultaniye (Köyceğiz), Karaada, Tavşanburnu (Bodrum) and Bozhöyük (Yatağan) (Çağlar et al., 2000).

Gökova region is suitable for the occureance of swarm activity. We can tell these characteristics as active faults, crustal fractures, volcanic rock outcrops, and hydro-thermal features. Our main aim is to determine the temporal features (frequency-magnitude distribution) of the earthquakes, to analyze waveform and spectrum characteristics of earthquakes in Gökova region and to give tectonic implications using these waveform data.

DATA AND METHODOLOGY

The seismicity in Gökova region is continously monitored using broadband stations operated by Kandilli Observatory Research Institute since 2004. Kalafat et al. (2005) installed a network of broad band stations called as Blue net to determine precise earthquake locations in that region. Before this date there was only a one component short period station (YER) in the region. This station was replaced with a broadband in July 2006.

The epicenter distribution of earthquakes located by the seismic network of KOERI (Kandilli Observatory and Earthquake Research Institute) during a period from the begining of 2004 to the end of 2005 are shown in Figure 5.

The locations of seismic events were determined by the program hypo71pc (Lee and Valdes, 1989) with a 1-D velocity depth

model (Kalafat et al., 1987).

In August 2004 earthquake series (3 August ML = 5.0; 4 August ML = 5.4; 4 August ML = 5.0) occured in Gulf of Gökova. Another earthquake sequence continued from 20 December, 2004 (Mw = 5.3) to 10 January, 2005 (Mw = 5.4). This activity continued for six months. 1558 seismic events were recorded at this period. Magnitudes range between 2.2 to 5.4 for these events. The most of the micro-earthquakes occured at depths between 1 and 30 km. Depth distribution of the events are seen in Figure 6.

The earthquakes are recorded digitally after the installation of MLSB (Milas) (installed in September 2003), DALT (Dalyan) (installed in August 2004), BODT (Bodrum) (installed in February 2005), DAT (Datça) (installed in October 2005) and YER (Yerkesik) (installed in July 2006) stations. All stations (MLS, DALT, BODT, DAT and YER) are located on limestone.

Sampling rate of the digital data is 20 samples per second before December 2005 and 50 sps after that.

The most common characterization of earthquake populations is the cumulative frequency–magnitude distribution that can be described by the Gutenberg–Richter relation (Gutenberg and Richter, 1956):

$$\log_{10} N = a - b \cdot M \qquad (1)$$

where N is the absolute number of earthquakes with magnitudes greater than or equal to a magnitude M. Frequency-magnitude distribution of the recent 2004 to 2005 earthquake swarms (1558 earthquakes) in Gökova region is examined in this study. We used the ZMAP software package for this examination (Wiemer, 2001). We also examined the digital waveforms and spectrum chracteristics of the earhquakes given in Table 1.

We used the fast Fourier transform (FFT) method of PITSA program (Programmable Interactive Toolbox for Seismological Analysis) (Scherbaum and Johnson, 1992) to calculate the normalized amplitude velocity spectra of seismograms.

RESULTS AND DISCUSSION

Waveforms and spectrum of the four events are seen in Figures 7, 8, 9 and 10 respectively. We gave the information for these four events in Table 1.

We found different behaviors when examined the waveforms and spectrum characteristics of the events occured in the Gökova region. We observed low frequencies on seismogram and spectrum for event 1 in Figure 7. Spectrum of this event has the frequency content restricted in a narrow band between 1 and 2.5 Hz.

Table 1. Location parameters used for waveform and spectral analyses of seismic events given in Figure 5 (filled big yellow circle).

Event number	Date (d:m:y)	Origin time (h:m:s)	Latitude (°N)	Longitude (°E)	Depth (km)	Magnitude
1	22.12.2004	20:29:16.8	37.06	28.19	3.0	3.6
2	25.09.2005	19:09:25.4	36.77	28.06	67.0	3.4
3	31.01.2007	23:13:46.0	36.97	27.80	11.0	3.7
4	21.05.2007	07:30:52.9	36.75	27.61	5.5	3.8

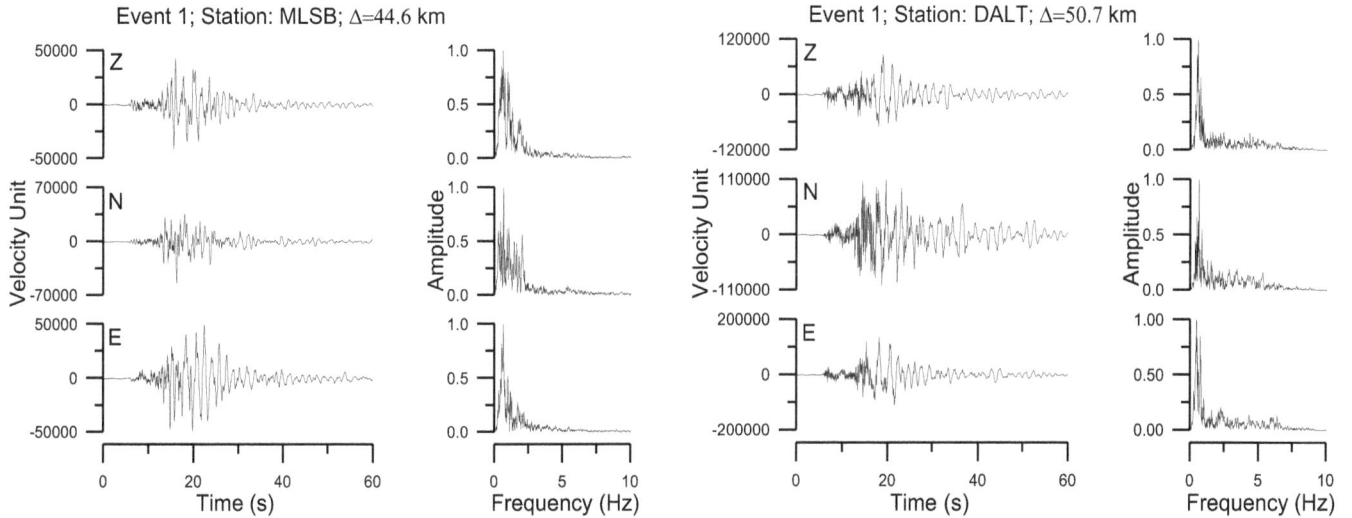

Figure 7. Three components of the velocity seismograms (MLSB, DALT) with a time window length of 60 s and its normalized amplitude spectrums for Event 1.

Source depth is 3 km for this event.

Whereas, we observed high frequencies on waveform and spectrum for events 2, 3 and 4 in Figures 8, 9 and 10. Event 2 has a 67 km focal depth. Hypocenter locations also give 60 to 70 km focal depth in this part of the region because of the Anatolian and Agean lithospheric border. The spectrum of MLSB and DALT for events 3 and 4 have lower frequency content compared to others (DAT, BODT and YER).

The frequencies of shallow earthquakes decrease when the seismic waves travel through attenuative medium. The fact that hot springs and volcanic rocks which attenuates high frequencies are observed in the area (Figure 5) can be linked to resulting low frequency spectrum found in this study.

Volcanic regions, particularly ones where shallow magma bodies and/or hydrothermal systems are present, frequently exhibit seismic swarm activity. Long Valley caldera (Hill et al., 2003), Campi Flegrei (Aster et al., 1992), Yellowstone (Waite and Smith, 2002), and the Socorro Magma Body have all experienced recent seismic swarm activity associated with vertical deformation.

The b-value parameter itself is often useful in understanding the causes of an earthquake swarm. For most tectonic regions of the Earth, $b \leq 1.0$ (Minakami,

1990). However, active volcanic areas can have much larger b-values, often with $b \geq 1.5$, because of increased crack density and/or high pore pressure. Examples include Mount Pinatubo, Philippines (Sánchez et al., 2004), Ito, Japan (Wyss et al., 1997), and Etna, Italy (Vinciguerra, 2002).

We utilize the maximum likelihood method (Weichert, 1980) to compute a, b value for this earthquake sequence (Figure 11). The b-value for this earthquake sequence is found as 1.73±0.08 (Figure 11). This high b-value is attributed to the presence of a high thermal gradient due to the emplacement of magmatic fluids, existence of hot springs and/or highly fractured heterogeneous media.

The b-value distribution for the Yellowstone volcanic region was determined as 1.5±0.05. This high value associated with the youthful 150,000-year old Mallard Lake resurgent dome (Farrel et al., 2009). Sánchez et al. (2004) obtained the frequency–magnitude distribution of earthquakes at Mount Pinatubo, Philippines measured by the b-value. They found that b-values are higher than normal (b = 1.0) and range between b = 1.0 and b = 1.8. This high b-value anomaly infered as increased crack density, and/or high pore pressure, related to the presence of nearby magma bodies.

We observed secondary phases on seismograms

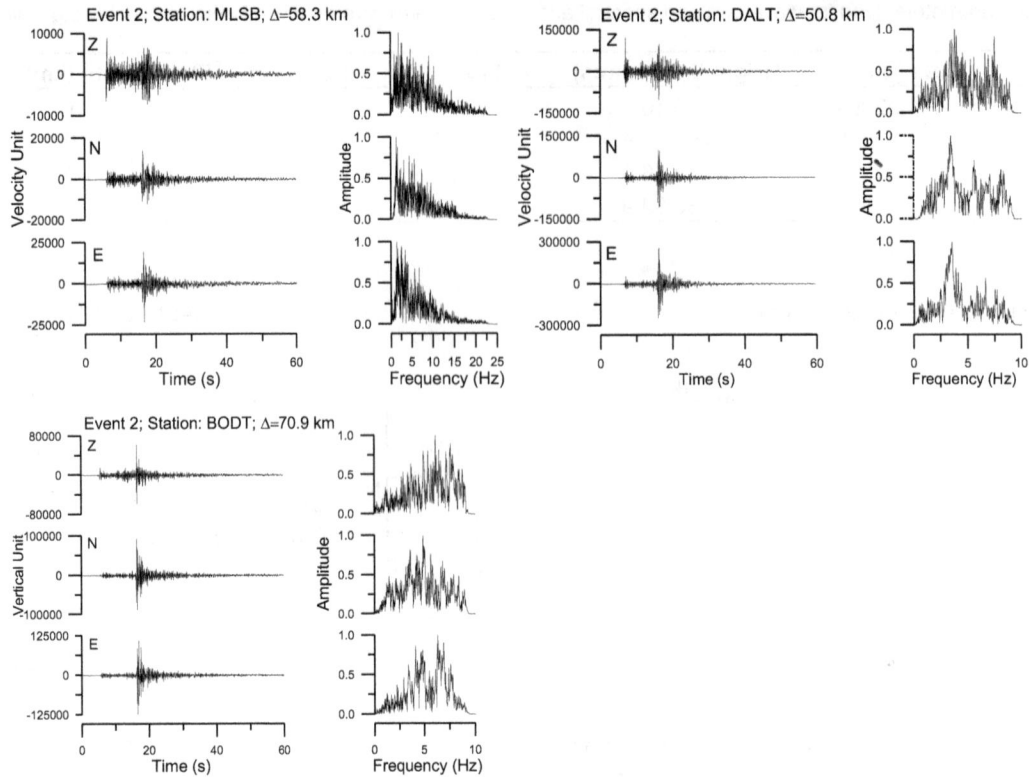

Figure 8. There components of the velocity seismograms (MLSB, DALT and BODT) with a time window length of 60 s and its normalized amplitude spectrum for Event 2.

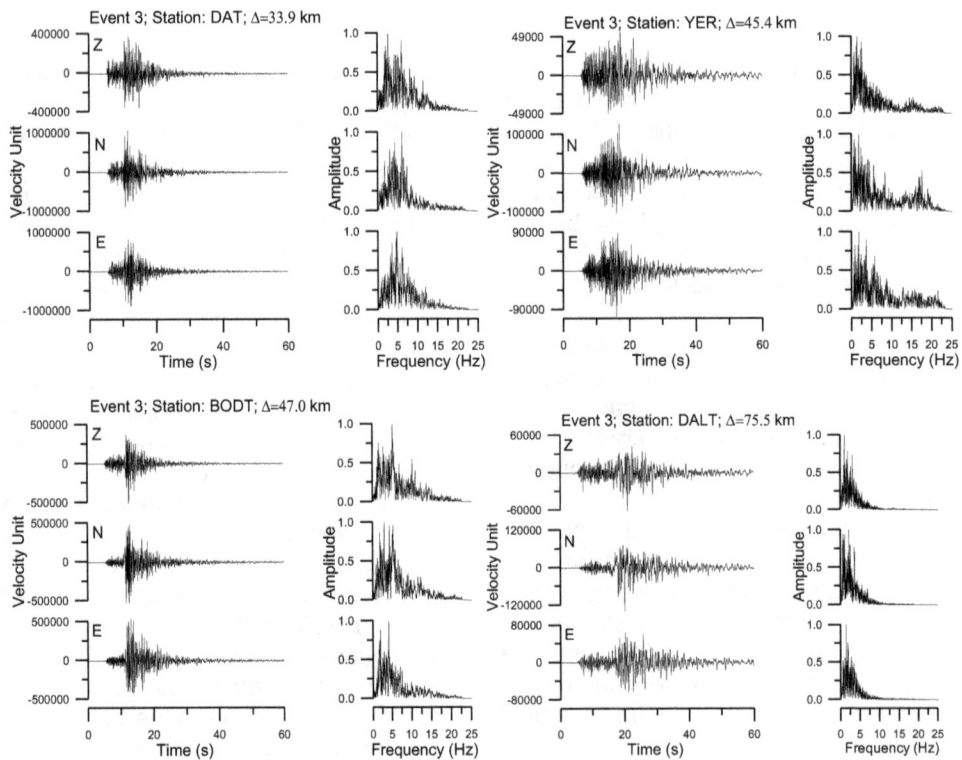

Figure 9. Three components of the velocity seismograms (DAT, MLSB, YER, BODT and DALT) with a time window length of 60 s and its normalized amplitude spectrum for Event 3.

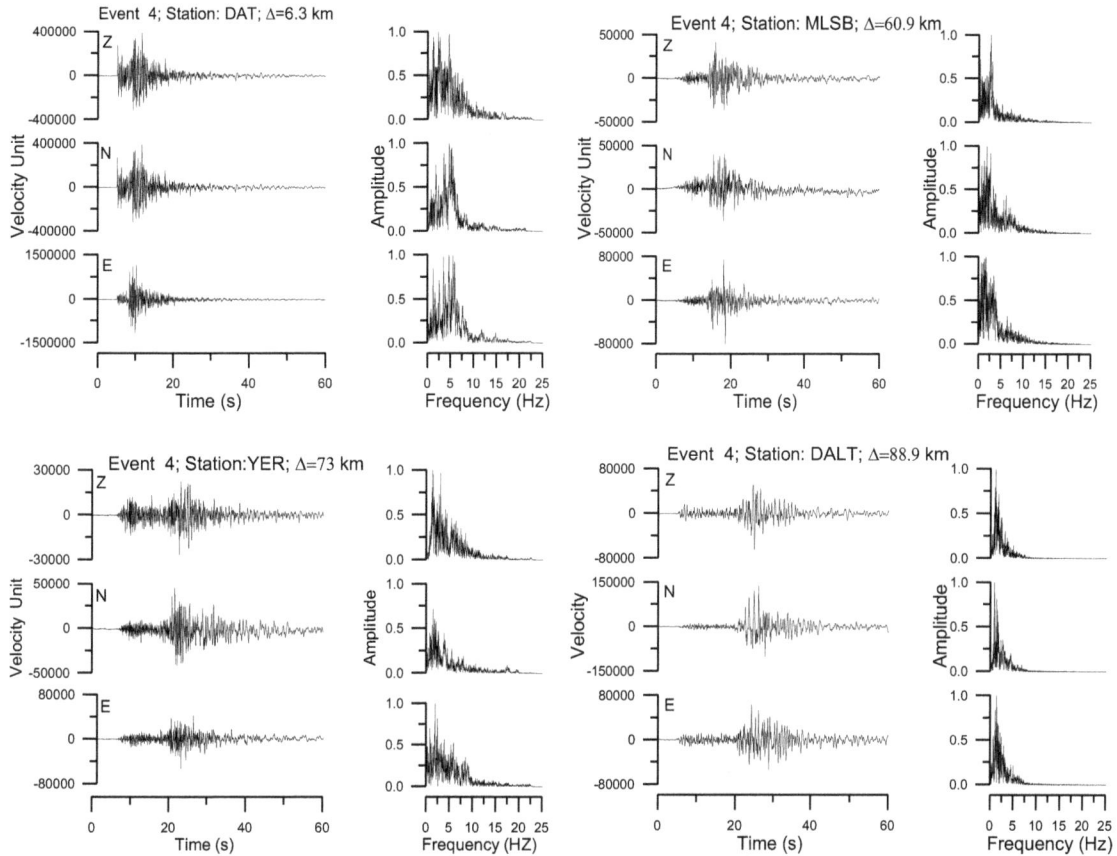

Figure 10. Three components of the velocity seismograms (DAT, MLSB, YER and DALT) with a time window length of 60 s and its normalized amplitude spectrum for Event 4. The event is not recorded at BODT station.

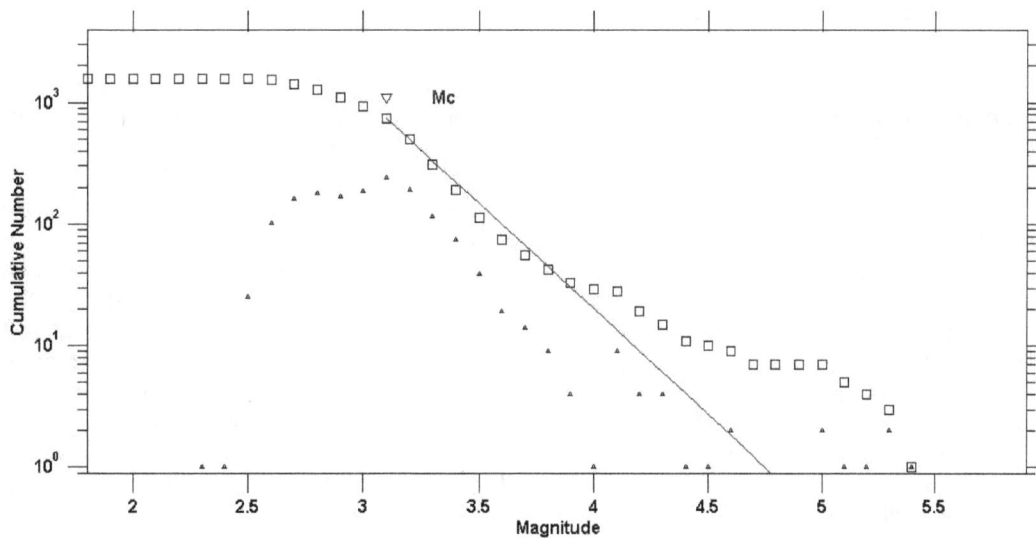

Maximum Likelihood Solution
b-value = 1.73 +/- 0.08, a value = 8.24, a value (annual) = 7.94
Magnitude of Completeness = 3.1

Figure 11. Plot of cumulative frequency of earthquakes as a function of magnitude for the Gökova earthquake sequence. The computed b value (line) obtained using the maximum likelihood method (Weichert, 1980) is 1.73±0.08. White triangle indicates data completeness magnitude, *M* 3.1.

a. Station: BODT; Date: 20 d: 0.8 m:2005 y;
Origin time: 02 h: 21 min : 26.76 s G M T;
M d = 3.0; D = 26.16 km

b. Station: BODT; Date: 22 d: 11 m: 2005 y;
Origin time: 03 h: 58 min : 54.15 s G M T;
M d = 2.2; D = 25.26 km

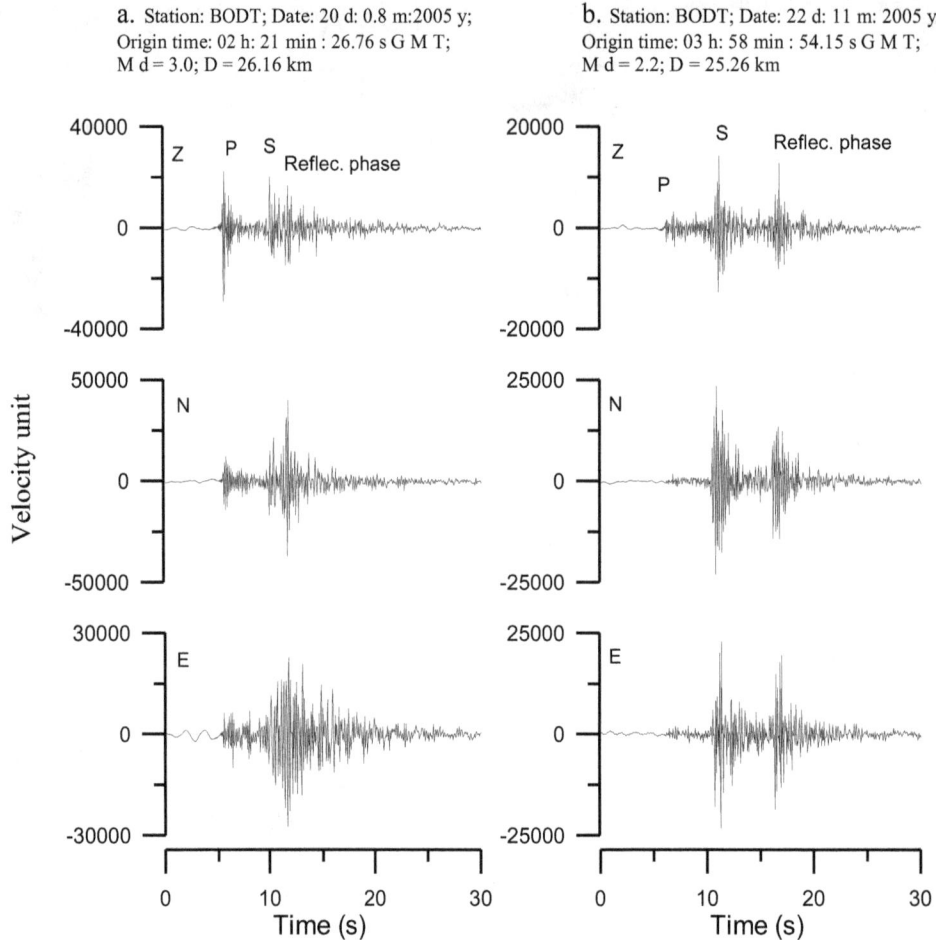

Figure 12. Microearthquake seismograms showing reflection phases. Seismograms are recorded by BODT which is a broadband station.

examined for the events in the region (Figure 12).

These phases arrive to BODT station ~ 1.9 to 7.3 s after the direct S-phase arrival. To help identify the secondary phases, we used travel time curve of events. The graph of travel times of direct and secondary phases versus distance (S-P interval) suggest that these secondary phases are reflections. In Figure 13, travel times for the two phases (direct and reflected S wave) from each earthquake are plotted with different symbols. These reflected phases are generally sharp and large amplitude in horizantal components. The large amplitudes of the reflected phases are explained by a large S-phase velocity contrast across the discontinuity and preferential downward radiation of S-wave energy from the earthquake foci (Sanford et al., 1973).

Sanford and Holmes (1961) first noted the presence of unusual secondary phases on microearthquake seismograms recorded at Socorro and suggested that the phases could be reflections. Sanford et al. (1973) attributed these phases to an interface between rigid and nonrigid crust. The fundamental characteristics of these waves are given in the paper of Sanford et al. (1973),

Sanford and Long (1965). Later, Sanford et al. (1977) concluded that this interface was the top of a sill-like magma body near 19 km depth and estimated its lateral extent by calculating the reflecting positions of SzS arrivals.

When Kurt et al. (1999) made multi-channel reflection study in Gökova bay, they did not see the continuation of the Datça fault in the deeper part of seismic section 11 (Figure 1). They interpreted that the hanging wall consist of Lycian Nappes at the bottom and basin fill at the top. They said that they observed strong reflections, due to the high acoustic impedance contrast where the fault plane is in contact with the basin fill at Gökova bay.

One of possibilities for the reflector is the detachment in fault plane and basin fill. Another possibility for the reflector at these depths is magma sources beneath Gökova region. There are volcanic rocks outcroped in the region. To determine the depth of reflecting discontinuity we used the S-wave velocity for crust and upper mantle as 3.37 and 4.64 km/s respectively (Atılganoğlu, 2007).

We calculated the reflection depth as ~17 to 18 km in Gökova region. To give more detailed charactertics of

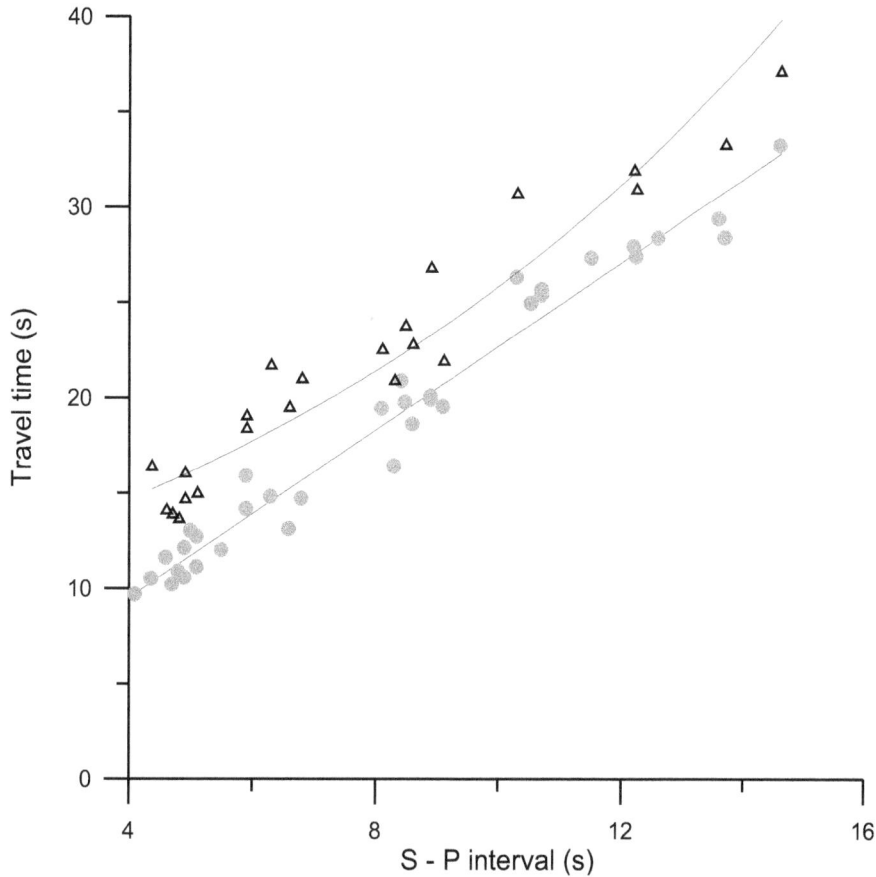

Figure 13. Direct S (filled circles) and reflectd phase (blank triangles) travel time versus S-P interval. Travel times for these phases obtained from the microearthquake events that was used in this study.

these reflected phases, further studies such as lateral velocity distribution and modelling are needed.

ACKNOWLEDGMENTS

We thank all members of Boğaziçi University Kandilli Observatory and Research Institute National Earthquake Monitoring Center of Turkey (NEMC) for making available the epicentral parameters. This work was supported by Boğaziçi University Research Fund within the scope of project BAP/SRP 6671. We thank Boğaziçi University Research Fund Commission and members. The generic mapping tools, GMT (Wessel and Smith, 1995) was used for mapping.

REFERENCES

Aldanmaz E (2002). Mantle source characteristics of alkali basalts and basanites in an extensional intracontinental plate setting, western Anatolia, Turkey: implications for multi-stage melting. Int. Geol. Rev. 44(5):440-457.

Atılganoğlu NB (2007). Bodrum istasyonun kaydettiği depremlerden zaman-uzaklık grafiğinin elde edilmesi, Lisans Bitirme Tezi, Sakarya Üniv. Sakarya.

Aster RC, Meyer R, De Natale G, Zollo A, Martini M, Del Pezzo E, Scarpa R, Iannaccone G (1992). Seismic investigation of the Campi Flegrei: a summary and synthesis of results, in Volcanic Seismol. Gasparini P, Scarpa R, Aki K (Editors), Vol. 3, Springer Verlag, Heidelberg, pp. 462-483.

Bozkurt E (2001). Neotectonics of Turkey – a synthesis, Geodinamica Acta. 14:3-30.

Çağlar I, Türk E, Akoğlu AM (2000). Türkiye sıcak su kaynakları,<http://www.geop.itu.edu.tr/~jeotermal/index1.html>

Çubukçu E (2002). Bodrum volkanizmasının petrolojik incelenmesi, Y. Lisans tezi, Hacettepe Uni. Ankara, Türkiye.

Ercan T, Günay E, Türcecan A (1984). Bodrum yarımadasındaki magmatik kayaçların petrolojisi ve kökensel yorumu. Türkiye Jeoloji Kurumu Bülteni 27:85-98.

Eyidoğan H, Akıncı A, Gündoğdu O, Polat O, Kaypak B (1996). Investigation of the recebt seismic activity of Gökova Basin. National Marine Geology and Geophysical Programme Workshop I, Proceedings, 8-9 February, pp. 68-71.

Farrell J, Husen S, Smith RB (2009). Earthquake swarm and b-value characterization of the Yellowstone volcano-tectonic system, J. Volcanol. Geother. Res. 188:260-276.

Genç ŞC, Karacık Z, Altunkaynak Ş, Yılmaz Y (2001). Geology of a magmatic complex in the Bodrum peninsula, SW Turkey. International Earth Science Colloquim on the Aegean Region, İzmir, Turkey, pp. 63-68.

Görür N, Sengör AMC, Sakınc M, Tüysüz O, Akkök R, Yigitbaş E, Oktay FY, Barka A, Sarıca N, Ecevitoglu B, Demirbag E, Ersoy S, Algan O, Güneysu C, Aykol A (1995). Rift formation in the Gökova region, southwest Anatolia: implications for the opening of the Aegean Sea.

Geol. Mag. 132:637-650.

Gutenberg B, Richter CF (1956). Earthquake magnitude, intensity, energy and acceleration. Bull. Seismol. Soc. Am. 46:105-145

Güleç N, Hilton DR (2006). Helium and heat distribution in Western Anatolia, Turkey: relationship to active extension and volcanism. In: Dilek Y, Pavlides S, editors. Post-collisional tectonics and magmatism in the eastern Mediterranean region, Geological Society of America, Special Paper 409:305-319.

Hill DP (1977). A model for earthquake swarms. J. Geophys. Res. 82:1347-1352.

Hill DP, Langbein JO, Prejean S (2003). Relations between seismicity and deformation during unrest in Long Valley Caldera, California, from 1995 through 1999, J. Volcanol. Geotherm. Res. 127:175-193.

Horálek J, Šflený J, Fischer T, Slancová A, Boušková A (2000). Scenario of the January 1997 west Bohemia earthquake swarm, Studia Geoph. et. Geod. II(44):Special Issue.

Kalafat D, Gürbüz C, Üçer SB (1987). Investigation of crustal and upper mantle structure at the western Turkey, Deprem Araştırma Bülteni 59:43-64.

Kalafat D, Öğütçü Z, Yılmazer M, Suvarıklı M, Güneş Y, Kekovalı K, Geçgel V (2005). Türkiye'de ilk defa gerçekleştirilen uydu bağlantılı genişbantlı alt ağlar : Mavi Ağ (Blue Net), Turuncu Ağ (Orange Net), Beyaz Ağ (White Net) Projeleri, Aktif Araştırma Grubu 9.Toplantısı ATAG9 Bildiri Özetleri Kitabı, Cumhuriyet Üniv. MF 22-24 Eylül 2005, Sivas, pp. 34-36.

Kurt H, Demirbağ E, Kuçu I (1999). Investigation of the submarine active tectonism in the Gulf of Gökova, southwest Antolia-southeast Aegean Sea, by multi-channe seismic reflection data, Tectonophysics 305:477-496.

Kurt H, Arslan M (2001). Bodrum (GB Anadolu) volkanik kayaçlarının jeokimyasal ve petrolojik özellikleri: fraksiyonel kristalleşmemagma karışımı ve asimilasyona ilişkin bulgular, Yerbilimleri. 23:15-32.

Lee WHK, Valdes CM (1989). User manual for HYPO71PC, IASPEI Software Library, 1, pp. 203-236.

Lees J M (1998). Multiplet analysis at Coso geothermal, Bull. Seismol. Soc. Am. 88:1127-1143.

Minakami T (1990). Prediction of volcanic eruptions, in Physical Volcanology, L. Civetta, P. Gasparini, G. Luongo, and A. Rapolla (Editors), Elsevier, Amsterdam, pp. 1-27.

Robert U, Foden J, Varne R (1992). The Dodocenase Province, SE Aegean: a model for tectonic control on potassic magnetism, Lithos. 28:241-260.

Sánchez JJ, McNutt SR, Power JA, Wyss M (2004). Spatial variations in the frequency-magnitude distribution of earthquakes at Mt. Pinatubo volcano, Bull. Seismol. Soc. Am. 94:430-438.

Sanford AR, Holmes CR (1961). Note on the July 1960 earthquakes in central New Mexico, Bull. Seismol. Soc. Am. 51:311-314.

Sanford AR , Long LT (1965). Microearthquake crustal reflections, Bull. Seismol. Soc. Am. 55:579-586.

Sanford AR, Alptekin O, Toppozada TR (1973). Use of reflection phases on microearthquake seismograms to map an unusual discontinuity beneath the Rio Grande rift, Bull. Seismol. Soc. Am. 63:2021-2034.

Sanford AR, Mott RP, Shuleski PJ, Rinehart EJ, Caravella FJ, Ward RM, Wallace TC (1977). Geophysical evidence for a magma body in the vicinity of Socorro, New Mexico, in The Earth's Crust: Its Nature and Physical Properties, J. G. Heacock (Editor), American Geophysical Monograph 20, pp. 385-404.

Scherbaum F, Johnson J (1992). Programmable Interactive Toolbox forSeismological Analysis (PITSA). In: Lee, W.H.K. (eds), IASPEI Software Library, USA,V 5, p. 269.

Stiros SC (2000). Fault pattern of Nisyros isand volcano (Aegean Sea, Greece): structural coastal and archaeol. evidence. In: McGuire, W. J., Grifaths, D. R, Hancock, P. L. & Stewart, I. S. (eds), The Archaeol. Geol.Catastrophes Geological Society, London, Special Publications 171, pp. 385-399.

Şaroğlu F, Emre Ö, Kuşçu I (1995). Türkiye diri fay haritası, MTA Int. Report.

Şengör AMC, Satir M, Akkök R (1984). Timing of tectonic events in the Menderes massif, Western Turkey: implications for tectonic evolution and evidence for Pan-African basement in Turkey. Tectonics 3:693-707.

Toda Shinji Stein RS, Sagiya T (2002). Evidence from the AD 2000 Izu islands earthquake swarm that stressing rate governs seismicity. Nature 419:58-61.

Tonarini S, Agostini S, Innocenti F, Manetti P (2005). d11B as tracer of slab dehydration and mantle evolution in western Anatolia Cenozoic magmatizm. TerraNova 17:259-264.

Uluğ A, Duman M, Ersoy Ş, Özel E, Avcı M (2005). Late quaternary sea-level change, sedimentation and neotectonics of the Gulf of Gökova: Southeastern Aegean Sea, Mar. Geol. 221:381-395.

Ulusoy I, Çubukçu E, Aydar E, Labazuy P, Gourgaud A, Vincent PM (2004). Volcanic and deformation history of the Bodrum resurgent caldera system (southwestern Turkey), J. Volcanol. Geotherm. Res. 136:71-96.

Vinciguerra S (2002). Damage mechanics preceding the September-October 1989 flank eruption at Mt. Etna volcano inferred by seismic scaling exponents, J. Volcanol. Geotherm. Res. 113:391-397.

Yılmaz Y, Genç SC, Gürer OF, Bozcu M, Yılmaz K, Karacık Z, Aktunkaynak S, Elmas A (2000). When did the western Anatolian grabens begin to develop? in Tectonics and Magmatism in Turkey and the Surrounding Area. eds Bozkurt E, Winchester JA, Piper JDA, Geol. Soc. London Special Publication 173:353-384.

Waite GP, Smith RB (2002). Seismic evidence for fluid migration accompanying subsidence of the Yellowstone caldera, J. Geophys. Res. 107:2177-2192.

Wessel P, and Smith WHF (1995). New version of the generic mapping tools (GMT)version 3.0 released, Trans., AGU, EOS, 76, p. 329.

Weichert DH (1980). Estimation of the earthquake recurrence parameters for unequal observation periods for different magnitudes, Bull. Seismol. Soc. Am. 70:1337-1346.

Wiemer S (2001). A software package to analyse seismicity: Zmap. Seismol Res Lett. 72:373-382.

Wyss M, Shimazaki K, Wiemer S (1997). Mapping active magma chambers by b-value beneath the off-Ito volcano, Japan, J. Geophys. Res. 102:20, 413-20, 422.

3

Derivation of the spatially flat Friedmann equation from Bohm-De Broglie interpretation in canonical quantum cosmology

Ch'ng Han Siong* and Shahidan Radiman

School of Applied Physics, Faculty of Science and Technology, Universiti Kebangsaan Malaysia 43600 UKM Bangi, Selangor D. E. Malaysia.

Friedmann equations play a central role in cosmology for describing the evolution of the universe. Initially, the Friedmann equations were derived by Alexander Friedmann in 1922. He derived the equations from Einstein's field equations for a universe which is spatially homogeneous and isotropic. The equations were again derived from Newtonian mechanics by Milne. However in this paper, we derive the spatially flat Friedmann equation from Wheeler-DeWitt equation. We apply the Bohm-de Broglie interpretation to the wave function of universe. In addition, we also set a condition to the wave function, where a large scale factor of the universe is needed.

Key words: Canonical quantum cosmology, Friedmann equations, Wheeler-DeWitt equation.

INTRODUCTION

Friedmann equations are important solutions of Einstein's field equations. They are obtained by plugging the Robertson-Walker metric into Einstein's field equations. It gives the description of evolution of the universe which is taken to be homogeneous and isotropic (Cosmological principle). We can obtain the information on how the universe evolves from Friedmann equations in which the evolution of the scale factor of universe is determined by the type of mass-energy in the universe. A Friedmann equation which is derived from tt - component of Einstein's field equations is given by (Islam, 1992):

$$\left(\frac{\dot{a}}{a}\right)^2 = \frac{8\pi G\varepsilon}{3c^2} - \frac{kc^2}{a^2}, \qquad (1)$$

where $a = a(t)$ is the scale factor, k is the curvature parameter, $\varepsilon = \varepsilon(a)$ is the mass-energy density of the universe and c is the speed of light. The curvature parameter, k can take the value $0, +1, -1$ which are for flat, closed and open universe respectively. In addition, Equation (1) is also called the first Friedmann equation.

There is another way of deriving the Friedmann equations. Milne (1934) had derived the equation without using general relativity (McCrea and Milne, 1934; Milne, 1934). It is totally based on Newtonian mechanics. However, in this paper, we will present an alternative way of deriving the spatially flat Friedmann equation (tt - component, $k = 0$). This is done by applying the Bohm-de Broglie interpretation to the solution of Wheeler-DeWitt equation. This solution is also called the wave function of the universe. The Bohm-de Broglie interpretation is typically applied in quantum cosmology (Kim, 1997; Pinto-Neto, 2000; Pinto-Neto and Colistete, 2001; Pinto-Neto and Sergio-Santini, 2003; Pinto-Neto, 2005; Shojai and Shirinifard, 2005; Shojai and Molladavoudi, 2007).

The Bohm-de Broglie interpretation is an alternative to the orthodox Copenhagen-interpretation. There are reasons that the Bohm-de Broglie interpretation is preferred over the Copenhagen interpretation in the context of quantum cosmology. The Copenhagen interpretation needs a measurement process which takes place outside the quantum system. However, quantizing the whole universe

*Corresponding author. E-mail: chng_hs@yahoo.com.

does not allow us to have a space outside it. Besides this, the Copenhagen interpretation is a probabilistic theory, which contradicts our belief that we have only one universe. In addition, Bohm-de Broglie interpretation is able to provide the precisely definable and continuously changing values of a variable in configuration space. In other words, it is causally determined. Hence, the Bohm-de Broglie interpretation is more appropriate to be applied in quantum cosmology in which the behaviour of evolution of the universe is studied.

Many physicists accept that quantum mechanics is a universal theory, which means that every physical system can be described or derived from quantum mechanics. Therefore, in principle, we should be able to recover the theory of Standard Cosmology (for example: Friedmann equation) from the Wheeler-DeWitt equation which is said to be the Schrödinger equation for the universe. Next, we present a spatially flat minisuperspace quantum cosmological model. Afterwards, we obtain the solution of the Wheeler-DeWitt equation. We then show that the spatially flat Friedmann equation can be obtained from the solution of Wheeler-DeWitt equation if we apply the Bohm-de Broglie interpretation to it. Finally, discussions and conclusions are given.

A SPATIALLY FLAT MINISUPERSPACE MODEL

We start by considering a homogeneous and isotropic spatially flat Robertson-Walker metric, which is given by:

$$ds^2 = -N^2 c^2 dt^2 + a^2 \left(dx^2 + dy^2 + dz^2 \right),$$ (2)

where $N = N(t)$ is the lapse function. The scale factor, $a = a(t)$ here takes the dimensions of length, while the coordinates x, y, z are dimensionless. Next, we introduce the Einstein-Hilbert action as follows (Maydanyuk, 2008; Khosravi et al., 2010; Norbury, 1998; Walecka, 2007):

$$S = \int \sqrt{-g} \left[\frac{Rc^4}{16\pi G} - \varepsilon \right] dx\,dy\,dz\,cdt,$$ (3)

where R is the Ricci scalar, g is the determinant of the metric, $\varepsilon = \varepsilon(a)$ is the energy density of the universe and furthermore it is a function of scale factor, a. Now, computing the Ricci scalar and determinant of the metric from Equation (2), we obtain them as follows:

$$R = \frac{6\ddot{a}}{N^2 c^2 a} - \frac{6\dot{N}\dot{a}}{N^3 c^2 a} + \frac{6\dot{a}^2}{N^2 c^2 a^2},$$ (4)

$$g = -N^2 a^6.$$ (5)

Next, we substitute Equation (4) and (5) into Einstein-Hilbert action (3) to obtain the following action:

$$S = \int Na^3 \left[\left(\frac{6\ddot{a}}{N^2 c^2 a} - \frac{6\dot{N}\dot{a}}{N^3 c^2 a} + \frac{6\dot{a}^2}{N^2 c^2 a^2} \right) \frac{c^4}{16\pi G} - \varepsilon \right] dx\,dy\,dz\,cdt$$ (6)

After integrating the first term in the bracket by parts with respect to t, Equation (6) becomes

$$S = V_o \int \left[-\frac{3a\dot{a}^2 c^3}{8\pi GN} - Na^3 c\varepsilon \right] dt.$$ (7)

In general, $\varepsilon = \varepsilon(a)$ can be written as follows:

$$\varepsilon = \frac{C_{(m)}}{a^m},$$ (8)

where $C_{(m)}$ is a constant and m is an integer. A value of m is chosen in Equation (8) depending on which type of energy that dominates the universe. Some types of energy density and their corresponding value of $C_{(m)}$ (Islam, 1992) are summarized in Table 1.

For any given value of t, the geometry of our universe is the same everywhere. Hence, the $V_o = \int dx\,dy\,dz$ in Equation (7) is constant and set equal to one by integrating over an appropriate compact region of space (Bojowald, 2011). Consequently, the Lagrangian from Equation (7) is:

$$L = -\frac{3a\dot{a}^2 c^3}{8\pi GN} - Na^3 c\varepsilon.$$ (9)

From this Lagrangian, we get the momentum conjugate to a as follows:

$$P_a = \frac{\partial L}{\partial \dot{a}} = -\frac{3a\dot{a}c^3}{4\pi GN}$$ (10)

Hence, the Hamiltonian is given by:

$$H = P_a \dot{a} - L = N \left[-\frac{2\pi GP_a^2}{3ac^3} + a^3 c\varepsilon \right].$$ (11)

The Hamiltonian constraint provides (for details see, Kolb and Turner, 1990)

Table 1. Three types of energy density and their values of $C_{(m)}$. Λ is the cosmological constant while O denotes present values.

Type of energy density, $\varepsilon(a)$	m	$C_{(m)}$
Radiation, ε_r	4	$C_{(4)} = \varepsilon_{ro}a_o^4$
Pressure less matter, ε_m	3	$C_{(3)} = \varepsilon_{mo}a_o^3$
Vacuum energy, ε_v	0	$C_{(0)} = \dfrac{\Lambda c^4}{8\pi G}$

$$-\frac{2\pi G P_a^2}{3ac^3} + a^3 c\varepsilon = 0. \tag{12}$$

So far, what we have worked out is purely classical. Next, we apply the canonical quantization procedure in Equation (12) to replace P_a by $-i\hbar\dfrac{d}{da}$, where i is an complex number and $\hbar$ is the reduced Planck constant. This procedure leads to the Wheeler-DeWitt equation which is written as follows:

$$\left\{ \frac{d^2}{da^2} + \frac{3C_{(m)}a^{4-m}c^4}{2\pi G\hbar^2} \right\}\psi = 0, \tag{13}$$

where ψ is the wave function of the universe.

BOHM-DE BROGLIE INTERPRETATION OF THE WAVE FUNCTION

The solution of Equation (13) is given by Polyanin and Zaitsev (2003):

$$\psi = a^{\frac{1}{2}}\left[C_1 J_{\frac{1}{6-m}}\left(\frac{2}{6-m}\sqrt{\frac{3C_{(m)}c^4}{2\pi G\hbar^2}}a^{3-\frac{m}{2}} \right) + C_2 Y_{\frac{1}{6-m}}\left(\frac{2}{6-m}\sqrt{\frac{3C_{(m)}c^4}{2\pi G\hbar^2}}a^{3-\frac{m}{2}} \right) \right] \tag{14}$$

where $J_{\frac{1}{6-m}}$ and $Y_{\frac{1}{6-m}}$ are Bessel functions of first and second kind, respectively, of order $\dfrac{1}{6-m}$. Now, let us choose $C_1 = A$ and $C_2 = -iA$ where A is a positive constant. Besides this, we also impose the condition:

$$\left(\frac{2}{6-m}\sqrt{\frac{3C_{(m)}c^4}{2\pi G\hbar^2}}a^{3-\frac{m}{2}} \right) \gg 1$$

as in Lea (2004) to get the

asymptotic expansion of the Bessel functions. Our current large scale universe satisfies this condition because the scale factor, a of a large scale universe is expected to be sufficiently large such that it obeys the following inequality: $a^{3-\frac{m}{2}} \gg \dfrac{6-m}{2}\sqrt{\dfrac{2\pi G\hbar^2}{3C_{(m)}c^4}}$. Hence now, we have $J_n(u) \sim \sqrt{\dfrac{2}{\pi u}}\cos\left(u - \dfrac{n\pi}{2} - \dfrac{\pi}{4} \right)$ and $Y_n(u) \sim \sqrt{\dfrac{2}{\pi u}}\sin\left(u - \dfrac{n\pi}{2} - \dfrac{\pi}{4} \right)$. Consequently, the wave function of the universe (14) becomes

$$\psi = A\sqrt{\frac{a^{\frac{m}{2}}(6-m)}{\pi a^2\sqrt{3C_{(m)}c^4/2\pi G\hbar^2}}}\cdot\exp\left\{ \frac{i}{\hbar}\left[\frac{-2}{6-m}\sqrt{\frac{3C_{(m)}c^4}{2\pi G}}a^{3-\frac{m}{2}} + \hbar\pi\left(\frac{1}{4} + \frac{1}{12-2m} \right) \right] \right\} \tag{15}$$

Now, let us make the following replacements:

$$R = A\sqrt{\frac{a^{\frac{m}{2}}(6-m)}{\pi a^2\sqrt{3C_{(m)}c^4/2\pi G\hbar^2}}} \tag{16}$$

and

$$S = \frac{-2}{6-m}\sqrt{\frac{3C_{(m)}c^4}{2\pi G}}a^{3-\frac{m}{2}} + \hbar\pi\left(\frac{1}{4} + \frac{1}{12-2m} \right), \tag{17}$$

where R and S are functions of the scale factor, a. Then, the wave function (15) is written (Bohm, 1952; Bohm et al., 1987; Holland, 1993) as follows:

$$\psi = R\exp\left(\frac{iS}{\hbar} \right). \tag{18}$$

To illustrate the Bohm-de Broglie interpretation, we substitute the wave function (18) into (13) and take the derivatives to obtain the following equations:

$$-\frac{2\pi G}{3c^3 a}\left(\frac{dS}{da}\right)^2 + a^3 c\varepsilon + \frac{2\pi G\hbar^2}{3c^3 a}\frac{1}{R}\frac{d^2 R}{da^2} = 0, \qquad (19)$$

$$\frac{d}{da}\left(R^2\frac{dS}{da}\right) = 0. \qquad (20)$$

Equation (20) is known as the continuity equation for probability, while Equation (19) is viewed as modified Hamilton-Jacobi equation where the last term on the left hand side is added to the usual classical Hamilton-Jacobi equation. This last term is regarded as an additional potential energy and called the quantum potential. It is responsible for the quantum effects. However, it has a very small value and effect here because the scale factor we consider is sufficiently large. Now from modified Hamilton-Jacobi Equation (19), we are to identify

$$\frac{dS}{da} = P_a. \qquad (21)$$

Equation (21) is called the guidance equation. Substituting Equation (10) into (21) and applying a gauge choice of $N = 1$, we obtain the following equation:

$$\frac{dS}{da} = -\frac{3a\dot{a}c^3}{4\pi G}. \qquad (22)$$

Then, after computing the derivative, we obtain

$$\left(\frac{\dot{a}}{a}\right)^2 = \frac{8\pi G}{3c^2}\left(\frac{C_{(m)}}{a^m}\right), \qquad (23)$$

where $\frac{C_{(m)}}{a^m} = \varepsilon$. Equation (23) is indeed the spatially flat Friedmann equation in which the value of m is dependent on the type of energy density that dominates the universe.

DISCUSSION AND CONCLUSIONS

We obtain the spatially flat Friedmann equation from canonical quantum cosmology by applying the Bohm-de Broglie interpretation to the wave function of the universe. The Bohm-de Broglie interpretation is able to provide the continuously varying values of variable. Solving Equation (23) would give us information on how the universe evolves with time. The condition $\left(\frac{2}{6-m}\sqrt{\frac{3C_{(m)}c^4}{2\pi G\hbar^2}}a^{\frac{3-m}{2}}\right) \gg 1$ is

taken into account when we derive the equation. However, some reference materials give different condition to obtain the asymptotic expansions, but all the conditions are applicable for the current late-time universe. However, what if the value of a is very small, such as in the early time of the expansion of universe?

Then, the condition might not be valid (it is certainly not valid for $a \to 0$). Can we still get the same equation of evolution? The answer is of course not. Thus, the spatially flat Friedmann equation is obtained when the universe we consider is in the big scale or, in other words, in quantum cosmology, the spatially flat Friedmann equation is only valid to describe the late-time universe. We argue that Einstein's field equations (obtained as Euler-Lagrange equation from Einstein-Hilbert action) are only applicable to the universe in the large scale.

ACKNOWLEDGEMENT

One of the authors, Ch'ng would like to thank the Ministry of Higher Education (MOHE), Malaysia for providing MY PhD scholarship.

REFERENCES

Bohm D (1952). A suggested interpretation of quantum theory in terms of hidden variables I and II. Phys. Rev. 85(2):166-193.

Bohm D, Hiley BJ, Kaloyerou, PN (1987). An ontological basis for the quantum theory. Phys. Rep. 144(6):323-375.

Bojowald M (2011). Canonical gravity and applications-cosmology, black holes, and quantum gravity. Cambridge University Press, New York. pp. 4-6.

Holland PR (1993). The quantum theory of motion. Cambridge University Press, United Kingdom.

Islam JN (1992). An introduction to mathematical cosmology. Cambridge University Press, United Kingdom, pp. 52, 64, 67, 95.

Khosravi N, Sepangi HR, Vakili B (2010). A cosmological viewpoint on the correspondence between deformed phase-space and canonical quantization. Gen. Relativ. Gravit. 42:1081-1102.

Kim SP (1997). Quantum potential and cosmological singularities. Phys. Lett. A 236:11-15.

Kolb EW, Turner MS (1990). The early universe. Addison-Wesley. pp. 451-464.

Lea SM (2004). Mathematics for physicists. Thomson Learning, United States of America. pp. 403-404.

Maydanyuk SP (2008). Wave function of the universe in the early stage of its evolution. Eur. Phys. J. C57:769-784.

McCrea WH, Milne EA (1934). Newtonian universe and the curvature of space. Q. J. Math. 5(17):73-80.

Milne EA (1934). A newtonian expanding universe. Q. J. Math. 5(17):64-72.

Norbury JW (1998). From Newton's laws to the Wheeler-DeWitt equation. Eur. J. Phys. 19:143-150.

Pinto-Neto N (2000). Quantum cosmology: How to interpret and obtain results. Braz. J. Phys. 30(2):330-345.

Pinto-Neto N, Colistete Jr R (2001). Graceful exit from inflation using quantum cosmology. Phys. Lett. A 290:219-226.

Pinto-Neto N, Sergio Santini E (2003). The accelerated expansion of the universe as a quantum cosmological effect. Phys. Lett. A 315:36-50.

Pinto-Neto N (2005). The bohm interpretation of quantum cosmology. Found. Phys. 35(4):577-603.

Polyanin AD, Zaitsev VF (2003). Handbook of exact solutions for ordinary differential equations. CRC Press, United States of America.

Shojai F, Molladavoudi S (2007). Quantum cosmology with varying speed of light and Bohmian trajectories. Gen. Relativ. Gravit. 39(6):795-813.

Shojai F, Shirinifard A (2005). Classical and quantum limits in bohmian quantum cosmology. Int. J. Mod. Phys. D 14(8):1333-1345.

Walecka JD (2007). Introduction to general relativity. World Scientific, Singapore.

4

Groundwater exploration using geoelectrical resistivity technique at Al-Quwy'yia area central Saudi Arabia

Mohamed Metwaly[1,2]*, Eslam Elawadi[1,3], Sayed S. R. Moustafal[1,2], F. Al Fouzan[4], S. Mogren[1] and N. Al Arifi[1]

[1]Department of Geology and Geophysics, Faculty of Science, King Saud University, Saudi Arabia.
[2]National Research Institute of Astronmy and Geophysics (NRIAG), Cairo, Egypt.
[3]Nuclear Materials Authority (NMA), Cairo, Egypt.
[4]King Abd Alaziz City for Science and Technology (KACST), Saudi Arabia.

Geoelectrical resistivity surveys were carried out in Al Quwy'yia area, located at the centeral part of Saudi Arabia, to map the acquifer and estimate the groundwater potentuality. The acquired vertical electrical sounding (VES) data sets have been collected along three longitudinal profiles trending East-West, perpendicular to the basment/sedimentary contact. The data sets have been analysed using 1D to obtain the initial figure out of the resistivity layers along the areas. Then, the data were inversion resistivity section using 2D inversion scheme. Information from two boreholes were incorporated during the processing to enhance the results and constrain the resistivity models with geological layers. The results revealed mainly two geoelectric layers represent mainly the basement and sedimentary rocks. The basement rocks dip generally east ward, where the sedimentary section increases in this direction. The depth to the basement is about 50 m in the western part of the area and can not be reached from the acquired data in the eastern part. The contact boundaring between the basment complex and sedimentary rocks can be determined. The static water table is coincident with the limestone rock of Khuff formation as indicated from the comparison between the individual resistivity models and the two wells located at the study area. The thickness of the aquifer is increasing in the north eastern direction where the possibility of the groundwater potentiality is increasing.

Key words: Vertical electrical sounding (VES), arid environment, Al Quwy'yia, 2D resistivity.

INTRODUCTION

Al Quwy'yia area is considered as one of the most promising areas for future sustainable developments in central part of Saudi Arabia. It is located at 160 km to the west of Riyadh city, the capital of Saudi Arabia (Figure 1). The area gains its importance as it is an essential stop for the internal pilgrimages during trips to holly Mecca. Because Al Quwy'yia area is located in the most arid region in the Arabian Peninsula, it has limited groundwater resources. The main charging groundwater resources are coming from the rare rainfall along the basement outcrops that bounds the area from the western side. The rain fall water is charging the aquifers in the Quaternary and the underlain calcareous deposits

of Khuff formation, which overlay the basement rocks particularly in the western part of the area. Water demands in Al Quwy'yia area is increasing due to the increasing of population and development activities. However, the groundwater studies of the area are rarely addressed, because it was thought that the area has not any groundwater potentiality and was out of the future development plan. However, most of the carried out work were dealing with the geological and mineral resources as the area is situated very close to the eastern part of Arabian shield (Senalp and Al-Duaiji, 2001). Moreover, there are some works, that have been done close to Al Quwy'yia area dealing with the groundwater recharging source along the shallow and deep aquifer in the eastern and central part of Saudi Arabia (Hoetzel, 1995). Also, Al-Amri (1996) applied the geoelectrical techniques in delineating the groundwater potentiality in the central part

---

*Corresponding author. E-mail: mmetwaly70@yahoo.com.

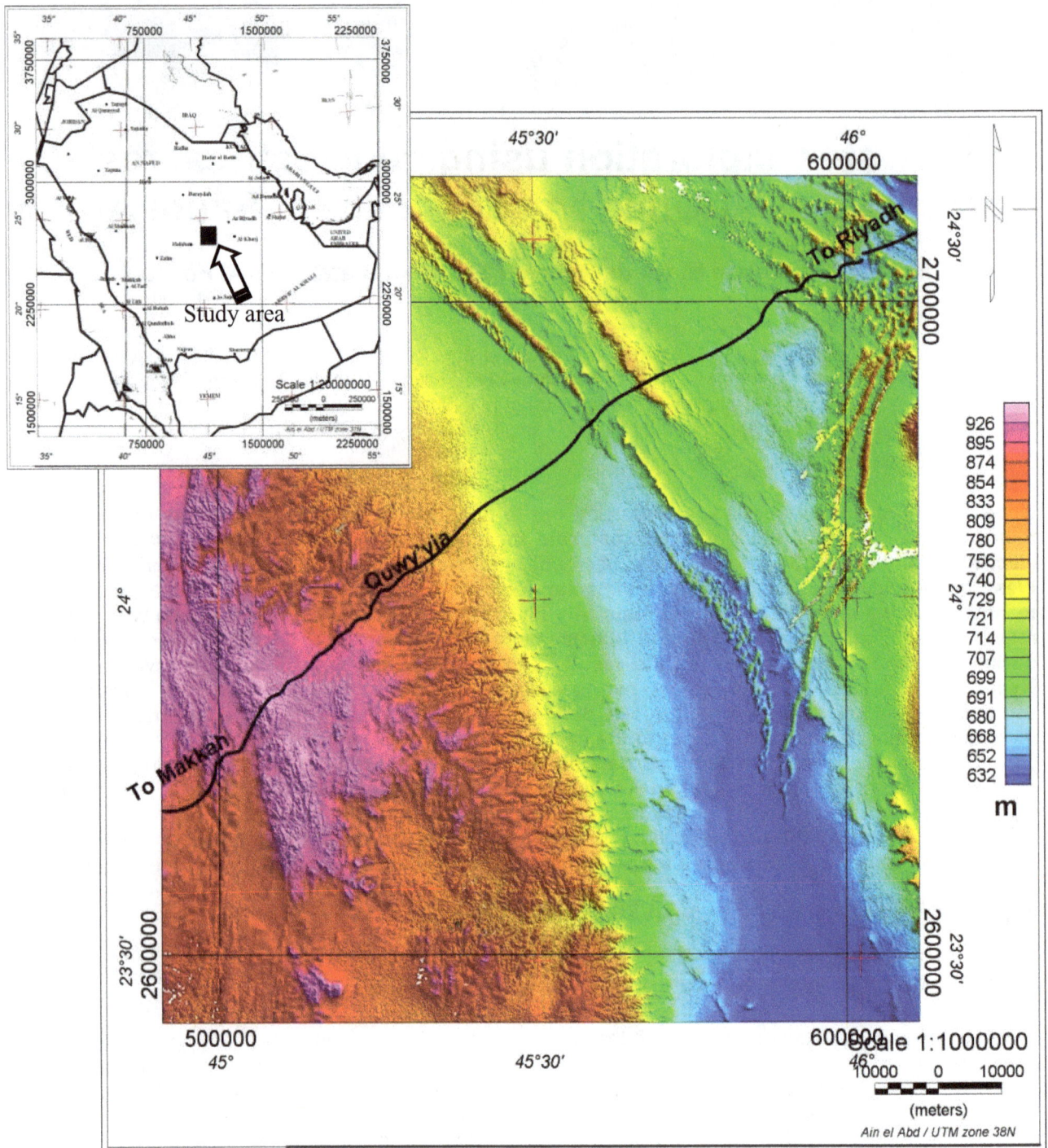

Figure 1. Location map of Al Quwy'yia area.

of Saudi Arabia. Zaidi and Kassem (2010) have applied the electrical resistivity tomography (ERT) in Diriyah area close to Riyadh, for estimating the depth of the water-bearing formations up to 70 m depth.

The present study aims to demonstrate and portray the shallow aquifer, applying surface electrical resistivity measurements in Al Quwy'yia area, along the contact

zone with the Arabian shield. In order to achieve this target, geoelectrical resistivity technique has been utilized as it is important exploring tool for studying and depicting the subsurface aquifer in arid areas (Asfahani, 2007; Chandra et al., 2010; Yadav and Singh, 2007). It is based on measuring the contrast in electrical conductivity of the different rock units which is varying according to the rock nature (density, porosity, pore size and shape), water content and its quality and temperature (Parasnis, 1997). The resistivity is more controlled by the water contents and its quality within the matrix of the formation than by the solid granular resistivity value itself. Therefore, the subsurface sedimentary succession may be subdivided into different geoelectrical units according to the different percentage of humidity.

Geological setting

The study area is part of a large plain extending north south over several hundred kilometers, between the Tuwaiq mountains in the east and the basement complex in the west (Figure 2). This plain slopes from the west (950 m elevation) to the eest (632 m elevation) in the east, whereas there are many dry wadis running from the west to the east. The south-west corner of the study area is made up of crystalline rocks with dark rounded hills dissected by white sandy wadis and separated by large sand dune areas. Going to the eastern side, the area is almost flat. The Khuff limestone exposures are dissected by a dense network of wadis, evidencing active erosion during rainy periods. Locally, most Al Quwy'yia area has been covered with Quaternary deposits, which are in the form of alluvium and sand dunes (Figure 2). The thickness of such deposites range from few meters to tens of meters (Figure 3). The Quaternary deposites are followed by calcareous units of Khuff formation (Permian), which is exposed as a hilly belt of 20 km wide at the eastern side of the basement. Going to west and south of Al Quwy'yia, the Khuff formation overlies directly the basement complex as indcated in wells (Qap2-1 and Qap1-1), located at the eastern side of Al Quwy'yia area (Figure 3). The Khuff formation comprises various types of limestone and dolomite, shales and siltstone, sandstone and marl.

ELECTRICAL RESISTIVITY DATA ACQUISITION

Most of the electrical resistivity techniques require injection of electrical currents into the subsurface via a pair of electrodes planted on the ground. By measuring the resulting variations in electrical potential at other pairs of planted electrodes, it is possible to determine the variations in resistivity (Dobrin, 1988; Ozcep et al., 2009; Alile et al., 2011). A conventional vertical electrical sounding (VES) survey was used for quantitative interpretation where the center point of the array remains fixed and the electrode spacing is increased for deeper penetration (Loke, 1999).

The ultimate aim of the resistivity survey is to determine the resistivity distribution with depth on the basis of surface measurements of the apparent resistivity and to interpret it in terms of geology or hydrogeology.

Nevertheless, when resistivity methods are used, limitations can be expected if ground inhomogeneities and anisotropy are present (SenosMatias, 2002). Due to large variation in resistivity of the formations, as well as inherent non-uniqueness in the interpretation techniques, the method sometimes results in mis-interpretation of the layer's parameters. The ambiguous interpretation, thus, often makes the results unreliable. It is thus necessary to interpret the soundings data taking parameters from other sources like geological and hydrological information into consideration (Kumar et al., 2007; Yilmaz, 2011) and/or applying the 2D inversion scheme to increase the data consistency. Such 2D inversion scheme reduces the uncertainty, which is common in the 1D inversion (Uchida, 1991).

Using the Syscal R2 acquisition system, operating with the Schlumberger electrode configuration, 16 VES data sets were recorded along three profiles passing through Al Quwy'yia area (Figure 2). For each VES, the current electrodes (AB/2) were varied from 3 to 1000 m and the potential electrodes (MN) were extended from 0.5 to 200 m in successive steps. Long steel electrodes (about 0.75 m) were used to optimise coupling between the electrodes and the ground particulary in the dry and friable sediments areas. At several locations, measurements were repeated or the current electrode positions were changed to improve the quality of the acquired resistivity data, which have been often checked during the data acquisition. Two VES (B2 and B3) have been carried out close to two boreholes (Qap2-1 and Qap 1-1). The geological data obtained from the wells were used in calibration process of the geoelectrical models and increase the constrains to minimize the uncertainties of the 1D inverted models (Figure 3).

Processing of VES data in 1D mode

Inspection of the acquired resistivity data curves (Figure 4) reveals certain properties that characterize the two distinct principal geological environments in the study area. The apparent resistivity curves are mostly of H-type, whereas the resistivity values are high at small AB/2 offsets due to the dry surface conditions. Then, the resistivity values decrease gradually due to the presence of aquifer saturated limestone rocks. With increasing AB/2 offsets, the apparent resistivity values increase again, probably due to the effect of bedrock.

Typical examples of 1D resistivity models obtained by iteratively inverting (IPI2win, 2000) code, which provides the opportunity to choose a set of equivalent solution and among them, select the best one with less fitting error between observed and calculated data. The routine is utilizing a least squares approach to minimize the difference between the input data and the theoretically derived curve. The quantitative interpretation has been applied to determine the thicknesses and resistivities of

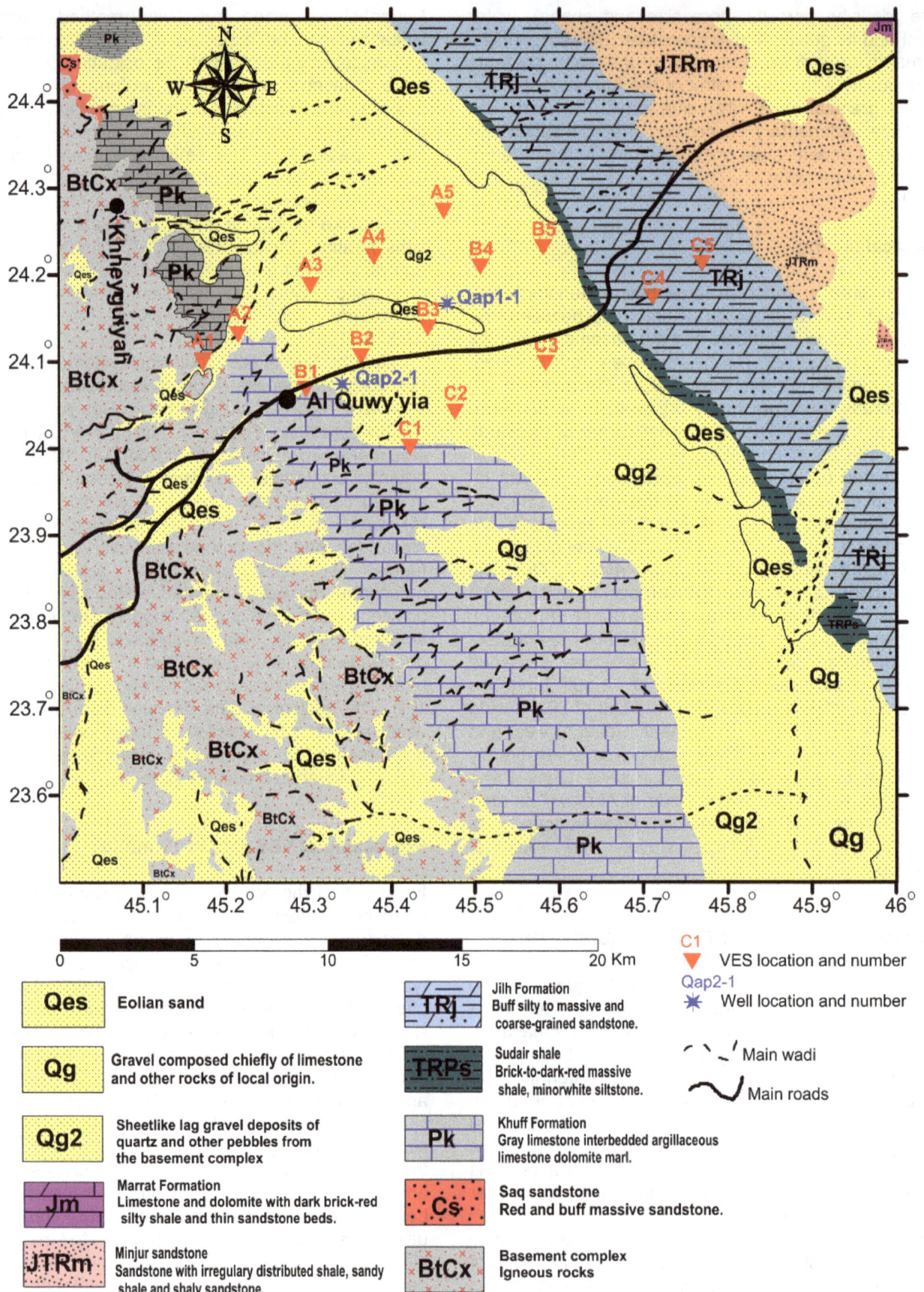

Figure 2. Location of VES and wells along the survyed site at Al Quwy'yia area superimposed on the geological map.

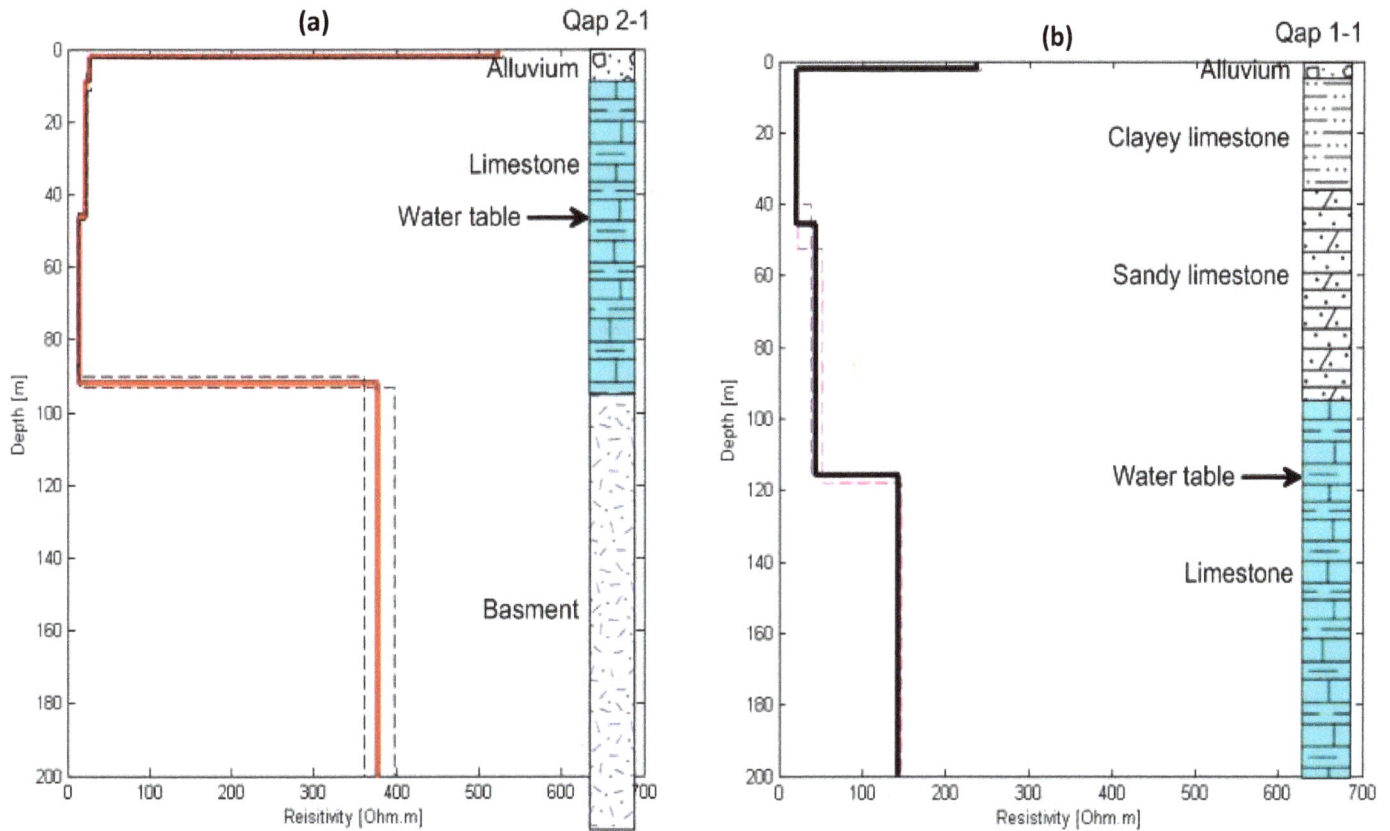

Figure 3. The layered resistivity and equivalent models constrained with the geological boundaries at two wells along profile B; (a) VES (2) and well (Qap2-1); (b) VES (3) and well (Qap1-1).

the different lithological units below each VES station (Figures 3 and 4). Resistivities are controlled primarily by the pore water conditions and ionic content rather than the lithological variation particularly in the sedimentary succession (Loke, 1999; Mohamed et al., 2011); therefore, there are wide ranges in resistivity for any particular subsurface matrix material. Accordingly, calibration of the resistivity model with the lithological layers is essential in such a case to reduce the non-uniqueness problem. During the processing of the 1D VES, it was essential to constrain the depths relative to the corresponding lithological boundaries of the two wells, whereas the resistivity values remain free during the inversion process. After some iterations, it was possible to achieve a good consistency between the resistivity model and the well lithological contacts explaining the changes in subsurface succession. The resistivity models in the two wells show a high resistivity at the shallow part due to the arid and dry condition of the alluvium layers. The thickness of this layer is less than 4 m. Then, the resistivity values decrease in both VES at the front of limestone layer in well (Qap2-1) and clayey sandstone, sandy limestone and limestone succession in well (Qap1-1). Then, resistivity increases sharply against the basement complex in well (Qap2-1), while the

maximum depth of penetration does not reach the basement in well (Qap1-1). Small changes in resistivity values in the second layer are referring to the variable amount of clayey and sand content in the limestone layer. The static water level has been marked in both wells and the saturated limestone layer is located at depth about 42 m in well (Qap2-1) and 118 m in well (Qap1-1) (Figure 3).

It is not possible to constrain the depth of each resistivity layers all over the surveyed site on the basis of borehole logs. It is, however, possible to derive trial-and-error models that match well VES data sets based on the behavior of the two models, concident with the two wells. The 1D models for all VES aquired along the three profiles have been ploted in Figure 4. Most of these models follow up the same characters obtained near the available well (Ministry of Water and Electricity, Saudi Arabia, Personal Comunication, 2011). However, there are some minor changes due to the different content of calyey matrials in limestone layer and the location of each VES relative to the basment of the Arabian scheild (Figure 2). The modeled layer for the VES along profile (B) are more systematic in comparison to other profiles. The corresponding equivalnce for each model have been ploted and only the minimum and maximum equivalences have been presented for simplicity (Figure 4). The

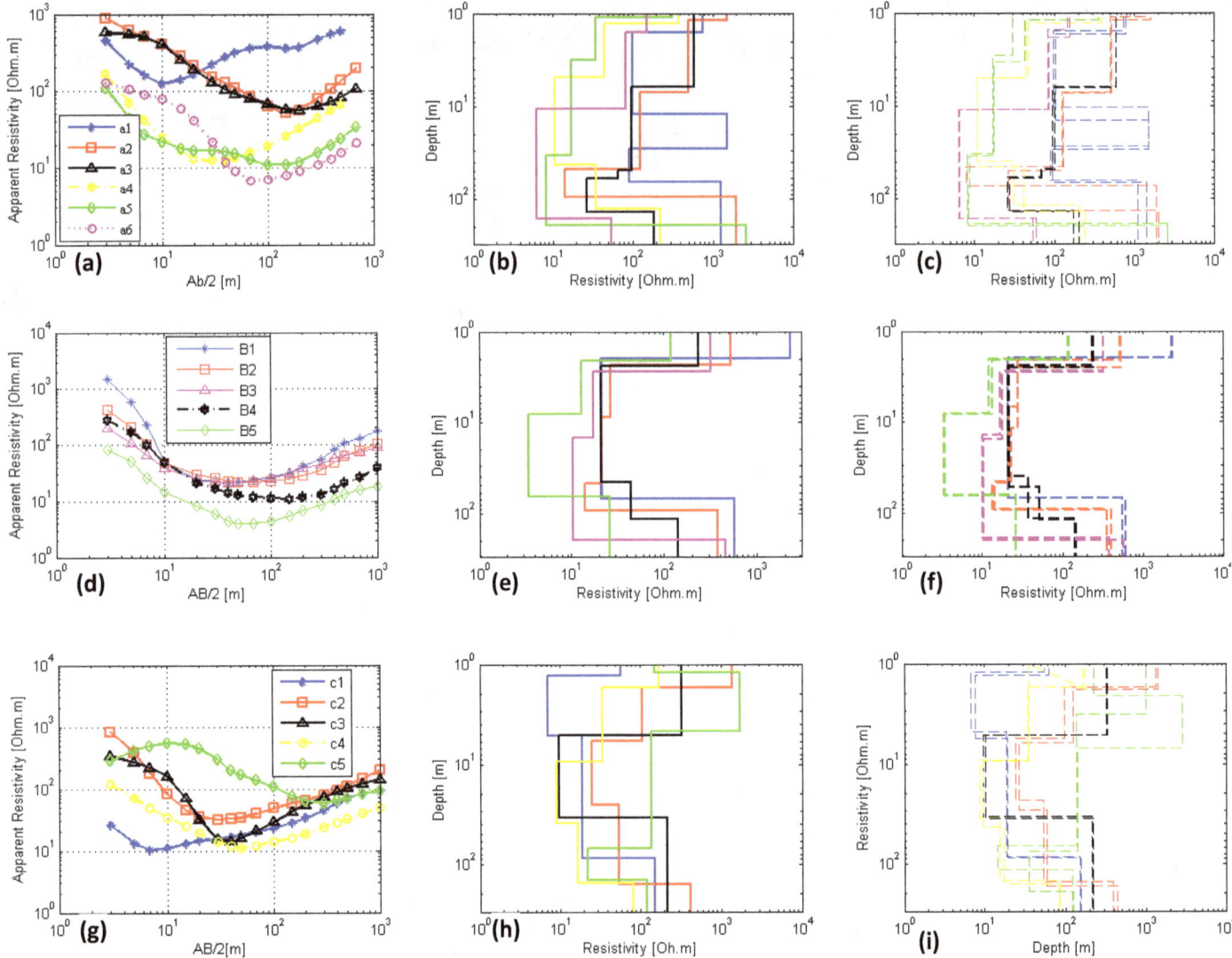

Figure 4. The acquired resistivity curves and the corresponding layered models with the equivalent mdoels along the study area; (a, b and c) at profile a; (d, e and f) at profile b; (g, h and i) at profile c.

equivalence models are consistent with the layer model with small deviation refering to the uncertianty of the 1D inversion process.

Processing of VES in 2D model

Since most of the VES data points were recorded along quasi-linear profiles, it is possible to process the acquired data in form of 2D subsurface models (Sasaki, 1989; Uchida, 1991). Applying the Uchida's (1991) algorithm, 2D resistivity models were derived for the available VES profiles. The algorithm is based on the Akaike Bayesian Information Criterion (ABIC) and utilize finite element calculation mesh (Sasaki, 1981; Tripp et al., 1984; Shima, 1990). The subsurface medium was represented by numerous rectangular blocks shown by the 2D mesh

(Figure 5). The numbers of blocks in the vertical and horizontal directions are functions of the number of VES sites and the extension of AB electrodes. Resistivity values were assigned to each block of the mesh individually. The model is obtained under the assumption that the data error and roughness (spatial derivatives of the parameters) are normally distributed with zero mean. The optimum smoothness is also obtained in the process of the likelihood maximization. For the least-squares inversion with smoothness regularization, we seek a model, which minimizes both the data misfit and model roughness.

In the first step of an inversion, Uchida's (1991) algorithm, outputs seven models based on different inversion parameters. The best of these models, according to the root-mean-square (RMS) misfit values and the smoothing factors, is then used as the initial model for

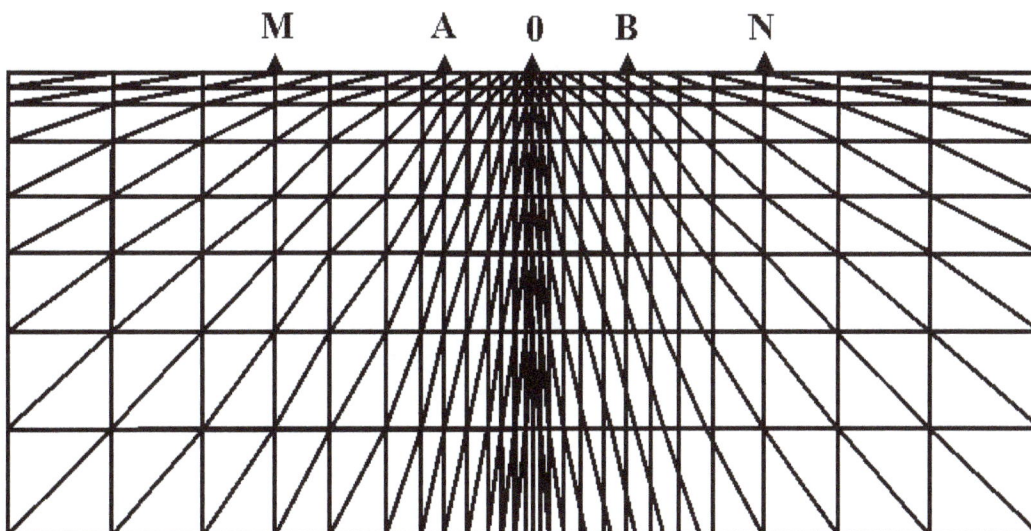

Figure 5. Sketch of the finite-element mesh used in the 2D inversion process. A and B are the current electrodes; M and N are the potential electrodes; O is the centre of the Schlumberger array.

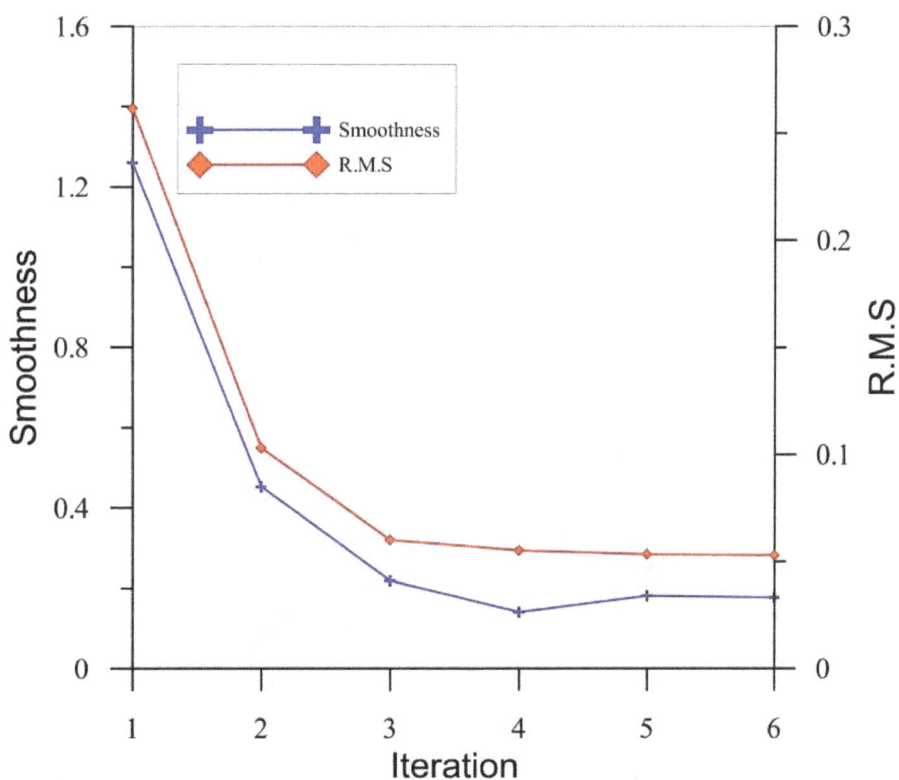

Figure 6. The inversion parameters (Smoothness and RMS) for profile (a).

the next iteration. Subsequently, the algorithm outputs seven new models. The process is repeated until the RMS misfits and smoothing parameters reach relatively stable minimum values. A typical example of the rate of convergence achieved by this procedure is presented in Figure 6. For profile (A), this figure shows that improvements in the RMS misfits and smoothnesses do not occur after the third iteration. For all VES profiles, convergence between the observed and predicted data was obtained by the third and fourth iteration.

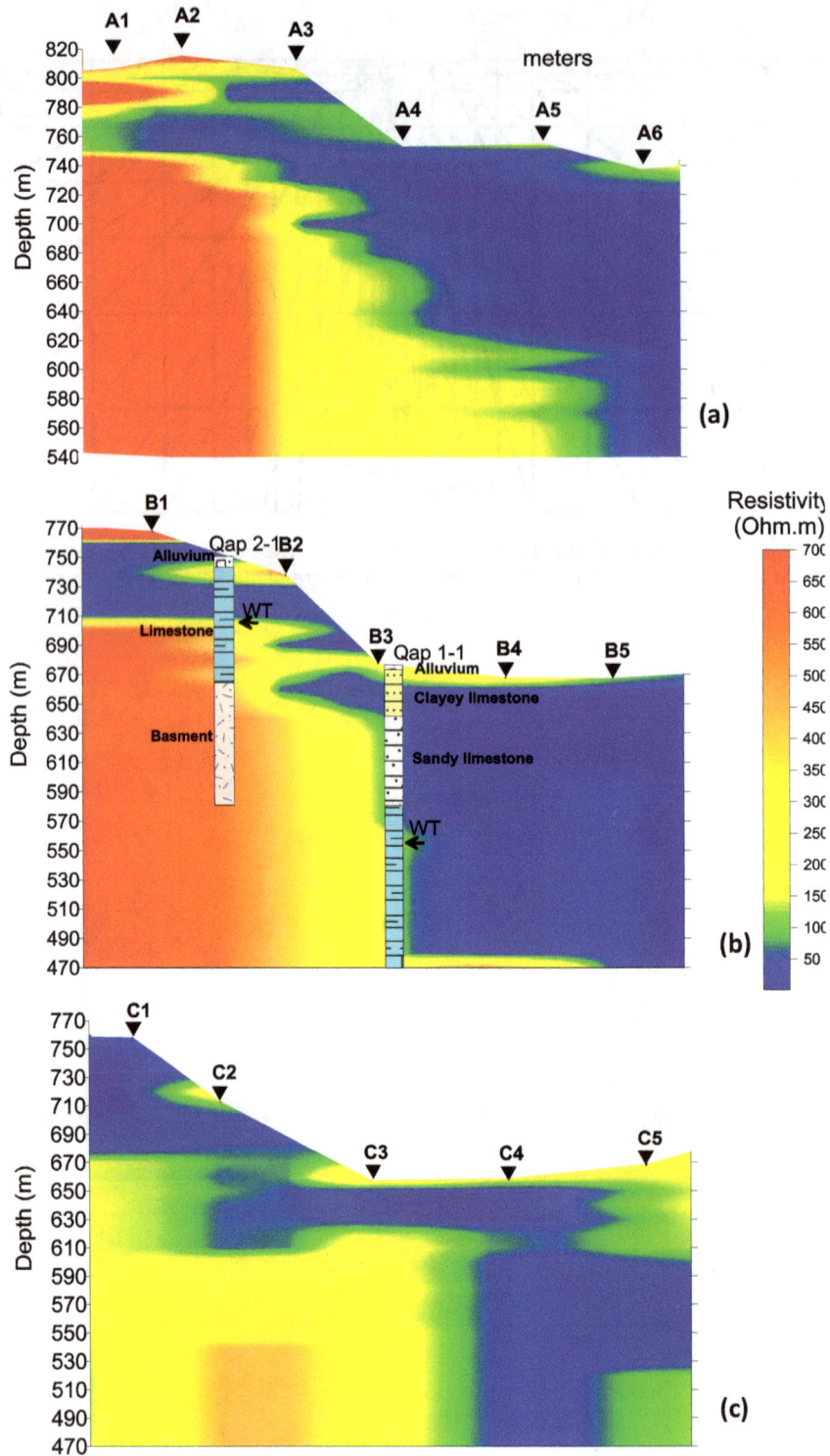

Figure 7. Geoelectrical cross sections deduced from 1D inversion models along profiles a, b and c.

INTERPRETATION

Based on the resultant 1D model of VES data, the electrical resistivity data are set of multi-layered models; each of them fit the observed field curve and describes the electrical properties of the subsurface medium. These models have been used to prepare 2D view of the electrical resistivity variation (Figure 7). Powerful geostatistical

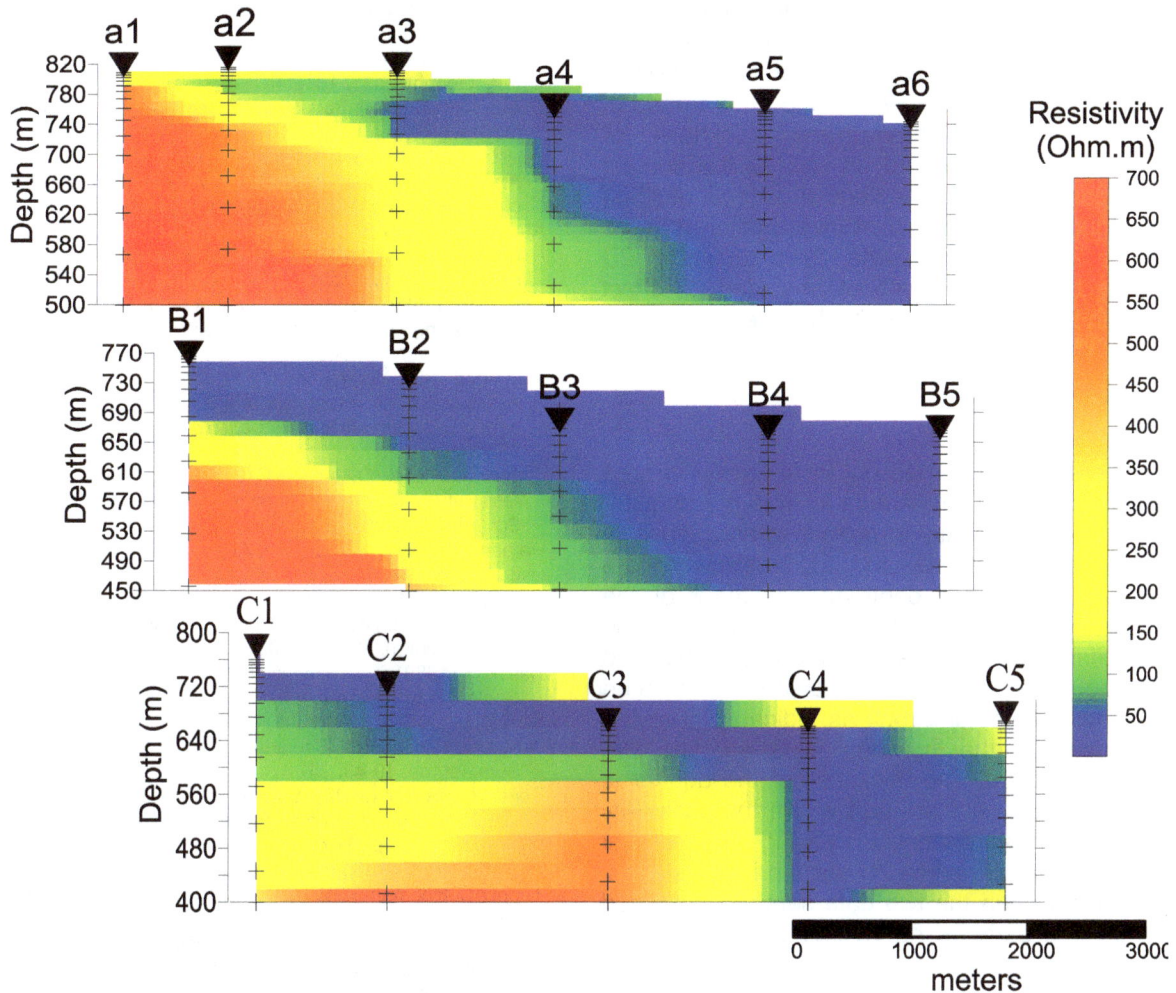

Figure 8. 2D Geoelectrical cross sections deduced from 2D inversion.

gridding (kriging) method has been applied to smoothly interpolate the inverted data. Linear color scale has been used to visualize the limited resistivity range (1 to 700 Ω/m). Generally, there are many features that can be interpreted from the resistivity cross sections. Firstly, underneath the most topographically elevated area in the south western side, the resistivity values are relatively high (> 500 ohm.m) and represent the occurrence of basement complex at shallow depth (Figure 7). This fact has been confirmed by the correlation between the well (Qap2-1) and the coincident VES (B2) (Figure 3a). Overlay the basement rocks, sedimentary cover exist with resistivity lower (<70 ohm.m) and variable thickness depends on the topography of the basement complex. Going to the eastern side, the topography is slightly flat and the resistivity values have relatively low values. This is due to the occurrence of limestone of Khuff formation, which has variable contents of clayey and sandy materials (Figure 7). The clay contents in the limestone rocks are increasing to the eastern direction as indicated

by the lower values of resistivity. It is noticeable that the boundary of basement can be identified in both resistivity sections (A and B), whereas it is located deeper in the southern west in profile (C). The static water table is coincident with the top boundary of limestone unit, giving an indication that the subsurface aquifer is increasing in depth and thickness eastward. However, there are small potentiality of groundwater underneath the elevated stations in the south western side due to the small thickness of the limestone unit and also, due to the moving of groundwater eastward with dipping of basement.

The 2D inversion results for the data sets have been presented in the form of 2D geoelectrical cross sections (Figure 8). The data which gave the lower RMS values have been presented and compared with the equivalent 1D sections based on the 1D VES inversions. Although, there are matching between the 2D and 1D inversion schemes in the general trends of increasing and decreasing resistivity values, some details can be observed only in the 1D inversion results particularly in the shallow parts of the sections. This is due to the difference of the inversion schemes where the 2D inversion scheme is concentrated on the general trend of the resistivity variations along the profile. However, the main trends of locating the basement trends and thickness of sedimentary cover are resolved as well as the contact between the basement and limestone sediments of Khuff formation (Figure 8). The depth of investigation is more deeper in the 2D inversion results. Such results of 2D inversion can be achieved without any kind of direct constrains with the boreholes like the 1D inversion.

Conclusion

Resistivity surveys were carried out in Al Quwy'yia area at the centeral part of Saudi Arabia to map the acquifer and groundwater potentuality in promising area close to Riyadh. The results of 1D and 2D resistivity data interpretation indicated the depth to basement at the south western part of the study area as well as the contact boundary between the basement complex and sedimentary rocks. The static water table is found to be matching with the limestone rock as indicated from the comparison between the individual VES and the two wells in the study area. The thickness of the aquifer is increasing in the north eastern part where the possibility of the groundwater potentiality is increasing. The result of 1D data set were confirmed by 2D section interpretation. The difference between the two data sets are observed only in the shallow parts as the 1D inversion is collected in dense manner and conistrained with the available geological information. Collecting more data sets using deep VES and time domain electromagnetic are recommended for more analysis of the acquifer in the eastern part of the study area.

ACKNOWLEDGEMENT

This work was supported financially by the National Plan for Science, Technology and Inovation (NPST) program, King Saud University, Saudi Arabia (Projact No. 09-ENV836-02).

REFERENCES

Al-Amri A (1996). The application of geoelectrical surveys in delineating groundwater in semiarid terrain, Case history from central Arabian Shield. M.E.R.C. Ain Shams Univ., Earth Sci. Servey, 10: 41-52.

Alile OM, Ujuanbi O, Evbuomwan IA (2011). Geoelectric investigation of groundwater in Obaretin -Iyanomon locality, Edo state, Nigeria. J. G. Mining Res., 3(1): 13-20.

Asfahani J (2007). Geoelectrical investigation for characterizing the hydrogeological conditions in semi-arid region in Khanasser valley, Syria. J.A. Environ., 68: 31-52.

Chandra S, Dewandel B, Dutta S, Ahmed S (2010). Geophysical model of geological discontinuities in a granitic aquifer: Analyzing small scale variability of electrical resistivity for groundwater occurrences, J. A. Geophys., 71: 137-148.

Dobrin MB (1988). Introduction to Geophysical Prospecting. New York: McGraw-Hill, p. 867.

Hoetzel H (1995). Groundwater recharge in an arid karst area (Saudi Arabia), Application of Tracers in Arid Zone Hydrology (Proceedings of the Vienna Symposium, August 1994). IAHS Publ. p. 232.

Kumar D, Ahmed S, Krishnamurthy NS, Dewandel B (2007). Reducing ambiguities in vertical electrical sounding interpretations: A geostatistical application. J. A. Geophys., 62: 16-32.

Loke MH (1999). Electrical imaging surveys for environmental and engineering studies. Apractical guide to 2D and 3D surveys: Austin, Texas, Advanced Geosciences Inc., p. 57.

Mohamed NE, Yaramanci U, Kheiralla KM, Abdelgalil MY (2011). Assessment of integrated electrical resistivity data on complex aquifer structuresin NE Nuba Mountains - Sudan. J.A.E. Sci., 60: 337-345.

Ozcep F, Tezel O, Asci M (2009). Correlation between electrical resistivity and soil-water content: Istanbul and Golcuk. Int. J. Phys. Sci., (4)6: 362-365.

Parasnis D (1997). Principle of Applied Geophysics, London: Chapman & Hall., p. 275

Sasaki Y (1981). Automatic interpretation of resistivity sounding data over two-dimensional structures, Geophysical Exploration of Japan (ButsuriTanko).34: 341-350.

Sasaki Y (1989). Two-dimensional joint inversion of magnetotelluric and dipole-dipole resistivity data. Geophys., 54: 254-262.

Senalp M, Al-Duaiji A (2001). Sequence stratigraphy of the Unayzah reservoir in central Saudi Arabia, Saudi ARAMCO of Technology, Summer, pp. 19-43.

SenosMatias MJ (2002). Square array anisotropy measurements and resistivity sounding interpretation. J. A. Geophys., 49: 185-194.

Shima H (1990). Two dimensional automatic resistivity inversion technique using alpha centres, Geophys., 55: 682-694.

Tripp AC Hohmann GW, Swift CM (1984).Two-dimensional resistivity inversion, Geophys., 49: 1708-1717.

Uchida T (1991). Two resistivity inversion for Schlumberger sounding, Geophs. Explor. of Japan (ButsuriTansa), 44: 1-17.

Yadav GS Singh SK (2007). Integrated resistivity surveys fordelineation of fractures for ground water exploration in hardrock area. J. A. Geophys., 62: 301-312.

Yilmaz S (2011). A case study of the application of electrical resistivity imaging for investigation of a landslide along highway. Int. J. Phys. Sci., 6(24): 5843-5849.

Zaidi KF, Kassem OM (2010). Use of electrical resistivity tomography in delineating zones of groundwater potential in arid regions: a case study from Diriyah region of Saudi Arabia, Arab J. Geosci., DOI: 10.1007/s12517-010-0165-7.

5

Geoelectrical studies for the delineation of potential groundwater zones at Oduma in Enugu State, Southeastern Nigeria

Austin, C. OKONKWO and Isaac, I. UJAM

Department of Geology and Mining, Enugu State University of Science and Technology, Enugu, Nigeria.

This work evaluates the use of geoelectrical method in the delineation of potential groundwater zones at Oduma in Enugu state, Southeastern Nigeria. Oduma lies within latitudes 6°02$'$ N to 6°07$'$ N and longitudes 7°35$'$ E to 7°41$'$ E with an area extent of about 102.6 km^2. The area is underlain by Awgu Shale group with its lateral arenaceous facie; Owelli Sandstone outcropping south of Oduma. Thirteen (13) vertical electrical soundings (VES) were carried out within the study area. Interpretated VES data shows predominance of Q and H curve type, indicating a fracture-shale subsurface. Contour maps of iso-resistivity, depth, transverse resistance, longitudinal conductance, aquifer transmissivity and hydraulic conductivity were constructed. Computed aquifer transmissivity from VES data values indicates a low yield aquifer. The latter was used to delineate the potential groundwater zones based on Gheorge aquifer transmissivity classifications. Comparisons of aquifer hydraulics estimated from geoelectrical sounding and the pump test analysis indicates a fairly good match. Three potential groundwater zones were delineated; the low, very low and negligible potential zones. The various contour maps and potential groundwater zone map will serve as a useful guide for groundwater exploration in the study area.

Key words: Resistivity, transmissivity, hydraulic conductivity, groundwater zones, aquifer yield.

INTRODUCTION

Oduma lies within latitude 6°02$'$ N to 6°07$'$ N and longitudes 7°35$'$ E to 7°41$'$ E with an area extent of about 102.6 km^2. It is located in Aninri local government area, Enugu state, southeastern Nigeria (Figure 1). The increasing population within Oduma and neigbouring towns has necessitated the high demand of groundwater development in the area. Cases of abortive water wells have been reported within Oduma and environs. Knowledge of groundwater zone is essential for a robust groundwater development program in the area. The natural flow of water through an aquifer is determined from the hydraulic properties of the aquifer. Hydraulic conductivity (k), transmissivity (T) and storativity (S) are

the aquifer properties. Transmissivity is the hydraulic conductivity multiplied by the saturated thickness of the aquifer. Predictions of these hydraulic properties are mainly from pumping test analysis. Now geophysical methods provide an effective technique for aquifer evaluation. Estimates of hydraulic properties from geoelectrical soundings have been made by several authors (Ezeh and Ugwu, 2010; Kelly, 1979; Urish, 1987). These parameters were estimated using empirical and semi-empirical relationships (Huntly, 1987; Koinski, 1981). Their study was aimed at characterizing the aquifers for optimum yield.

In the present study, an attempt has been made by

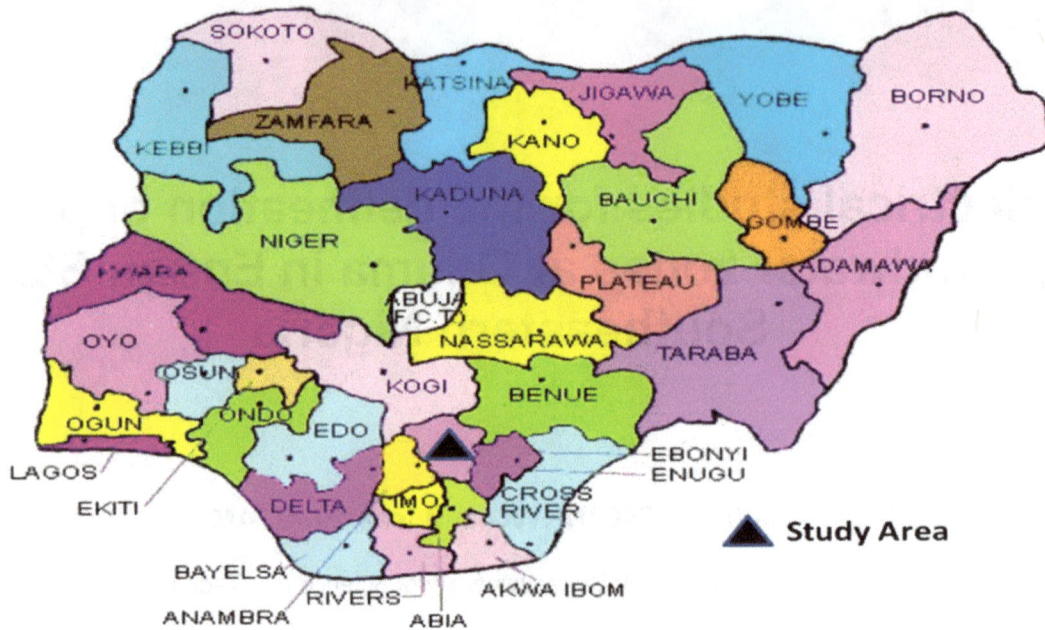

Figure 1. Map of Nigeria showing the study area (World Gazette, 2011).

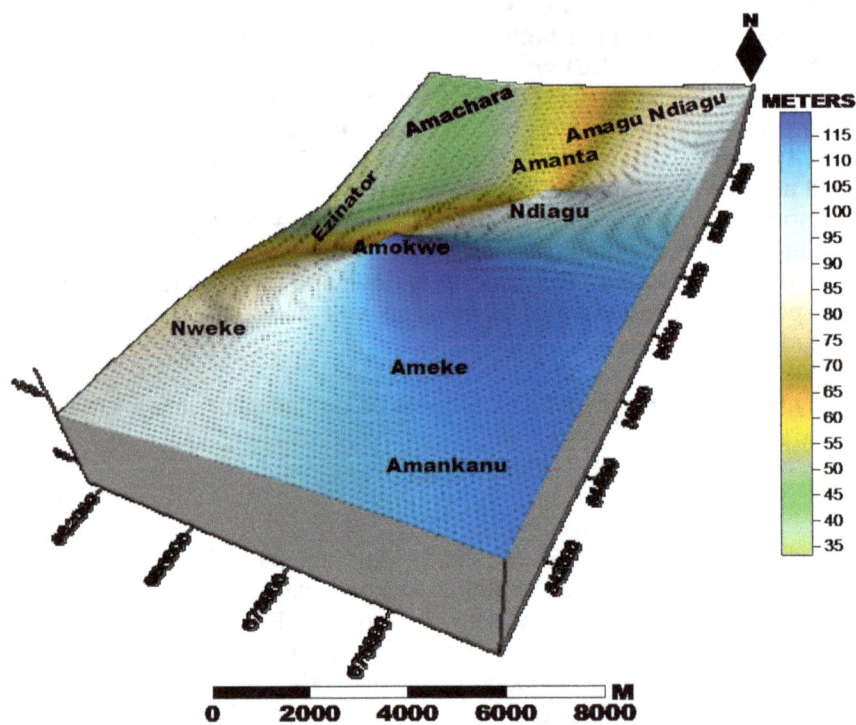

Figure 2. Surface map of the study area.

using the distribution of these hydraulic parameters to delineate the groundwater potential zone at Oduma and environs. Modeled estimates of the hydraulic parameters from geoelectrics were compared with data from pumping test to have a better picture of groundwater potential zone.

Study area

Physiography

The study area is fairly a lowland topography (Figure 2). Amokwe community is about the highest in the area, with

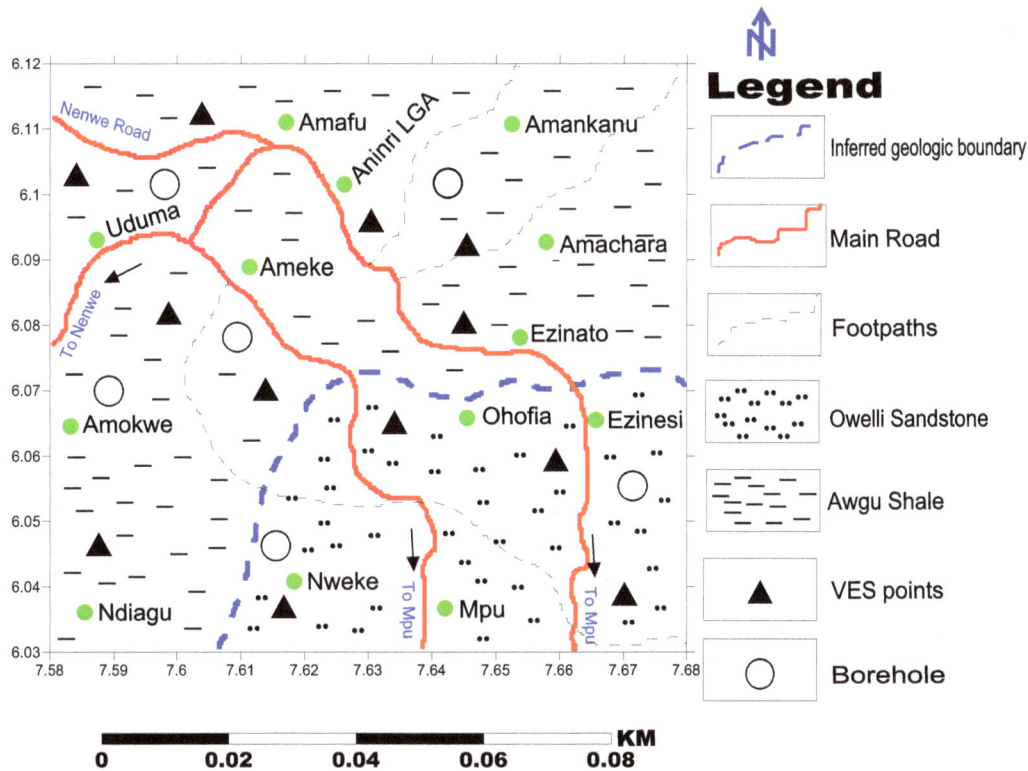

Figure 3. Geologic map of the study area showing VES stations and borehole points.

an elevation of 120 m above sea level (ASL). The lowland is indicative of the cultural land use system in the area. As the area is predominately rice farm terrain as a result of stagnant water, the flat topography is also controlled by the subsurface geology of the area.

Geology

The study area falls within the geologic complex called, the Lower Benue Trough. It is underlain by Awgu Shale unit which is coniacian in age, with an arenaceous facies (Owelli Sandstone) development to the south of Oduma (Figure 3). The unit consists of bluish grey, well bedded shales with occasional intercalations of fine-grained, pale yellow, calcareous sandstones and shaly limestones (Reyment, 1965). It is about 900 m thick and gently folded.

Hydrogeology

The study area falls within the Cross River Basin, which is hydrogeologically a problematic groundwater basin (Offordile, 2002). This is as a result of poor yield and saliferous groundwater. More than 90% of the basin is underlain by cretaceous rocks of the Asu River, Ezeaku, Awgu, Nkporo and Mamu Formations, with the oldest, the

Asu River Formation, underlain by the basement complex rocks. With the exception of Awgu and Ezeaku formation, all these rock units are very poor aquifers. The sandstones within the Awgu formation are thin and generally limited in extent and as a result, give poor yields. Aneke (2007) proposed an exploration strategy for exploiting the groundwater from the fractured shaley units which are the main water bearing units in the study area.

THEORY AND METHODS

The electrical resistivity method is utilized in diverse ways for groundwater exploration (Zohdy, 1976; Choudhury et al, 2001; Frohlich and Urish, 2002). Electrical surveys are usually designed to measure the electrical resistivity of subsurface materials by making measurements at the earth surface. Current is introduced into the ground by a pair of electrodes, while measuring the subsurface expression of the resulting potential field with an additional pair of electrodes at appropriate spacing.

Data acquisition and interpretation

A total of thirteen vertical electrical sounding (VES) was acquired within and outside the study area (Figure 2). Some were stationed very close to existing boreholes, for correlation purposes. The Schlumberger electrode configuration was used with maximum current electrode separation ranging from 400 to 600 m. The equipment used for the fieldwork was the versatile Ohmega resistivity meter.

Table 1. Interpreted model geoelectric parameters and curve types from the study area.

S/N	Location	VES No	NL	ρ_1	ρ_2	ρ_3	ρ_4	ρ_5	ρ_6	ρ_7	ρ_8	T_1	T_2	T_3	T_4	T_5	T_6	T_7	Curve type
1	Nkwo Amorji	1	7	5	2	2	6	10	10	38	-	0.6	1.2	1.2	2.0	7.0	42.0	-	HA
2	Nawu Ezinesi	2	7	50	35	15	6	5	8	125	-	0.8	0.4	1.8	12.0	15.0	28.0	-	QQA
3	Ezinator	3	7	15	5	3	5	8	15	80	-	0.8	1.4	2.8	18.0	17.0	18.0	-	QAA
4	Nweke	4	7	380	220	100	20	5	105	1152	-	0.8	1.2	2.5	10.5	25.0	20.0	-	QQA
5	Ameke	5	7	1205	2785	12	14	23	4	35	-	0.8	1.1	2.6	20.5	31.0	69.0	-	KAH
6	Amaorji	6	7	280	225	18	9	7	6	18	-	0.8	1.7	2.0	5.5	32.0	93	-	QQH
7	Amankanu	7	7	750	520	40	18	4	2	85	-	1.0	1.5	5.0	14.5	43.0	63.0	-	QQH
8	Enugu Agu	8	7	780	120	18	25	24	11	55	-	0.8	1.7	4.3	18.2	31.0	64.0	-	QKH
9	Ndiagu	9	7	1002	3452	275	13	23	90	20	-	0.8	1.2	5.5	18.5	30.0	69.0	-	KHK
10	Amokwe	10	6	165	12	6	10	2	11	-	-	0.8	1.7	2.5	40.0	19.0	-	-	QK
11	Amanta	11	6	46	13	8	14	21	8	-	-	0.8	1.2	4.0	14.0	40.0	-	-	QA
12	Amachara	12	7	14	9	13	10	14	20	8	-	0.8	1.7	1.5	11.0	25.0	45.0	-	HHK
13	Amagu Ndiagu	13	5	105	75	22	2	250	-	-	-	0.8	1.4	7.8	15.0	-	-	-	QH

After acquiring the data, the measured field resistance (R) in ohms was converted to apparent resistivity (ρ_a) in ohm-meter by multiplying resistance (R) by the geometric factor (K). A log-log graph plot of apparent resistivity (ρ_a) against current electrode distance (AB/2) was plotted for each VES station to generate a sounding curve. Using the conventional partial curve matching technique, in conjunction with auxiliary point diagrams (Orellana and Mooney, 1966; Koefoed, 1979; Kellar and Frischknecht, 1966), layer resistivities and thickness were obtained, which served as a starting point for computer-assisted interpretation. The computer program RESOUND was used to interpret all the data sets obtained. From the interpretation of the resistivity data, it has been possible to compute for every VES station, the longitudinal conductance(S).

$$S = h_i / \rho_i \qquad (1)$$

And transverse resistance(R)

$$R = h_i . \rho_i \qquad (2)$$

Where h_i and ρ_i are thickness and resistivity of the aquiferous layer. These parameters R and S are known as the Dar-zarrouk variable and Dar-zarrouk function, respectively (Maillet, 1947). Further quantitative analysis for aquifer hydraulics in the study area are based on Equations 1 and 2 using analytical relationship of Niwas and Singhal (1981). They showed that: in areas of similar geologic setting and water quality, the product kσ (hydraulic conductivity) remain fairly constant.

Interpretation

The form of curves obtained by sounding over a horizontally stratified medium is a function of the resistivities and thicknesses of the layers as well as the electrode configuration (Zohdy, 1976). The resistivity curve type associated with the study area from VES 1-13 include: HA, QQA, QAA, QQA, KAH, QQH, QKH, KHK, QK, QA, HHK, and QH curve types respectively (Table 1). The first dominant curve type is Q. This is indicating a shaly terrain. The H curve type is the second dominant. This also indicates fractured shale horizons which are targets for groundwater exploration.

RESULTS AND DISCUSSION

Geoelectrical sounding

Contour maps of the apparent resistivity, the isopach, the depth, the longitudinal conductance, the transverse resistance, the transmissivity and the hydraulic conductivity of the aquiferous horizon have been constructed using the results of the resistivity sounding interpretation. Apparent resistivity variation (Figure 4) indicates a high resistivity to the southeast and southwest with low resistivity to the north, around Amokwe and Nawu Ezinesi. Aquifer depth variation is a function of topography. A NW-SE trend variation predominates (Figure 5). The isopach map also show similar trend (Figure 6). The distribution of the aquifer transverse resistance and longitudinal conductance computed from the VES interpretation is shown in Figures 7 and 8 respectively. Maximum values of transverse resistance are observed around Ndiagu-Nweke-Amachara axis. Aquifer transmissivity (Figure 9) does not show similar trend, with highest value of 11m^2/day at Amachara, indicating a low permeability aquifer (Ekwe et al, 2010) and very low potential (Ezeh, 2012). The longitudinal conductance shows a thick resistive horizon at Amankanu and also in a NW-SE trend. Hydraulic conductivity computed from VES interpretation (Figure 10) show an aquifer with a poor yield, practically depicting a shaly terrain (Figure 12).

Borehole data

Aquifer parameters from pumping test analysis were also acquired. They are transmissivity (Figure 13), hydraulic conductivity (Figure 14) and aquifer yield (Figure 15). Contour maps for the former were also produced.

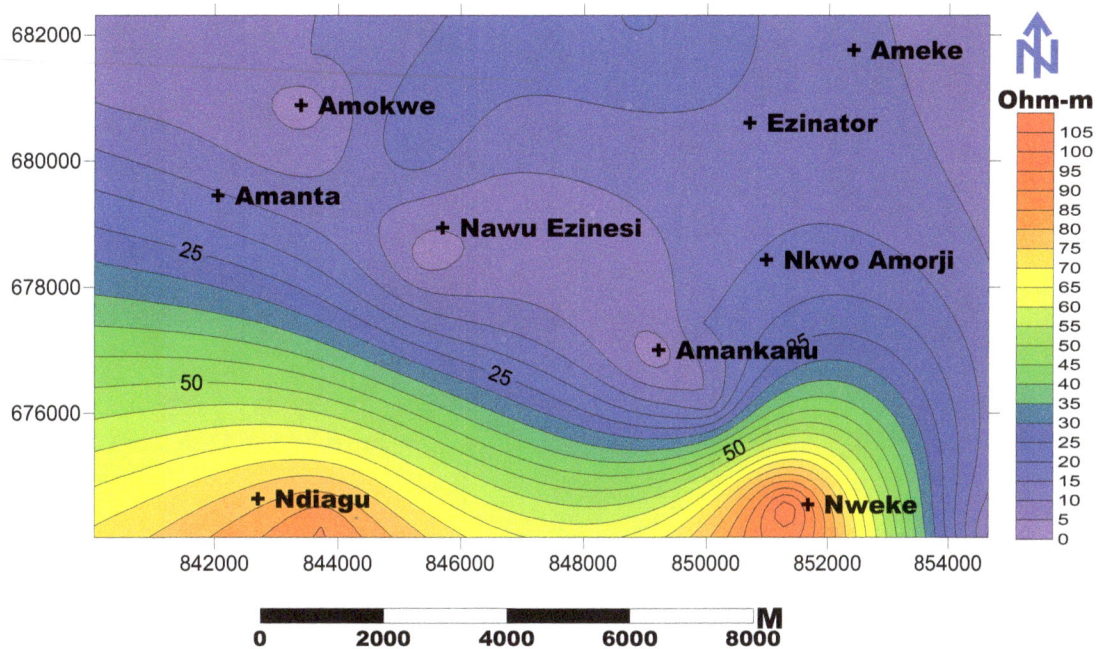

Figure 4. Iso-apparent resistivity map of aquiferous horizon in the study area.

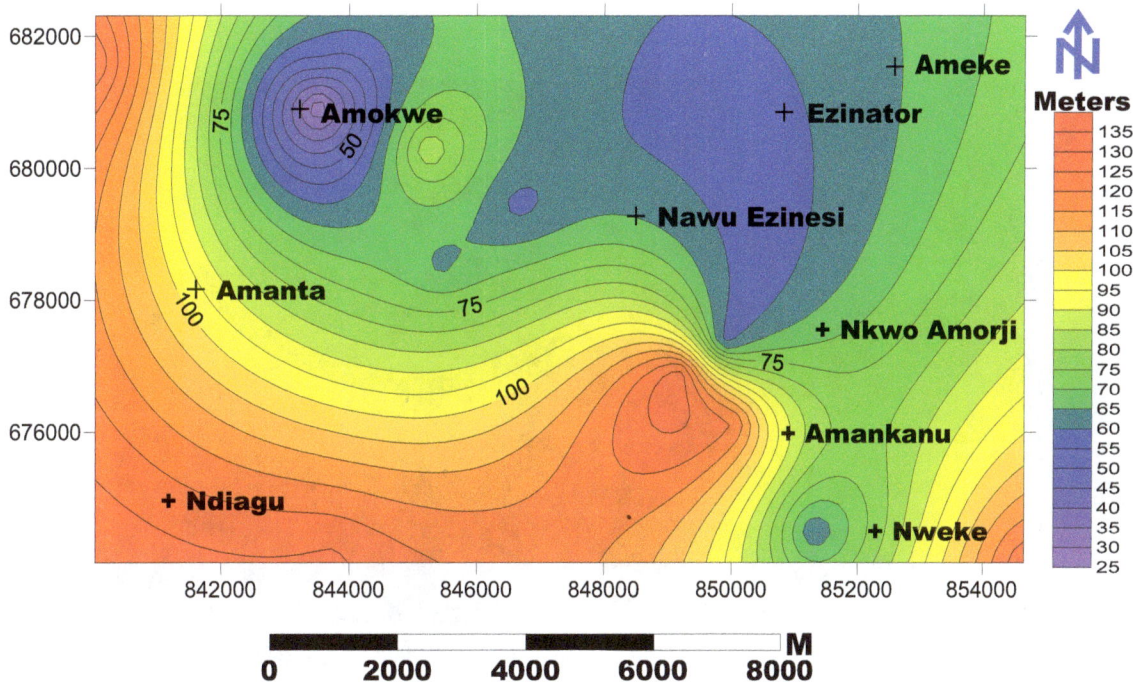

Figure 5. Aquifer depth map of the study area.

Comparisons of aquifer hydraulics estimated from geoelectrical sounding and the pump test analysis indicates a fairly good match. Estimated aquifer transmissivity (Figure 9) around Nweke, Amokwe, Amachara and Ndiagu fairly matches aquifer transmissivity from pump test data. Similarly the estimated hydraulic conductivity from geoelectrical sounding also fairly matches hydraulic conductivity from pumping test data in the study area. The aquifer yield (Figure 15) depicts the true picture of the study area as a

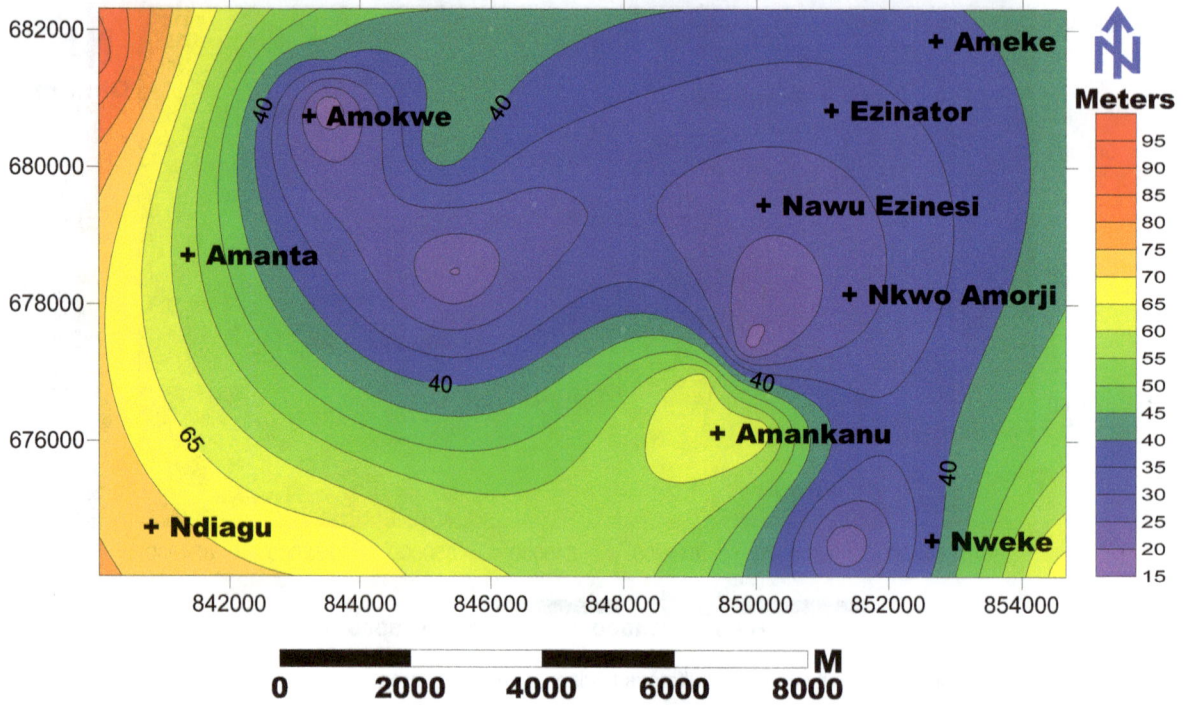

Figure 6. Isopach map of the aquiferous horizon in the study area.

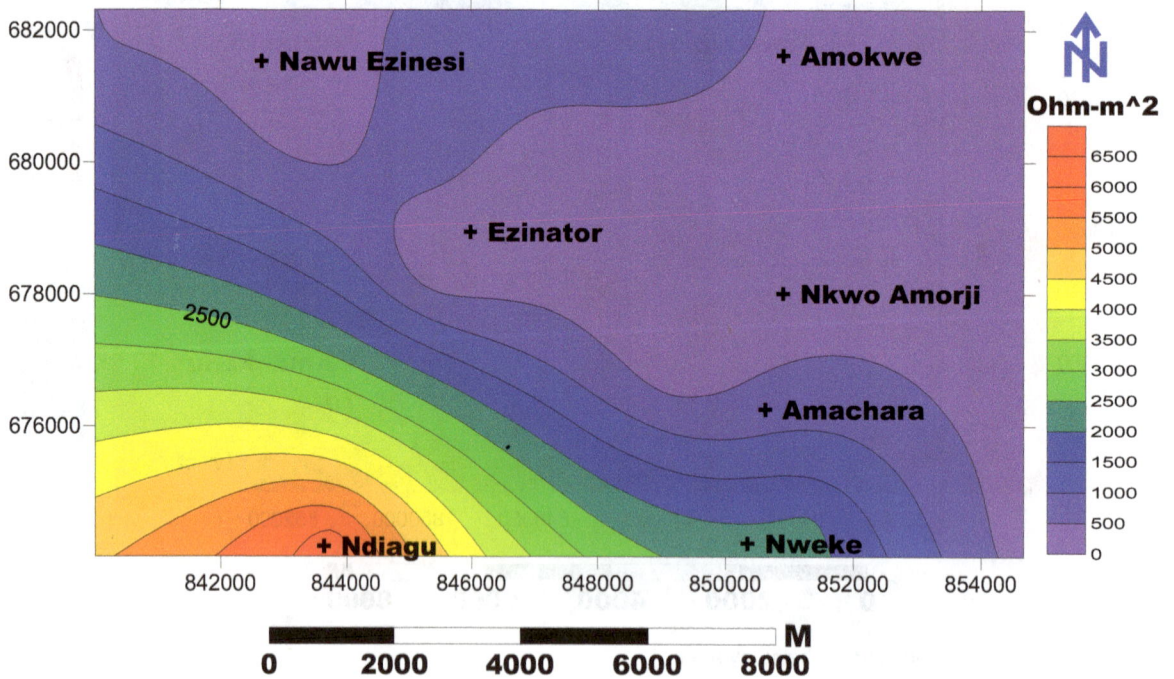

Figure 7. Transverse resistance map of the study area.

low permeability area. The highest aquifer yield in the area is about 6.20 m³/h at southeast corner near Nweke village.

Groundwater potential evaluation

The groundwater potential zones was delineated (Figure

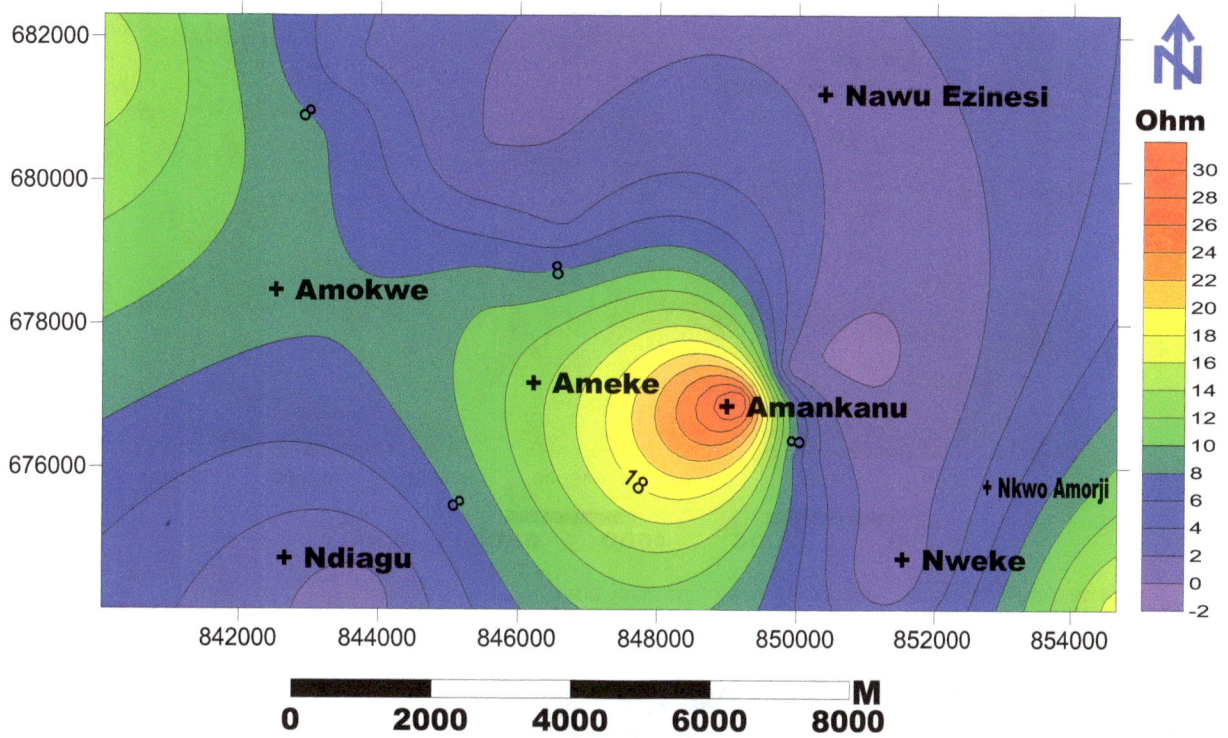

Figure 8. Longitudinal conductance map of the study area.

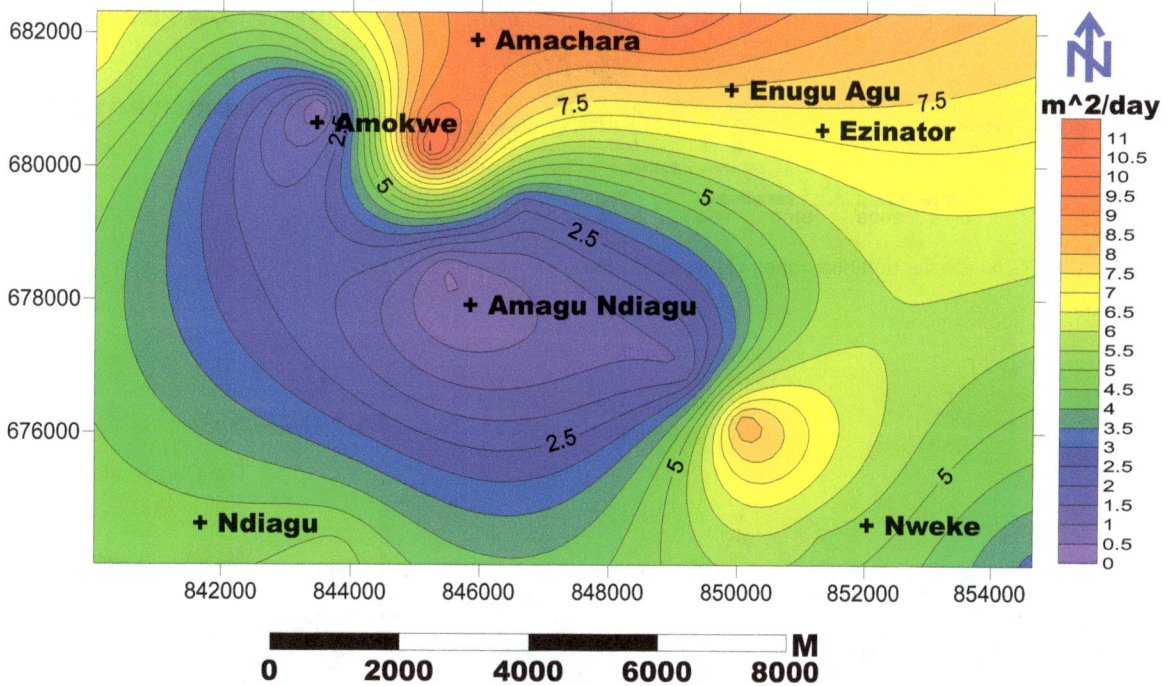

Figure 9. Aquifer transmissivity map of the study area.

11) based on Gheorghe (1978) aquifer transmissivity classifications. Groundwater potential is a function of complex inter-relationship between geology, physiography, groundwater flow pattern, recharge and

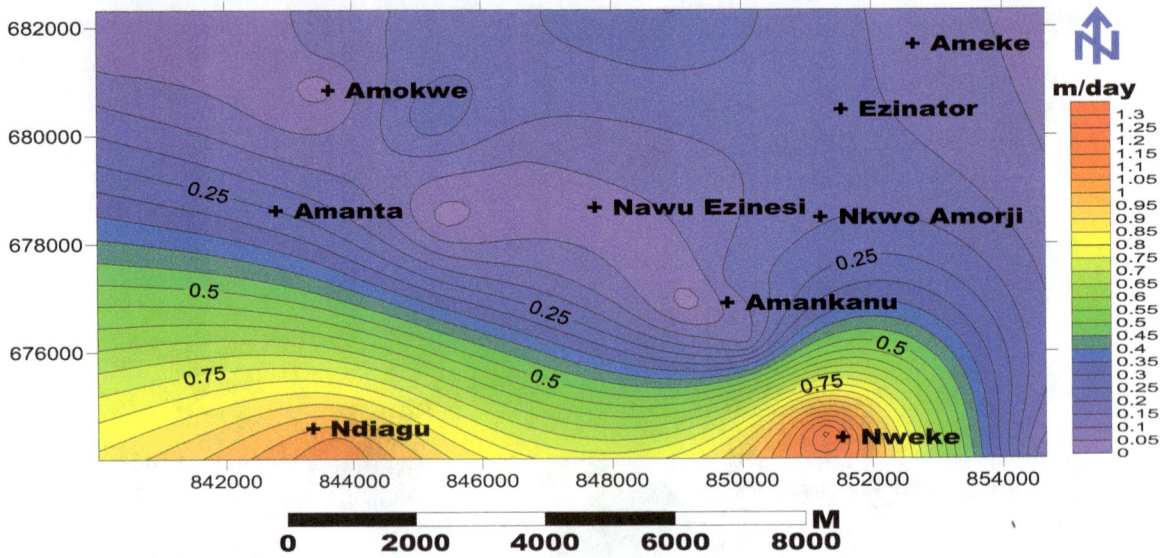

Figure 10. Aquifer hydraulic conductivity map of the study area.

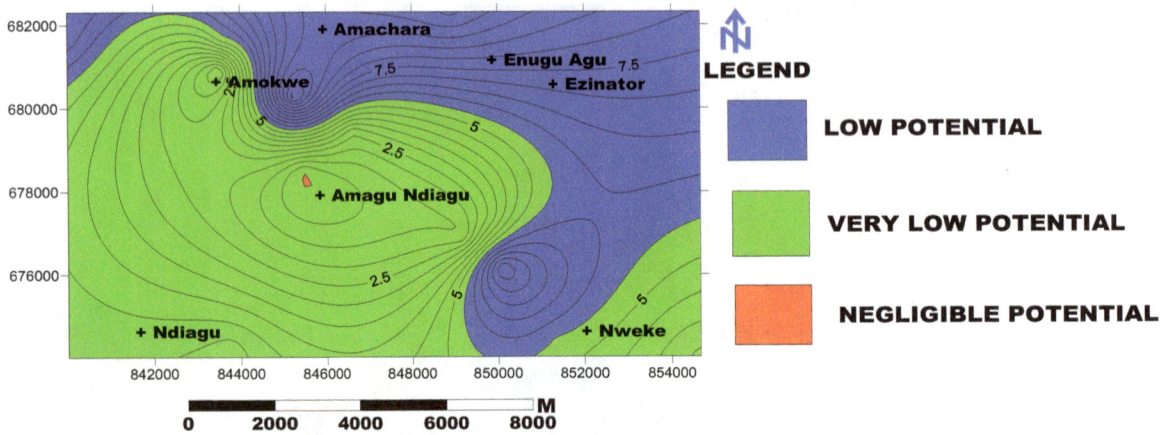

Figure 11. Groundwater potential zones of the study area.

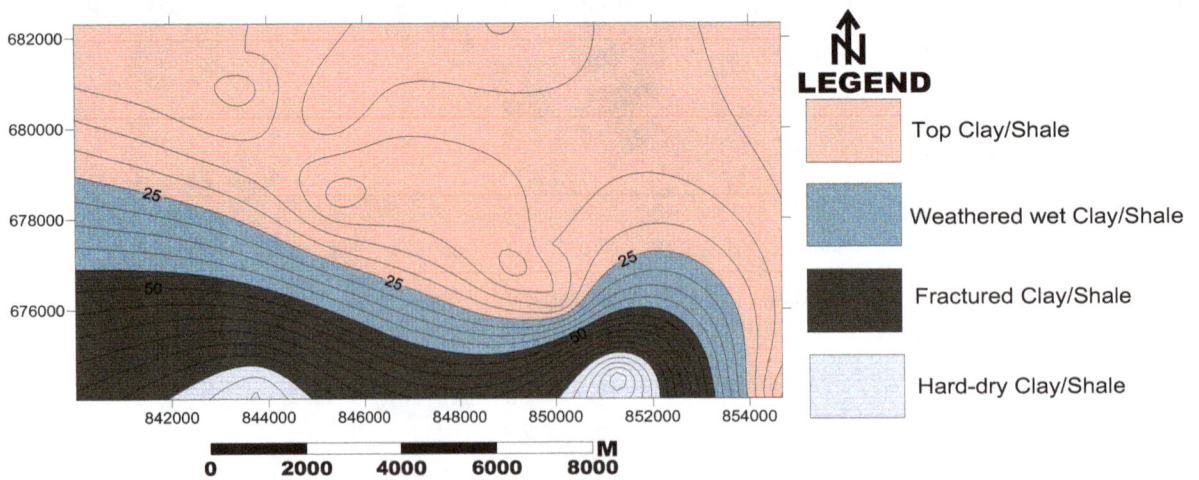

Figure 12. Possible geoelectric layer distribution.

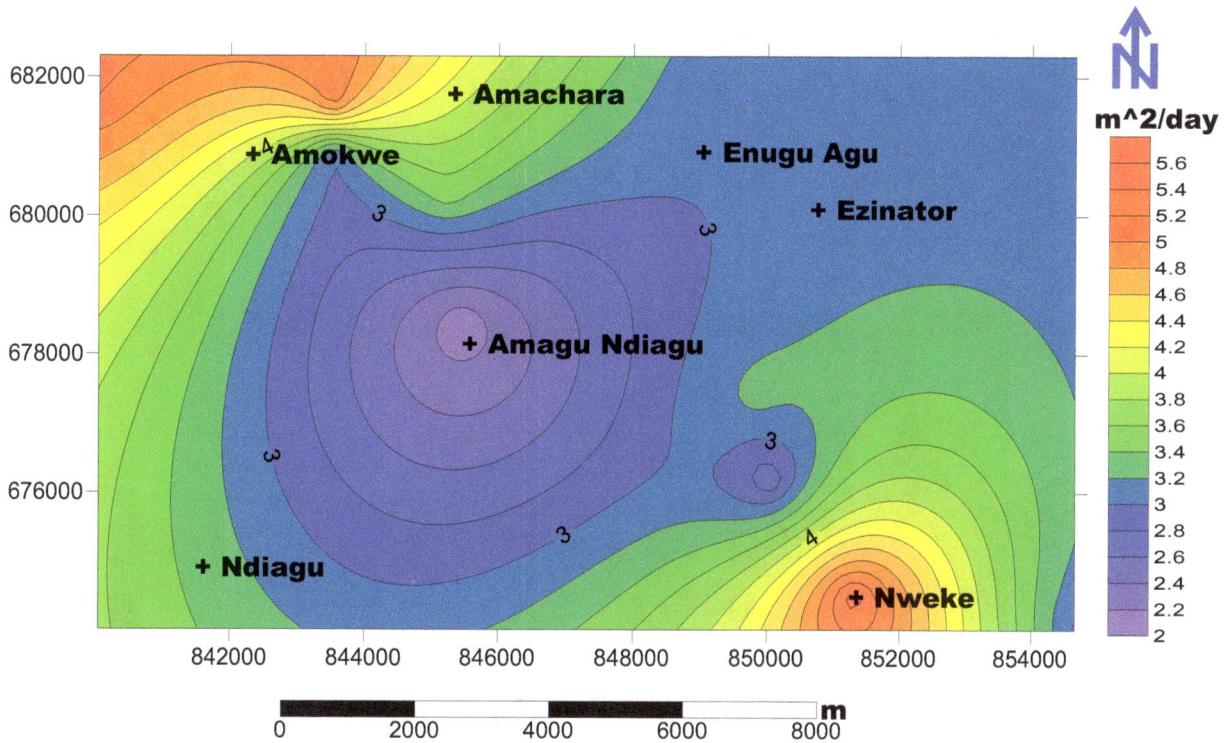

Figure 13. Aquifer transmissivity map (from pumping test analysis) of the study area.

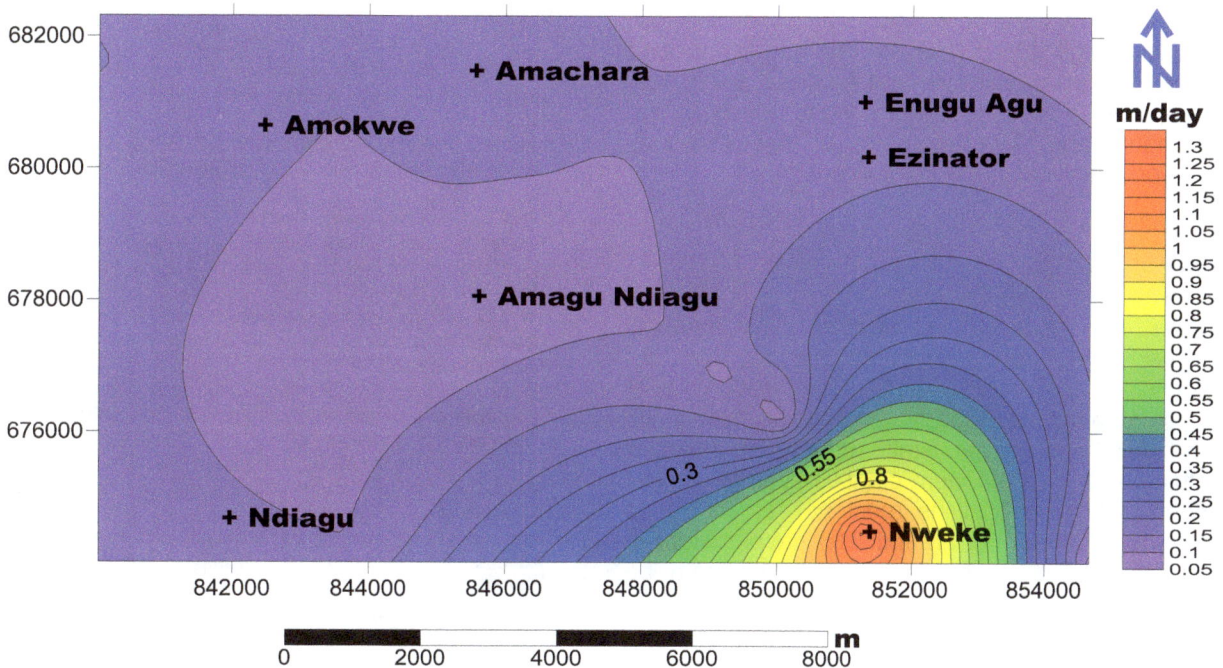

Figure 14. Aquifer hydraulic conductivity map (from pumping test analysis) of the study area.

discharge processes (Ezeh, 2012). The present evaluation of the groundwater potential of the study area has been based on aquifer geoelectrical parameters obtained from VES interpretation results. Three potential

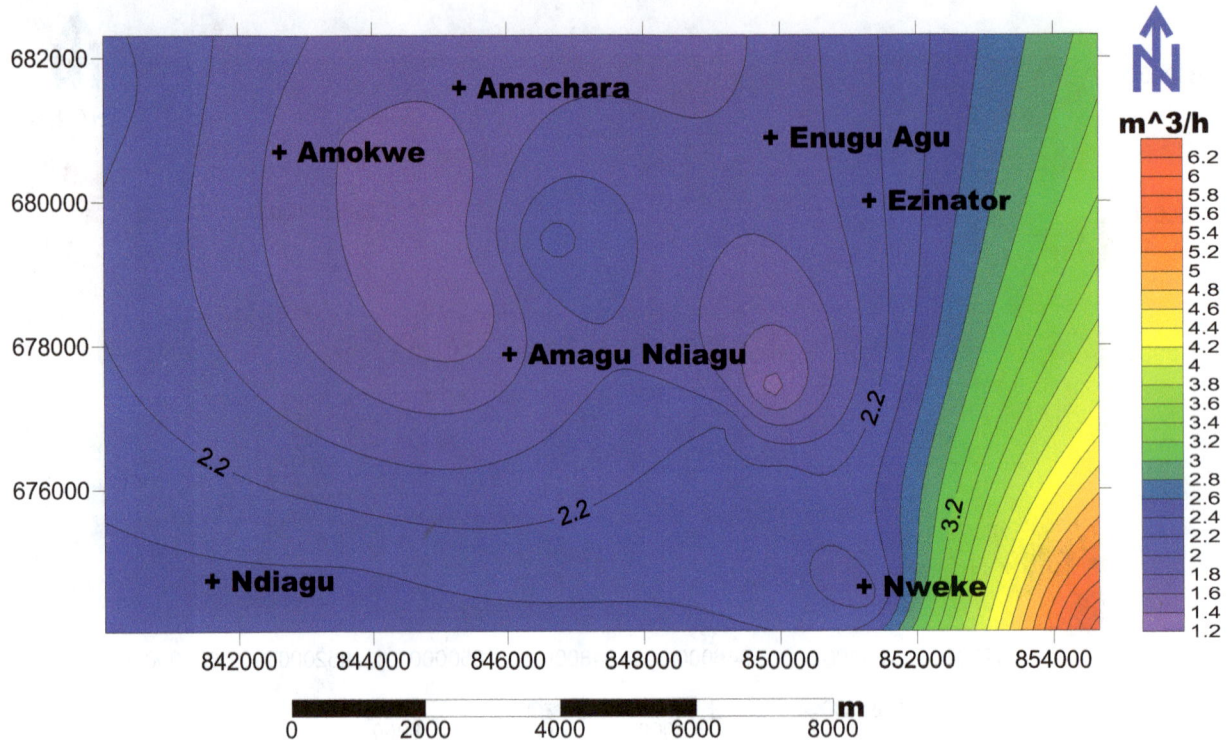

Figure 15. Aquifer yield map (from pumping test analysis) contour map from the study area.

groundwater zones were delineated. The zones are low, very low and negligible potentials. The country around Amagu Ndiagu, Ndiagu, Nweke and Amokwe are of very low potential while areas to the northeast around Enugu Agu, Ezinator and Amachara are of low potential. Negligible potential was quite insignificant.

Conclusion

Based on the geoelectrical studies, the potential groundwater zones were delineated; the low, very low and negligible potential zones. Computed aquifer hydraulics parameters indicate a low aquifer yield. However, this should not stall further groundwater exploration in the area but more detailed hydrogeological and geophysical investigations must be carried out to determine good point(s) for groundwater development. A depth greater than 60 m but less than 100 m may be recommended.

ACKNOWLEDGEMENT

The authors are grateful to EssBee Geotechnical ventures limited for the resistivity data processing. Many thanks to Mr. Emmanuel Enang, principal geologist Felgralinks Nigeria limited, for his qualitative role during the data acquisition stage.

REFERENCES

Aneke BC (2007). Water resources of the Awgu Shale group, Enugu state, southeastern Nigeria. Ph.D Thesis, University of Nigeria, Nsukka.

Choudhury K, Saha DK, Chakraborky P (2001). Geophysical study for saline water intrusion in a coastal alluvial terrain. J. Appl. Geophys. 46:189-200.

Ekwe AC, Nnodu IN, Ugwumbah KI, Onwuka OS (2010). Estimation of aquifer hydraulic Characteristics of low permeability formation from geosounding data: A case study of Oduma Town, Enugu State. J. Earth Sci. 4(1):19-26.

Ezeh CC (2012). Hydrogeophysical studies for the Delineation of potential groundwater zones In Enugu State, Nigeria. Int. Res. J. Geol. Min. 2(5):103-112.

Ezeh CC, Ugwu GZ (2010). Geolectrical sounding for estimating groundwater potential in Nsukka L.G.A. Enugu State, Nigeria. Int. J. Phys. Sci. 5(5):415-420.

Frohlich RK, Urish D (2002). The use of geoelectrics and test wells for the assessment of Groundwater quality of a coastal industrial site. J. Appl. Geophys. 46:261-278.

Gheorge A (1978). Processing and synthesis of hydrogeological data. Abacus press, Tunbridge, Kent. 136 pp.

Huntly D (1987). Relations between permeability and electrical resistivity in granular aquifers. Groundwater 24:466-474.

Kelly WE (1979). Geoelectric sounding for estimating aquifer hydraulic conductivity. Groundwater 50(6):420–425.

Keller GV, Frischknecht FC (1966). Electrical methods in geophysical prospecting. Pergamon Press.

Koefoed O (1979). Geosounding principles, 1 Elsevier, Amsterdam.

Koinski WK (1981). Geolectric soundings for predicting aquifer properties. Groundwater 19(2):163-171.

Maillet R (1947). The fundamental equations of electrical prospecting. Geophysics 12:529-556.

Niwas S, Singhal DC (1981). Estimation of aquifer transmissivity from Dar-zarrouk parameters in porous media. J. Hydrol. 50:393-399.

Offordile ME (2002). Groundwater study and development in Nigeria. 2nd Edition. Mecon geology And Engineering Services Ltd. Jos, Nigeria. 453 pp.

Orellana E, Mooney HM (1966). Master tables and curves for vertical electrical sounding over Layered structures. Interscience, Madrid.

Urish DW (1987). Electrical resistivity-hydraulic conductivity relationships in glacial outwash aquifers. Water Resour. Res. 17(5):1401-1408.

World Gazette (2011). Colourful map of Nigeria with 36 states. www.world-gazette.com

Zohdy AAR (1976). Application of surface geophysical (electrical methods to groundwater Investigations) techniques of water resources investigations of the United States Geological Survey pp. 5-55.

6

Evaluation of performance of z-component of Nigerian seismographic stations from spectral analysis

Kadiri Umar Afegbua[1] and F. O. Ezomo[2]

[1]Department of Earthquake Seismology, Centre for Geodesy and Geodynamics, P. M. B. 11, Toro, Bauchi State, Nigeria.
[2]Department of Physics, Faculty of Physical Sciences, University of Benin Ugbowo, Benin City, Edo State, Nigeria.

Preliminary noise analysis has been carried out on the Z-components of the Nigerian National Network of Seismographic Stations (NNNSS) using data collected from the five operational stations (Toro, Kaduna, Nsukka, Awka and Ife) in 2012. The results of the analysis from noise spectra using SEISAN software clearly showed that the noise levels at the respective stations are high and the average noise levels are above the high noise model of Peterson curves. The possible sources of noise are cultural, actual earth vibrations and instrumentation. The study is intended to quantify the amount of noise present in the existing NNNSS and compare with Peterson noise curves, and understand the noise characteristics in order to adopt better practices in the day-to-day seismic stations' operations in Nigeria.

Key words: National network of seismographic stations (NNNSS), noise, spectral analysis, signal to noise ratio.

INTRODUCTION

In seismology and other fields that deal with signal, regarding noise as an undesirable component of the signal. Conventionally, noise is described as a disturbance in the signal which does not represent part of a message from a specified source (Sherrif, 1991).

Seismologists collect data on seismic background noise for assessing the suitability of sites for temporary or permanent seismic recordings. The stations that form the National Network of Seismographic Stations (NNNSS) are permanent seismic stations and installed with broadband seismometers. Site quality requirements depend on the task of seismic observations on their resolution, dynamic range, bandwidth and frequency range (Bormann, 1998). Till now, noise data are collected with a wide range of instruments, both analog and digital of different bandwidth, resolution and transfer functions. Accordingly, noise appearance in seismic records,

amplitude- and frequency-wise, differs and the various kinds of noise spectra derived there vary too (Alguacil and Havskov, 2010). Apart from the noise spectra adopted in this study, another possible way to investigate noise level of a seismic station is to obtain velocity power spectra of ambient seismic noise at noisy and quiet conditions for each station based on their different geological setting (Aki and Richards, 1980) and when the frequency spectrum of the seismic signal of interest differs significantly from that of the superposed seismic noise, band-pass filtering can help to improve the signal-to-noise ratio (SNR) (Bormann, 2009 NMSOP). The velocity power spectra of ambient seismic noise and different ways of improving Signal to Noise Ratio (SNR) are not the purpose for this study.

Recorded seismic signals always contain noise and it is important to be aware of both the source of the noise

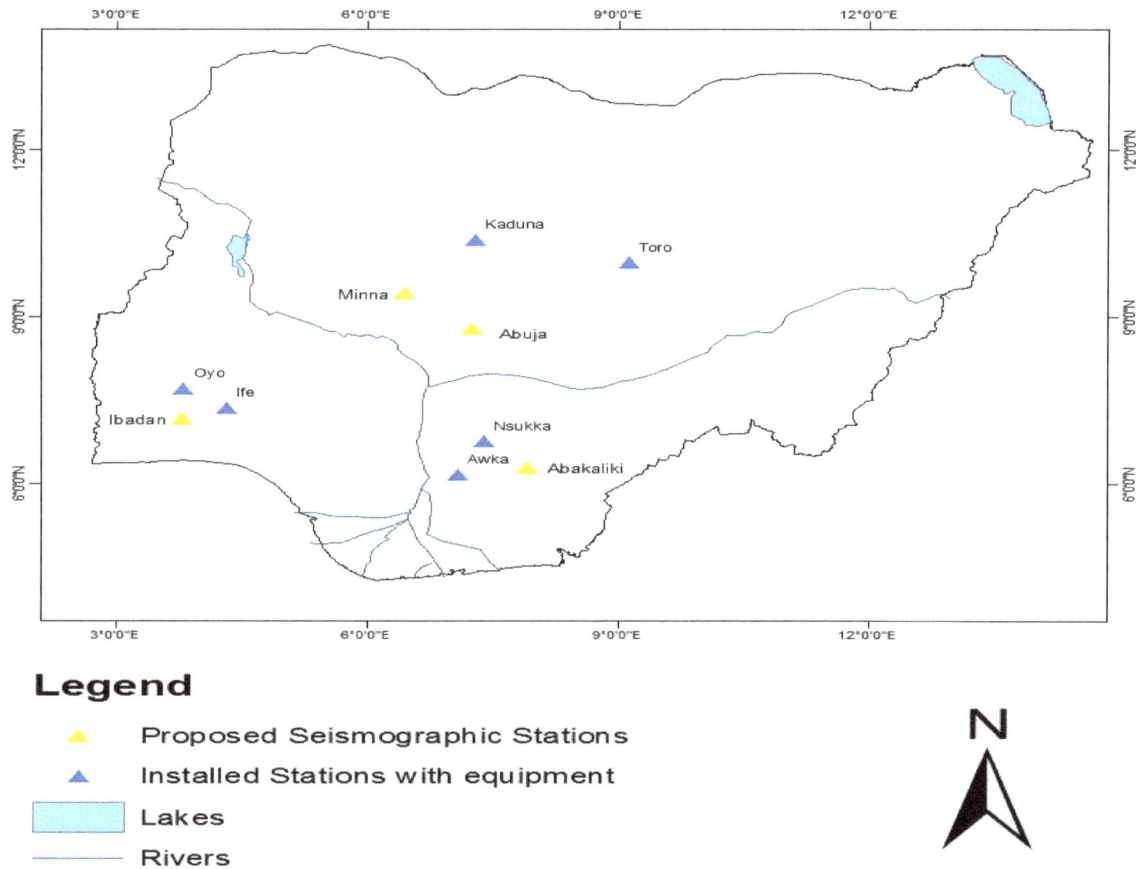

Figure 1. Map of Nigeria showing locations of existing and planned stations.

and how to measure it (Alguacil and Havskcov, 2010). Noise can have two origins: Noise generated in the instrumentation and 'real' seismic noise from earth vibrations. Normally, the instrument noise is well below the seismic noise although most sensors will have some frequency band where the instrumental noise is dominating (e.g. an accelerometer at low frequencies) (Havskov and Ottemoller, 2008).

Ambient seismic noise defines vibrations of the ground caused by sources such as tides, water waves striking the coast, turbulent wind, effects of wind on trees or buildings, traffic or human based noise (Bonnefoy-Claudet et al., 2006). Ambient seismic noise basically has two different origins-cultural and natural representing the microtremors and microseisms, respectively (Alguacil and Havscov, 2010). Origins of microtremors are mainly cultural from the actions of human beings. Other sources are rain, traffic, wind, industrial noise in the urban areas, geologic noise. Cultural noise are seen as high frequency noise surface waves greater than 1.0 Hz and attenuate within several kilometers in distance and depth (Alguacil and Havskcov, 2010). So it has strong significant lower noise levels in boreholes and deep tunnels. The noise at low frequency is typically higher on the horizontal

component than on the vertical component due to the difficulty of stabilizing the station for small tilts and noise level vary greatly between different sites and different frequencies (Alguacil and Havskcov, 2010).

The Peterson (1993) noise curves and noise spectral level (Figure 5) has been used to evaluate noise levels for the IRIS station BOCO. Seismic noise at the station BNG (Bangui, Central Africa) had been evaluated (Bormann, 2009) and compared to the new global seismic noise model by the Peterson (1963) noise curves. Variations of seismic background noise in South Korea have been investigated using power spectral analysis (Sheen et al., 2009). These so-called Peterson curves (Figure 5) have become the standard, by which the noise levels at seismic stations are evaluated. The curves of Figure 5 represent upper and lower bounds of a cumulative compilation of representative ground acceleration power spectral densities determined for noisy and quiet periods at 75 worldwide distributed digital stations (Havskov and Ottemoller, 2008).

The Centre for Geodesy and Geodynamics (CGG) has been operating the NNNSS located in triangulations (Figure 1) across Nigeria. The properties of the respective stations are shown in Table 1. Other

Table 1. Location of NNNSS (Modified after Akpan and Yakubu, 2010).

S/N	Station name	Latitude	Longitude	Elevation (m)	Geologic foundation	Instrumentation
1	Oyo	07°53.131'N	03°57.078'E	295	Granite	Seismograph: DR4000 recorder Seismometer: SP400 medium period seismometer
2	Ibadan	07°27.251'N	03°53.520'E	193	Gneiss	No instrument installed
3	Ile-Ife	07°32.800'N	04°32.815'E	289	Gneiss	Seismograph: DR4000 recorder Seismometer: EP105 broadband seismometer
4	Awka	06°14.561'N	07°06.693'E	50	Shale and siltstone	No instrument installed
5	Nsukka	06°52.011'N	07°25.045'E	430	Sandstone	Seismograph: DR4000 recorder Seismometer: EP105 broadband seismometer
6	Abakiliki	06°23.453'N	08°01.474'E	82	Sandstone	No instrument installed
7	Abuja	08°59.126'N	07°23.380'E	432	Granite	No instrument installed
8	Toro (Central)	10°03.303N	09°07.089'E	882	Gneiss	No instrument installed
9	Kaduna	10°26.101'N	07°38.484'E	668	Granite	Seismograph: DR4000 recorder Seismometer: SP400 medium period
10	Minna	09°30.702'N	06°26.411'E	203	Granite	No instrument installed

Table 2. Information about response files parameters of the Entec sensors/Recorders at NNNSS (Courtesy CGG, Toro).

Stations	Free period (s)	Damping rate	Generator constant (V/m/s)	Digitizer sensitivity	Sampling rate	Amplifier gain
Kaduna	16	0.7	2000	419,430C/V	40	0.0
Nsukka	30	0.7	2000	"	"	"
Toro	60	0.7	2000	"	"	"
Awka	16	0.7	2000	"	"	"
Ife	60	0.7	2000	"	"	"

information regarding the equipment at the respective stations is shown in Table 2. The Awka, Kaduna, Nsukka, Ife and Toro are currently operational while installation of equipment at Abakilike, Oyo, Minna, Abuja, Ibadan stations would be completed soon. The sensors at Nsukka, Awka and Ife are placed in a vault of about 6-10 m deep in University communities with surrounding residential buildings, while the sensors at Kaduna and Toro stations are placed on the surface of a basement in relatively quiet environments and are about 100 – 600 m from light vehicular movement, and some surrounding trees. The NNNSS has been collecting data since 2008.

While some stations are located few kilometers away from the Atlantic Ocean and some fast running streams in the Southern part of the country with surrounding human settlements, others are located farther away in the north with sparsely populated hamlets, no streams apart from surrounding trees.

Although, geologic foundation for each site was established before sitting of the five operational stations (Figure 1), no noise analysis was conducted at the site to ascertain the noise level and possibly identifying the sources of the noise. The study is intended to quantify the amount of noise present in the existing NNNSS data

Figure 2. A Eentec EP 105 sensor installed and insulated at Toro Station.

Figure 3. A DR 4000 Eentec recorder at Toro Station.

and compare with Peterson (1993) noise curves, since this simple comparison had been adopted to investigate the performance of seismic stations around the world. The spectra figures obtained from this study can be compared with those of Peterson's and see if they are behaving in line with the global standard (Figures 2 to 5 and Tables 1 and 2).

METHODOLOGY

The data in miniSEED format obtained from the five stations were used for this study. Average of an hour-long Z-component data for six month from each station and their respective noise spectra were plotted using spectral analysis method (Alguacil and Havskcov, 2010). With digital data like NNNSS's data, it is possible to make spectral analysis, and thereby get the noise level at all frequencies

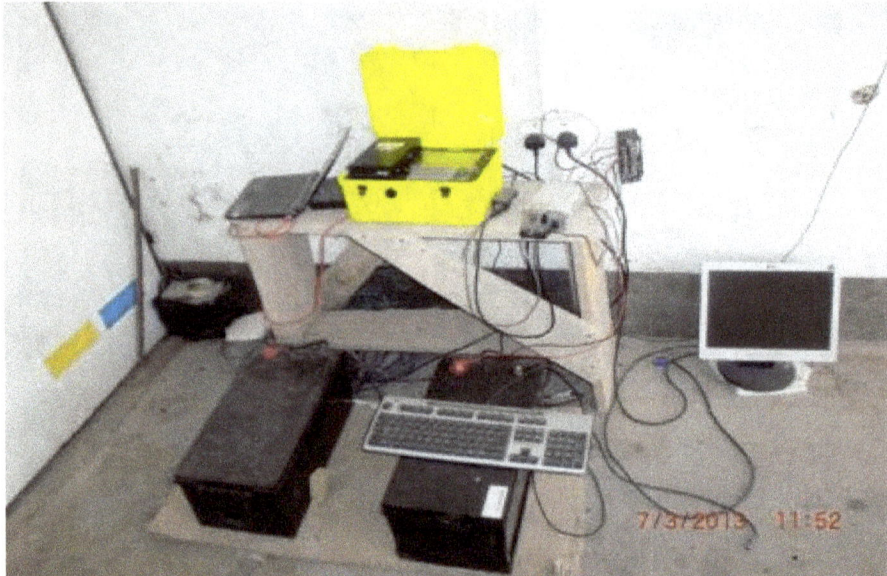

Figure 4. A typical set up of NNNSS stations; this one at Toro station showing recorder, batteries connected to solar panels installed outside, computer monitor to temporary store data before they are downloaded and processed.

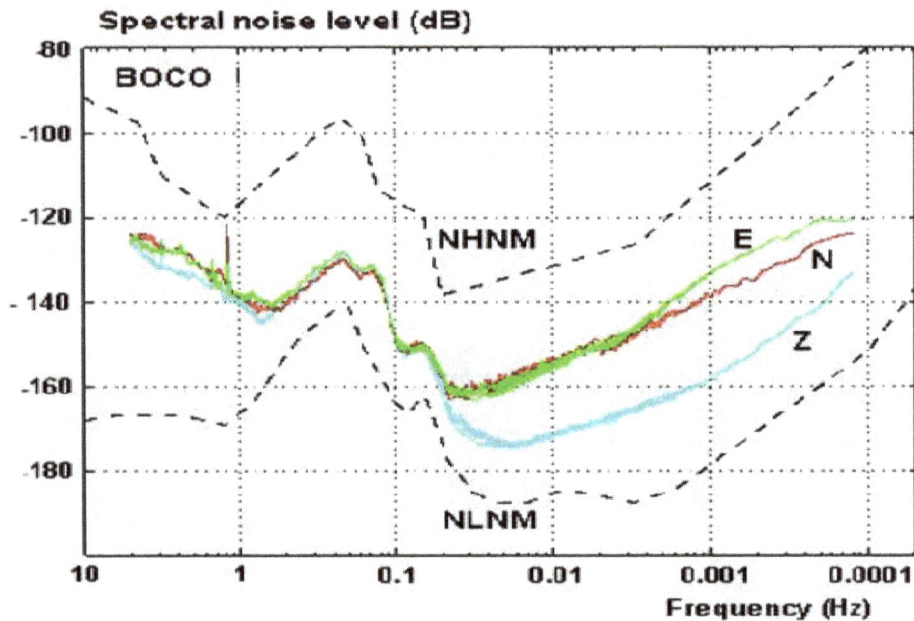

Figure 5. New global high (NHNM) and low noise models (NLNM) of Peterson (1993) and noise curves and noise spectral level for the IRIS station BOCO, which was considered a good station from this figure. The Peterson high and low noise models are shown with dashed lines (Havskov and Ottemoller, 2008; Alguacil and Havscov, 2010). However, other techniques like noise analysis using Pascal Quick Look Extended package and others can also be used to test for performance of seismic stations that give the Power Spectral Densities.

in one simple operation ((Alguacil and Havskcov, 2010; Chapman et al., 2006). The noise spectra is represented as the noise power density acceleration spectrum $Pa(\omega)$, noise level is thus, calculated as:

$$\text{Noise Level} = 10 \log [Pa(\omega)/(m/s^2)^2/Hz] \qquad (1)$$

It is also possible to relate the power spectra to amplitude measurements (Bormann, 2009). Approximate relationship can be calculated between the noise power density $N(dB)$ given in dB and the ground displacement d in meters:

$$d = f-1.5/39 * 10N(dB)/20 \quad (3.17) \qquad (2)$$

$$N(dB) = 20log(d) + 30log(f)+32 \qquad (3)$$

Where f is the average frequency of the filter and dB is relative to 1 $(ms^{-2})^2/Hz$ (Alguacil and Havscov, 2010). Using Equation 2 or 3, noise level in dB and at various frequencies range can be computed.

Although, it can be used to realize other results, the spectral analysis is commonly used to make the correction of attenuation and instrument displacement spectrum and determine the flat spectral level and corner frequency from which the seismic moment, source radius and stress drop can be calculated (Ottemoller et al., 2012).

To determine moment, source radius and stress drop using spectral analysis, spectral option (Spec) is used with (d) for displacement, (v) for velocity, and (a) for acceleration or (r) for raw spectrum. However, instead of selecting d, v, a or r in Spec program (SEISAN), just press the same characters in upper case to make power spectrum and noise spectrum which is the interest of this study (Ottemoller et al., 2012).

Specifically, this study considered the seismic background noise which is often displayed as acceleration power spectral density in dB relative to $((1m/s**2)**2)/Hz$. Instead of selecting d, v or a as mentioned, pressing n instead will show the Peterson (1993) new global high and low noise models superimposed on the observed spectrum as showed in the figures. As it is applicable when determining moment, source radius, and stress drop of an event from analysis, no attenuation correction is done when doing noise spectra (Ottemoller et al., 2012). Although, an hour-long window was used for the spectral analysis, the resulting spectrum according to Ottemoller et al. (2012) can be normalized using the following relation:

$$P = |F^{DFT}|^2 * \frac{\Delta t^2}{T} * 2 \qquad (4)$$

where P is the Peterson power spectrum, F^{DFT} is the discrete Fourier transform, Δt is the sample interval and T is the length of the time window. The factor 2 comes from the fact that only the positive frequencies are used so only half the energy is accounted for. The total power is proportional to the length of the time window since the noise is considered stationary, so by normalizing by T, the length of the time window should not influence the results. This noise option is a good and straightforward method of checking the noise characteristics of a given seismic station and compare it to global standards. This is exactly what we have demonstrated in this study, with limitations like plotting noise spectra in longer window and obtaining power spectra densities directly, which could be improved upon on in future.

RESULTS

There are several software that can give better results that we obtained here, but within the limit of spectral analysis technique in SEISAN software that was handy for this work, the noise spectra were plotted here. One hour long data collected on April 1, 2012 from Kaduna; July 1, 2012 from Nsukka; June 28, 2012 and June 30, 2012 from Awka; July 1, 2012; and June 3, 2012 from Toro and Ife stations respectively as compared to data collected for the six months were used, since there was no much deviation from month to month. It is pertinent to point out that observation of pattern or consistency of noise on the Z-component of the five stations (Figure 1) was conducted between January and July 2012 and average day time/night time noise as observed are represented in Figures 14 to 23. It is also assumed that the seasonal variation of noise is addressed in the analysis since the data convered six months. However, more detailed study that will clearly reflect seasonal variations in noise would be undertaken in future. The SEISAN software can plot power spectra but has limitation in the length of window for data spectral figures as one hour-long data were accommodated and averaged to get general overview of noise pattern. The amplitude and phase responses for the five stations are showed in Figures 6 to 10. It is a prerequisite to create the response files within CAL directory of SEISAN before performing noise spectral analysis. The response files give information about the instruments and the need or not for correction for instrument response. Figures 14 to 23 are the noise spectra for Z-component of the five stations and these spectra were compared with the Peterson (1993) curves and as showed in Figure 5).

DISCUSSION

Figures 6 to 10 are the amplitude and the phase response graphs from the Z-component of each station. These response files were created in the CAL directory of SEISAN software to test for the instrument information prior to the plotting of the noise spectra. Figure 11 shows noise traces from the five stations (from top bottom: Awka, Kaduna, Nsukka, Toro and Ife). PQLII was used toremove the mean and other trends from the traces before plotting the figures. It was clearly observed that Nsukka is more noisy followed by Awka while Toro is less noisy. Figure 12 shows noise spectra plotted with PQLII on the same window, having Nsukka with more amplified cultural noise and least on Toro. In this case, the noise spectra were not compared with Peterson (1993). Overlay of the noise traces in Figure 13 resulted in Nsukka (Green) and Awka (Purple) overshadowing other three stations of Ife, Kaduan and Toro. Figures 11 to 13 were plotted to demonstrate dominance of background noise of some stations over the others.

Ocean generated microseisms are constant source of energy and ambient seismic noise is dominated by two peak of microseism at 7.0 and 14 s period (Friedrich et al., 1998). A peak, which is known as microseismic peak or double frequency peak, takes place around 7.0 s.

At Ife Station (Figures 14 and 15), the primary and secondary microseismic peaks are distinct at 0.05 and 0.3 Hz, respectively. The observed cultural noise was

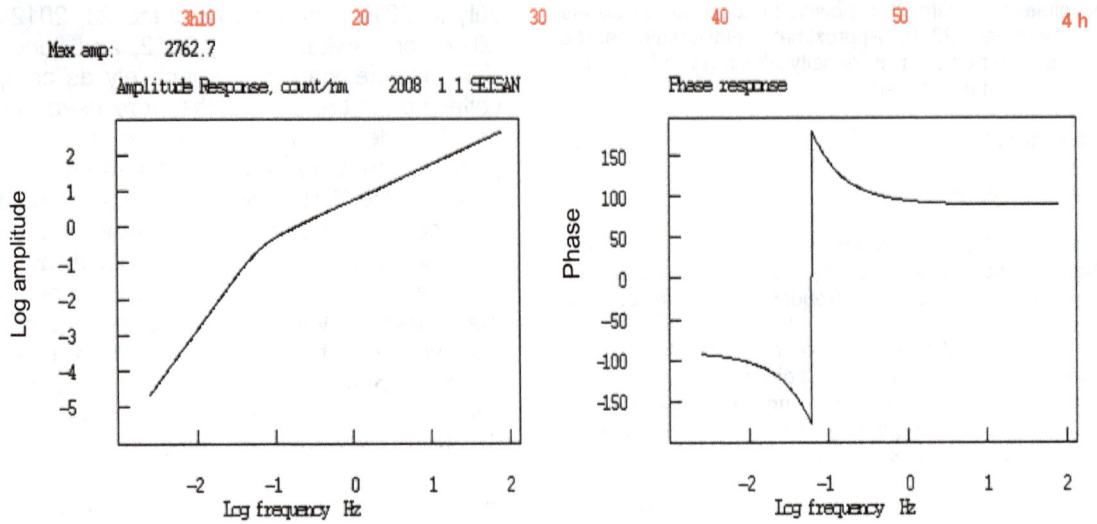

Figure 6. Amplitude and phase responses from Ife Station (Z-component).

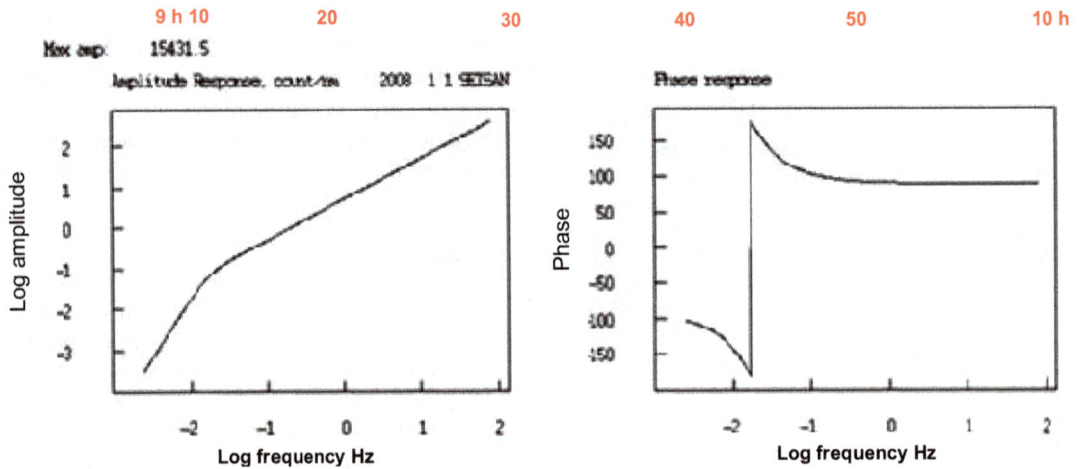

Figure 7. Amplitude and phase responses from Toro Station (Z-component).

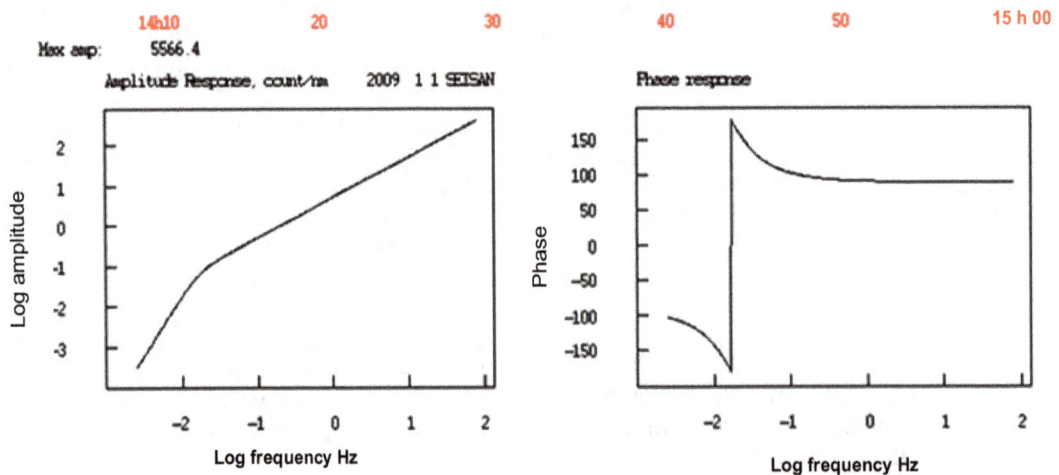

Figure 8. Amplitude and phase responses from Kaduna Station (Z-component).

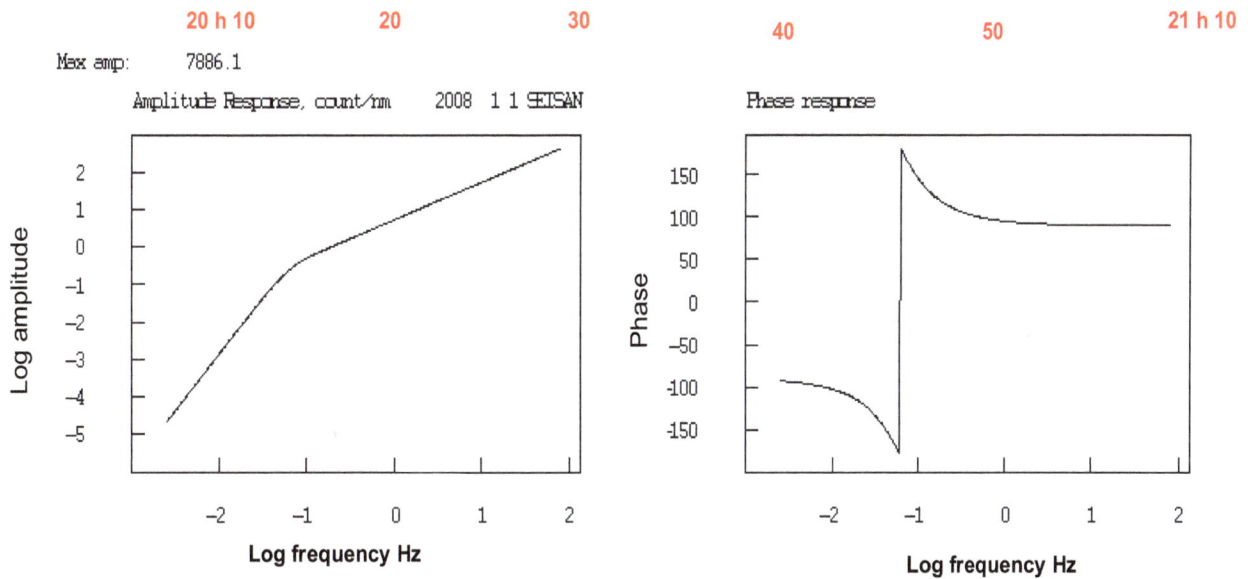

Figure 9. Amplitude and phase responses from Nsukka Station (Z-Component).

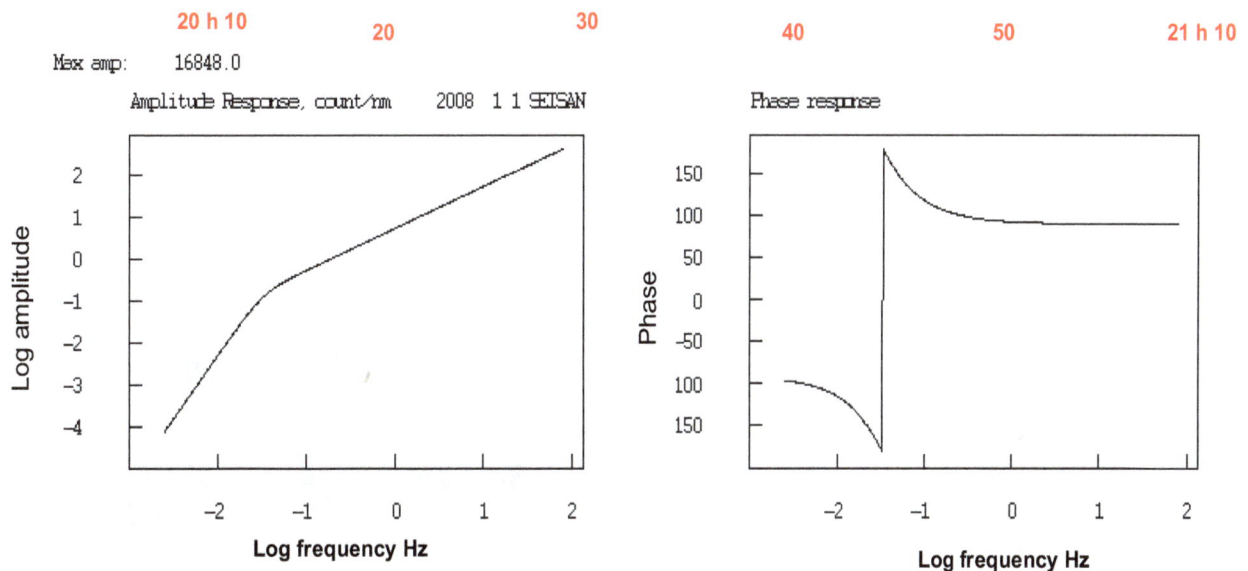

Figure 10. Amplitude and phase responses from Nsukka Station (Z-component).

slightly higher during the night time than in the day time, which is likely due to wind or rainfall or other sources of noise during the night the data was generated. The average noise spectra are good compared to the Peterson (1993) noise spectra curves. Between 0.1 to 2.0 Hz, both spectra assumed the shape of Peterson curve which represents the ground noise and station resolving power. Ife station is located in a quiet environment apart from surrounding trees, some running streams, and it is several kilometers away from the Atlantic Ocean in the

southern part of Nigeria.

At the Nsukka Station (Figures 16 and17), the noise spectra are good compared to the shape of Peterson (1993) noise curves but cultural noise for both day time (10 am-12 noon) and night time (12 mid-night to 2 am) are very high likely due to wind and vehicular traffic, as the station is located far away from human settlements but about one kilometer from highway. Looking at Figure 16 and 17, it could be seen that there is no significant noise variation between day and night periods as high

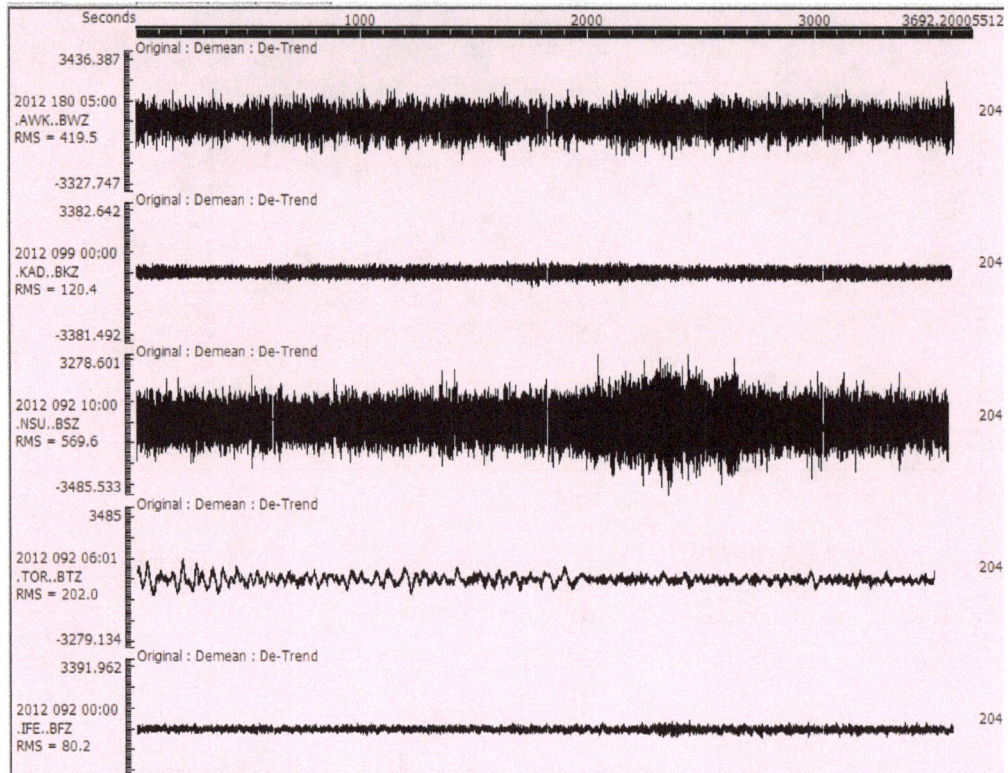

Figure 11. Showing noise traces (from top: Awka, Kaduna, Nsukka, Toro and Ife). No filters were applied, but the mean and other trends were removed.

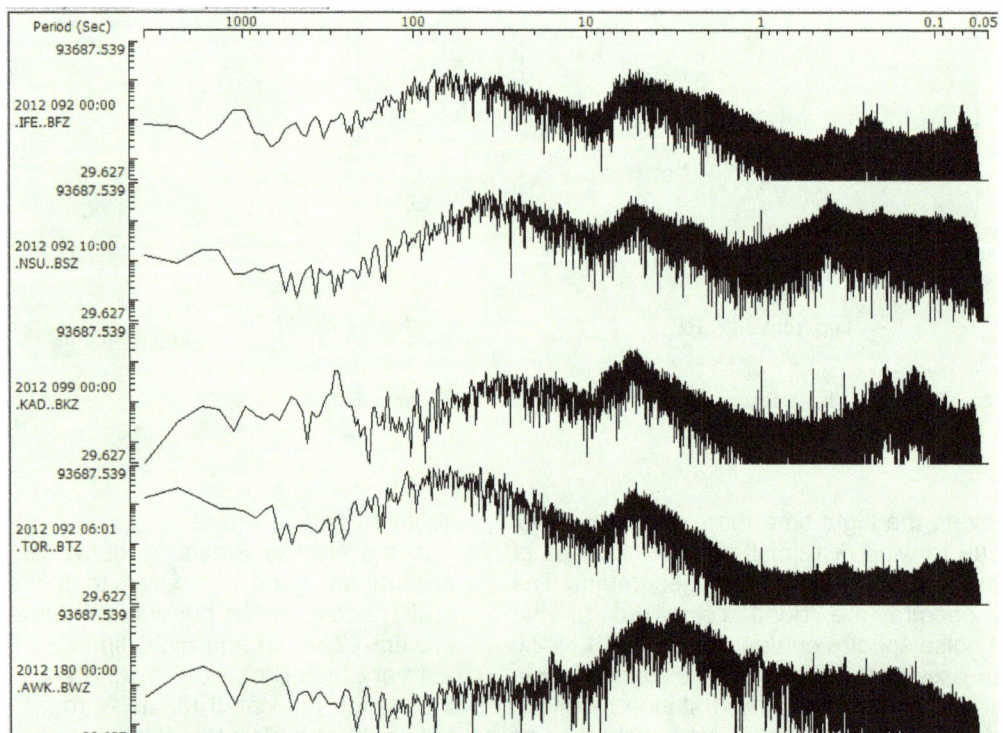

Figure 12. Noise spectra of (from top Ife, Nsukka, Kaduna, Toro and Awka). The spectra were here not compared with Peterson (1993) noise models.

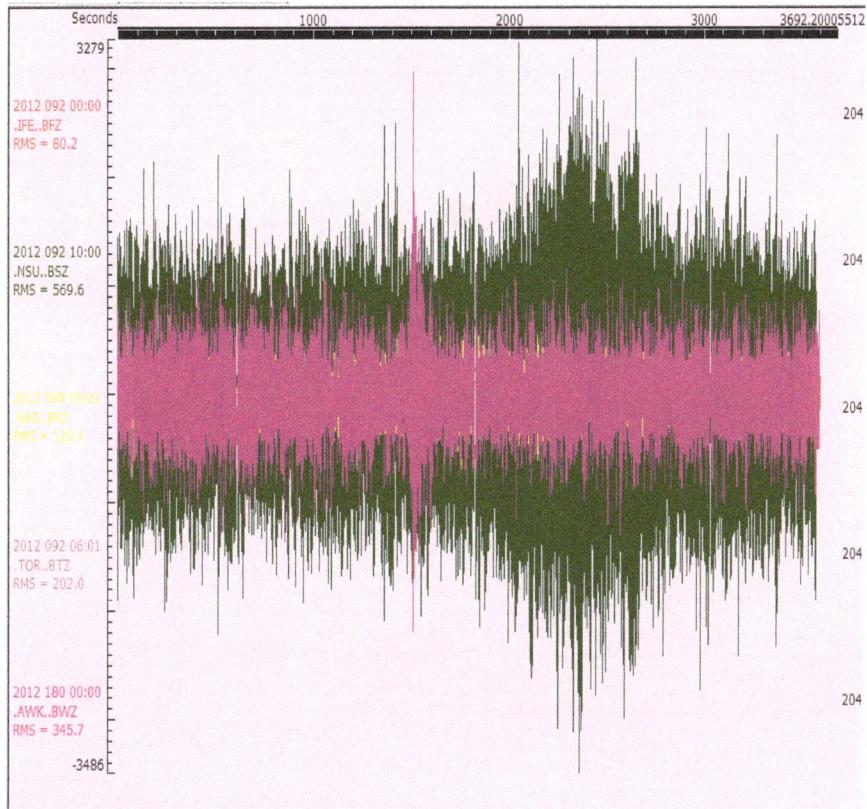

Figure 13. Overly of the Z-component noise traces (from the top to bottom: Ife, Nsukka, Kaduna, Toro and Awka).

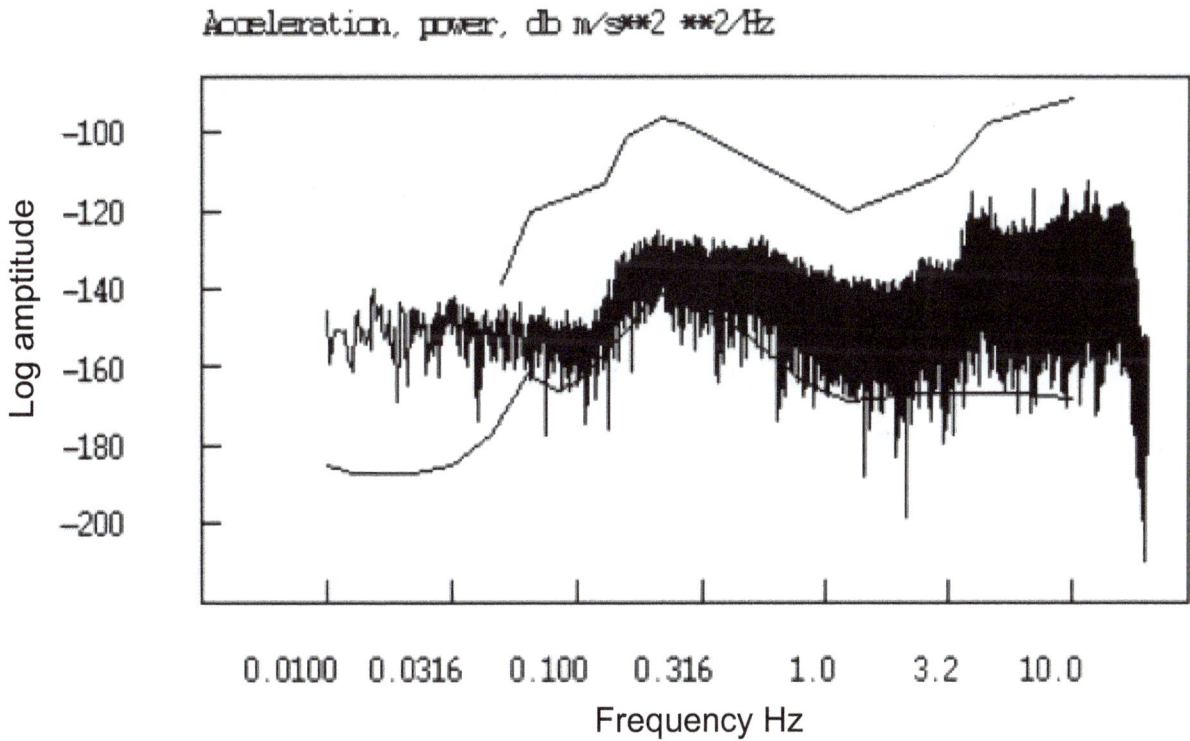

Figure 14. Noise variation between 12:00am – 4:00am (Ife Station).

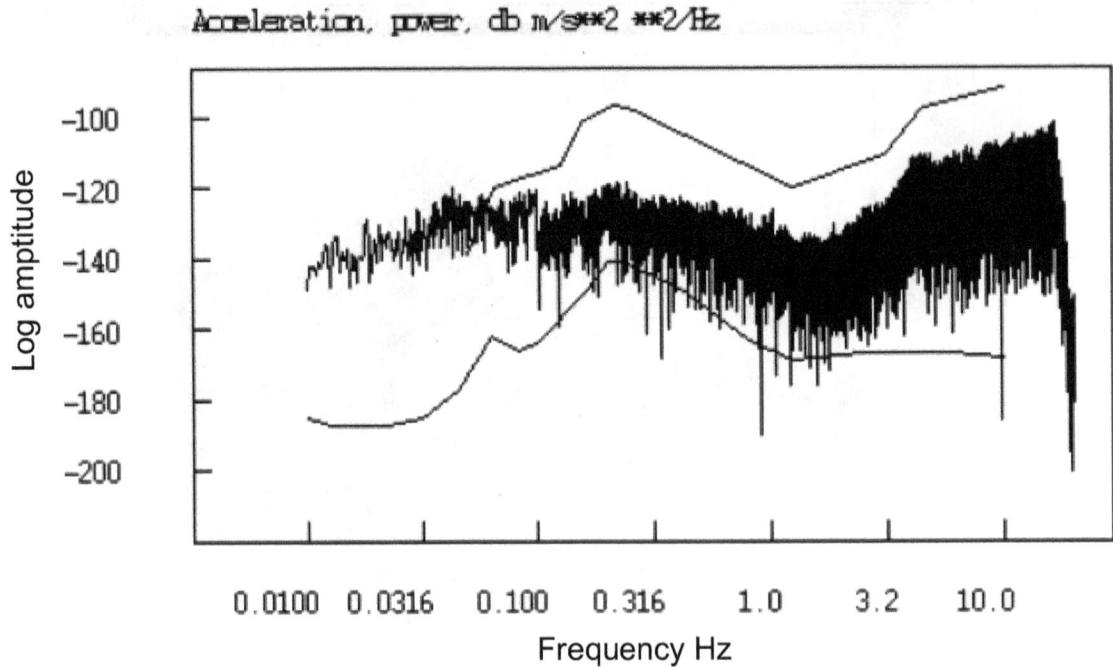

Figure 15. Noise variation between 11:00am-3:00pm (Ife Station).

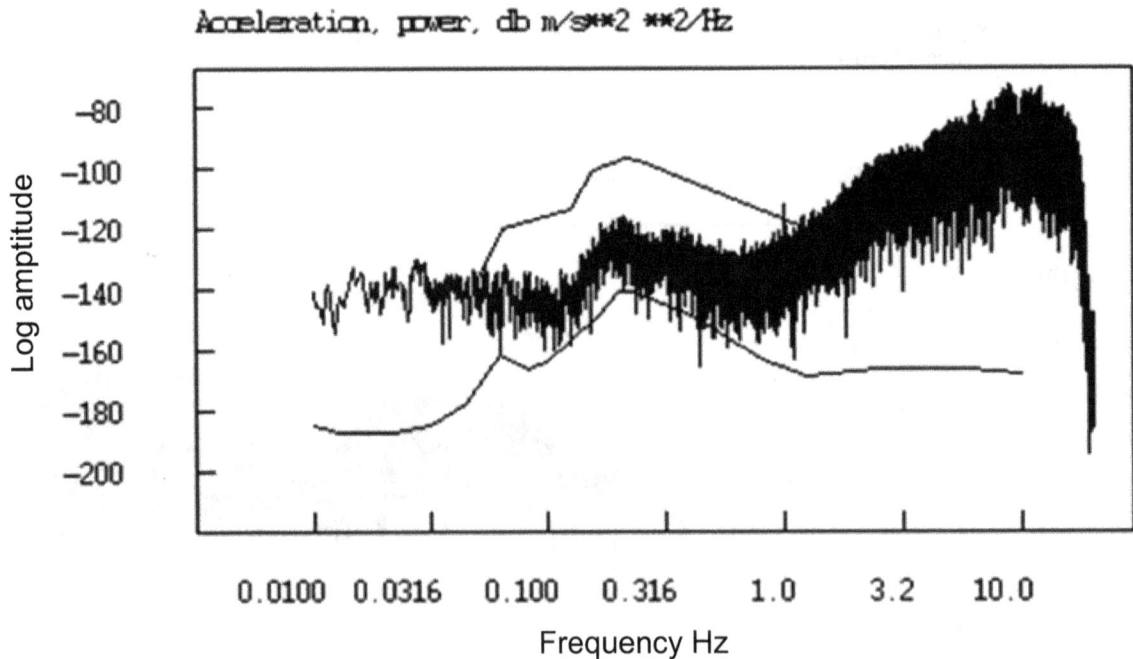

Figure 16. Noise variation between 10am -2pm (Nsukka Station).

level of noise was observed at the Z-component at high frequencies, that is, from 1 Hz and above and it is higher than the average Peterson noise curves (1993). At lower frequencies below 0.15 Hz, it was observed that the spectrum had almost the same shape as the Peterson curves which represents the ground noise and thus the resolving power of the station in that frequency range. The high noise level at high frequencies may indicate the

Acceleration, power, db m/s**2 **2/Hz

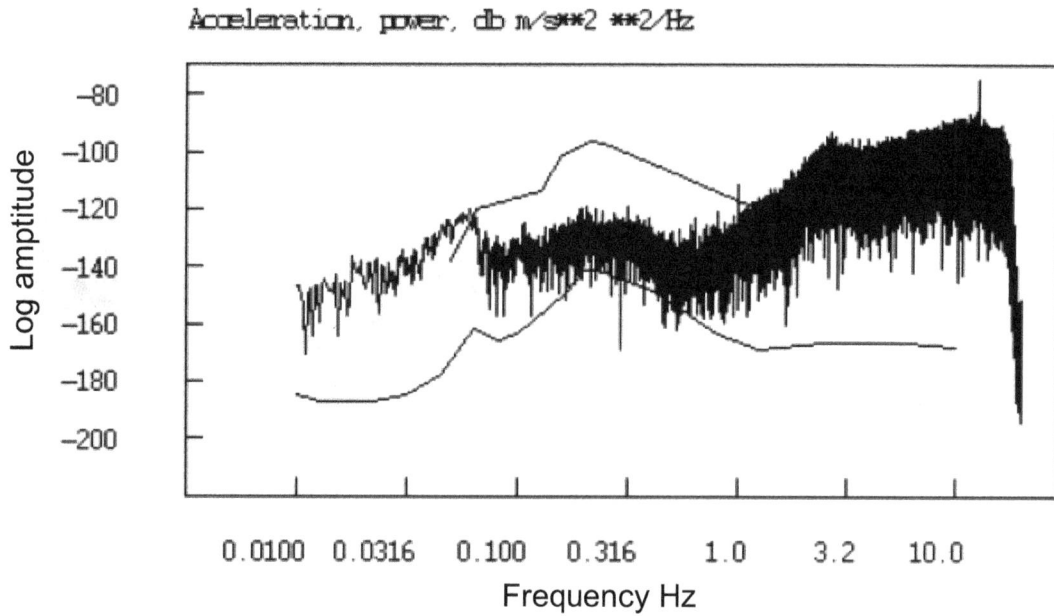

Figure 17. Noise variation from 12am-2am (Nsukka Station).

Acceleration, power, db m/s**2 **2/Hz

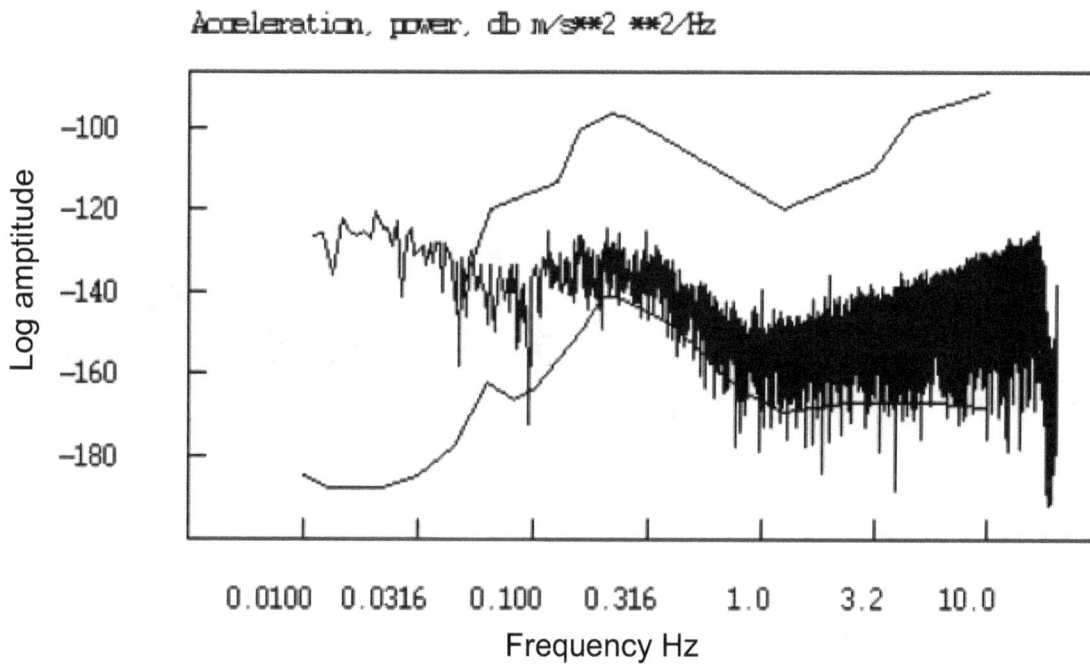

Figure 18. Noise Variation between 12am -3am (Toro Station).

contribution of cultural noise and on the lower periods (<1 s) maybe related to instrument self noise.

At Toro Station (Figures 18 and 19), although observed cultural noise at high frequencies is low on the average compared with Peterson (1993) noise curves, the noise is higher during the day time than during the night. Also, higher noise was observed at low frequencies between 0.01-0.1 Hz which is likely due to instrument self-noise. Since the sensor of this station is on solid, immobile rock hosting other monitoring equipment like Global Positioning System (GPS) etc., and with scanty settlements few meters away from the station, the likely sources of noise are wind, vehicular traffic, and human activities. Construction of a vault for the sensor would

Acceleration, power, db m/s**2 **2/Hz

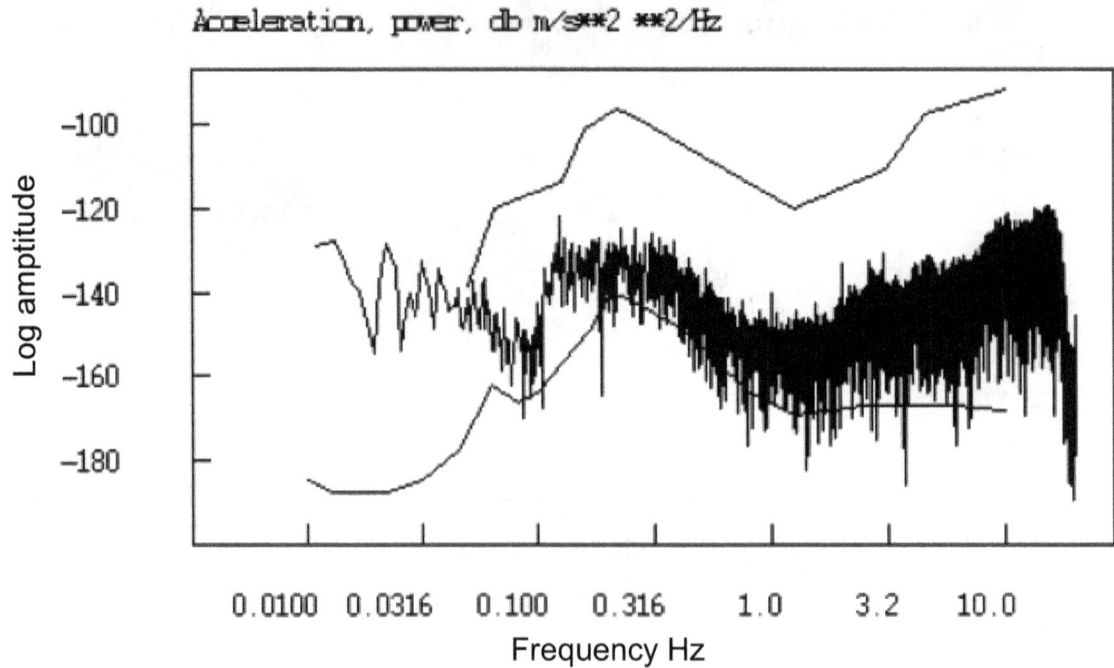

Figure 19. Noise Variation between 10am-2pm (Toro Station).

Acceleration, power, db m/s**2 **2/Hz

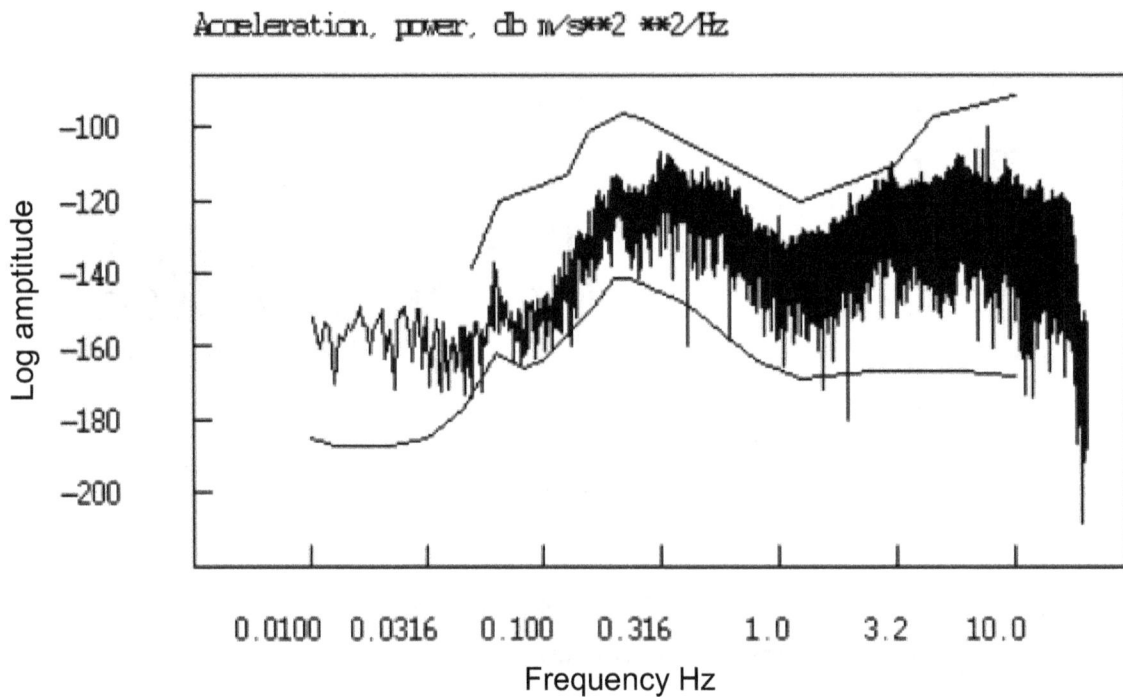

Figure 20. Noise variation between 9pm-12am (Awka Station).

likely minimize the observed noise.

At Awka Station (Figures 20 and 21), the noise spectra of the Z-component has the same shape as the Peterson curves at frequency band 0.1-10 Hz which apparently

represents the ground noise of the station and of course the station's resolving power. The noise is low at lower frequencies between 0.01-0.1. The fairly high noise observed at higher frequencies above 1.0 Hz may be as

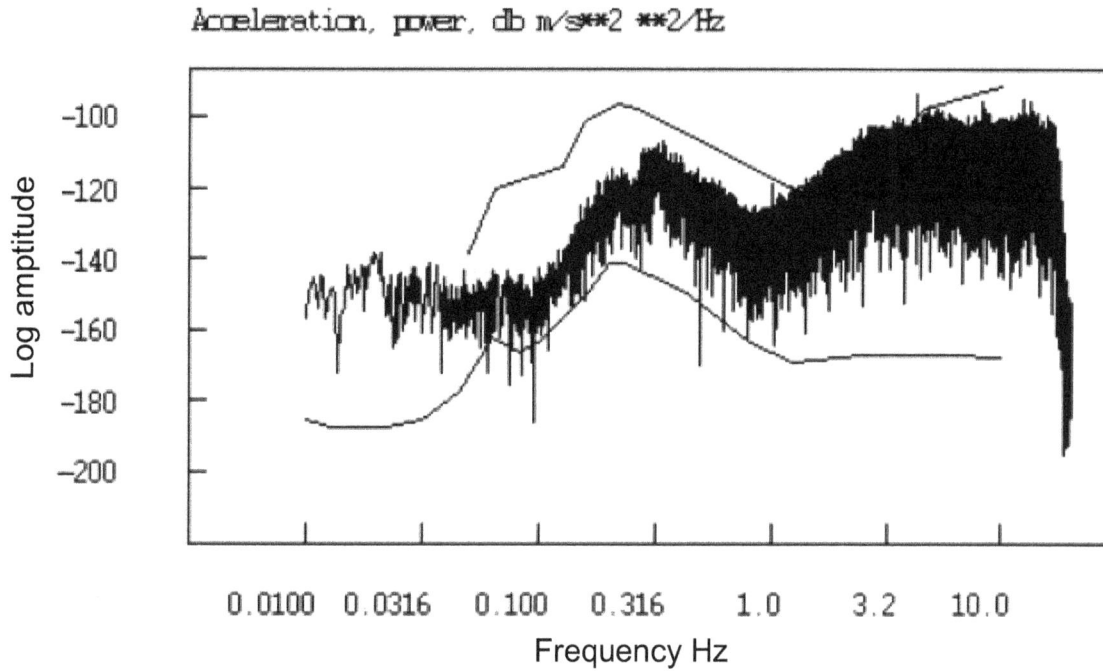

Figure 21. Noise Variation between 11am-2pm (Awka Station).

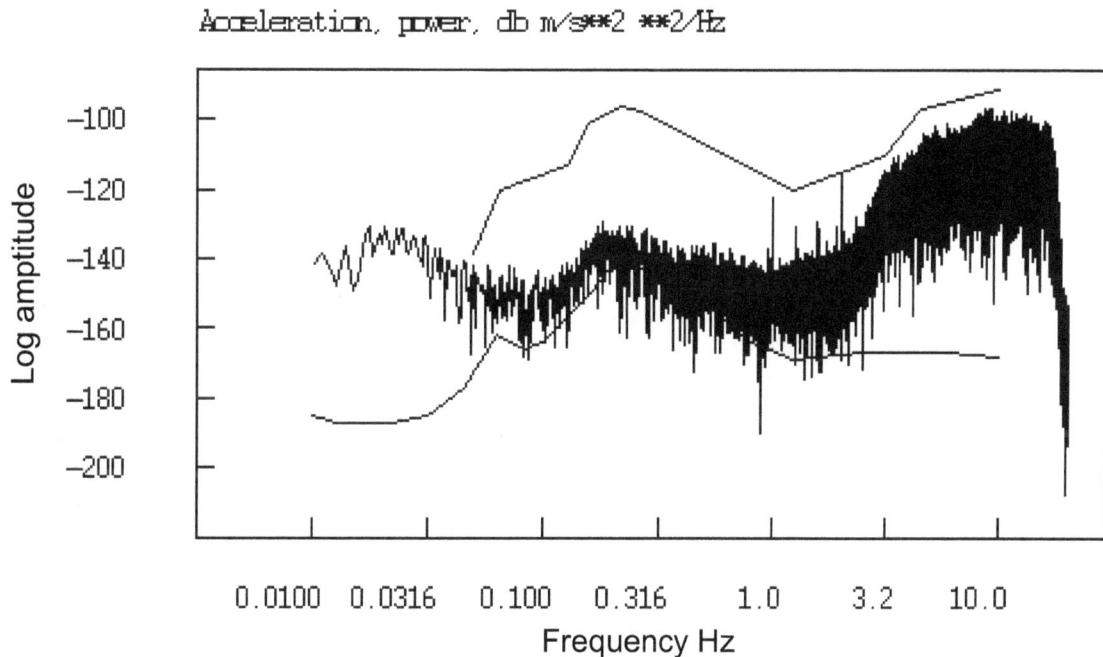

Figure 22. Noise variation between 9pm-1am (Kaduna Station).

a result of the contribution of cultural noise from the busy surroundings, it is a surprise that the noise is slightly higher during the night time than in the day time period. This may be due to windy condition or even rainfall, but the clear microseismic peak at 0.3 Hz is a clear indication that the station is a good station.

There is no significant noise variation at Kaduna Station (Figures 22 and 23) noise spectra are within acceptable frequency limits but noise is high at high frequencies between 3.2 and 6.0 Hz high noise level was

Acceleration, power, db m/s**2 **2/Hz

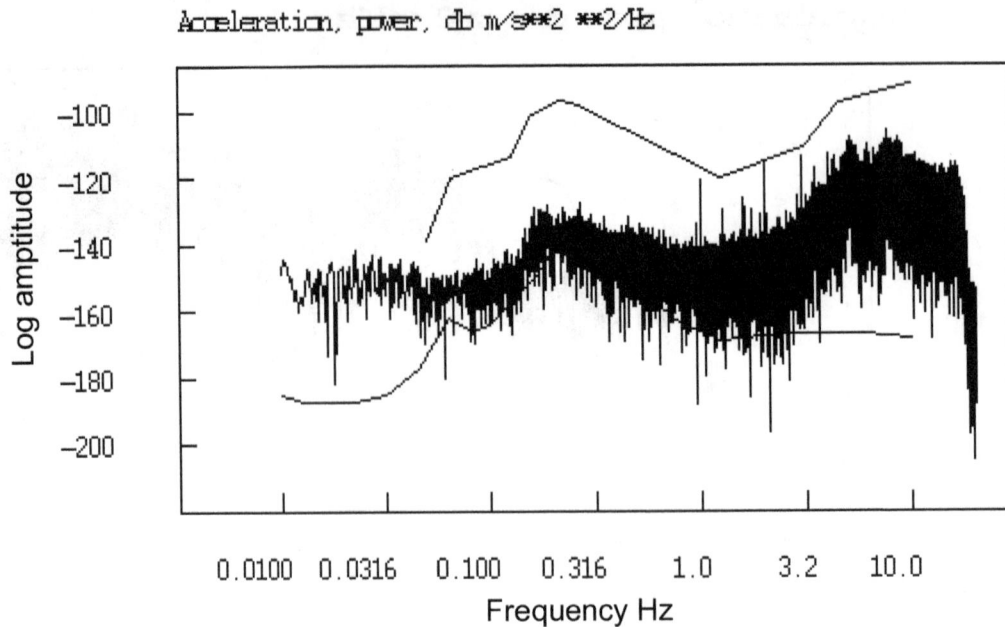

Figure 23. Noise variation between 9am-1pm (Kaduna Station).

also observed at lower frequency below 0.3 Hz. Since this station is located far away from busy settlements and heavy traffic, it is likely that the observed high noise at the Z-component is due to the contribution of cultural noise from human activities around the site, traffic and wind, noise from surrounding trees and instrumental noise.

Conclusion

Noise spectral analysis was conducted for all the NNNSS (Figure 1) and the various noise spectra (Figures 14-23) obtained were compared with Peterson noise curves and as observed in (Figure 5). The observed high levels of noise at the stations are likely due to a number of factors ranging from contribution of cultural noise to instrument self-noise. Since noise analysis was not carried out before the construction of these stations (Figure 1), it is highly encouraged to perform a more comprehensive noise analysis in order to understand the noise characteristics and the isolation of those stations must be provided in order to obtain better data quality and high S/N ratio. Also, the stations with observed higher noise level like Nsukka needed to be better installed and insulated. The orientation and leveling of the instruments should be checked and possibly instrument response and calibration files checked. Since this is a preliminary study, there is the need to undertake a site assessment to ascertain to a reasonable degree, SOURCES of seismic noise at each Station using recommended distances by Willmore (1979).

Abbreviations: NNNSS, Nigerian national network of seismographic stations; **CGG,** centre for geodesy and geodynamics; **SNR,** signal to noise ratio.

REFERENCES

Akpan OU, Yakubu TA (2010). A Review of Earthquake Occurrences and Observations in Nigeria. Earthq. Sci. 23(3):289-294.

Aki K, Richards PO (1980). Quantitative seismology. Freeman and Co., New York. pp. 613-614.

Alguacil G, Havskov J (2010). Instrumentation in Earthquake Seismology. 3:77-85.

Bonnefoy-Claudet S, Cotton F, Bard PY (2006). The nature of noise wavefield and its applications for site effects studies. A literature review. Earth Sci. Rev. 79:205-227.

Bormann P (1998). Conversion and comparability of data presentation on seismic background noise. J. Seismol. 2(1).

Bormann P (2009). Seismic Signal and Noise. New Manual of Seismological Observatory Practice. 1:213-236.

Chapman M, Liu E, Li X (2006). The Influence of fluid sensitive dispersion and attenuation on AVO analysis. Geophys. J. Int. 167(1):89-105.

Friedrich A, Klinge K, Krüger F (1998). Ocean-generated microseismic noise located with the Gräfenberg array. J. Seism. 2:47-64.

Havskov J, Ottemoller L (2008). Processing Earthquake Data. pp. 64-68.

Havskov J, Ottemoller L (2008). Computer exercises in processing earthquake data using SEISAN and introduction to SEISAN. pp. 46-50.

Peterson J (1993). Observations and Modeling of Seismic Background Noise, U.S. Geological Survey Open File Report. 93(322):94.

Peterson (1963). FDSN Station Book http://www.fdsn.org/station_book/G/BNG/bng.g_allyr.gif). And Bormann, New Manual of Seismological Observatory Practice.

Sheen DH, Shin JS, Kang TS (2009). Seismic noise level variations in South Korea. Geosci. J. 13(2):183-190.

Sherrif RE (1991). Encyclopedic Dictionary of Exploration Geophysics. Third Edition: Society of Exploration Geophysic. P. 387.

7

Traffic accident in India in relation with solar and geomagnetic activity parameters and cosmic ray intensity (1989 to 2010)

P. L. Verma

Department of Physics, Government Vivekan and P. G. College Maihar Satna M. P. India.

We studied the relation between deaths due to traffic accident in India and geomagnetic activity indices: Kp, Ap, Dst and solar activity (SA) parameters, sun spot number, solar flare index (FI), coronal index (CI) and yearly average of cosmic ray intensity, observed during the period of 1989-2011. It is seen that rate of death by traffic accident are well correlated with these parameters. We have found large negative correlation with correlation coefficient of -0.84,-0.79 between rate of death by traffic accident and Kp and Ap indices; large positive correlation with correlation coefficient 0.79 rate of death by traffic accident and Dst index. Further good medium correlation with correlation coefficient of -0.56,-0.60.-0.56 has been found between rate of death by traffic accident and yearly average of sunspot number, solar flare index and coronal index respectively. A positive correlation with correlation coefficient of 0.69 between rate of death by traffic accident and yearly average of cosmic ray intensity was also obtained.

Key words: Sunspot numbers, solar flare index, coronal index, cosmic ray intensity, Ap index, Kp index, Dst index, traffic accident death.

INTRODUCTION

Solar activity (SA), geomagnetic activity (GMA) and cosmic ray activity (CRA), respectively are major constituents of space weather that affects our daily life-navigation (Ptitsyna et al., 1996) as well as human health in space and in the Earth (Breus, 2003). Over the last years, many studies have been carried out concerning the possible effect that solar and geomagnetic activity might have on human physiological state (Stoupel, 2002; Dorman et al., 2001; Cornelissen et al., 2002; Dimitrova et al., 2004). Cosmic ray intensity (CRI) and GMA variations can influence not only the performance and reliability of space-borne or ground-based technological systems, but also human life (Cornelissen et al.,

2002;Dzvonik et al., 2006; Stoupel et al., 2007).

These results refer not only to the possible influence of GMA disturbances on the human cardiovascular state through variations of physiological parameters such as heart rate (HR) and arterial diastolic and systolic blood pressure (Dimitrova et al., 2009), but also on the central and vegetative nervous system through changes of the human brain's functional state and the psycho - emotional state (Babayev et al., 2006). Some studies revealed that the most significant effects on myocardial infarctions, brain strokes, and traffic accidents were observed on the days of geomagnetic field disturbances accompanied with Forbush decreases (FDs)

(Zhadin, 2001; Ptitsyna et al., 1998) and especially during the declining phase of FD (Villoresi et al., 1994; Dorman, 2005). At the same time, it was shown that very low GMA could also adversely affect human cardio-vascular system (Stoupel et al., 2004, 2005, 2007; Stoupel, 2006) and that is why it is suggested that the role of environmental physical factors becoming more active in low GMA, like CR (neutron) activity, should be an object of further studies (Stoupel, 2006).

In the last decades, many scientists have worked on the impact of space weather parameters, through the geomagnetic field, on different diseases (Dorman et al., 2001; Stoupel, 2002; Gmitrov and Ohkubo, 2002; Gmitrov and Gmitrova, 2004; Dimitrova et al., 2004). It has been revealed that cardiovascular, circulatory, nervous and other functional systems react under changes of geophysical factors (Kay, 1994; Watanabe et al., 1994; Persinger and Richards, 1995; Gurfinkel et al., 1995; Cornelissen et al., 2002). It has long been claimed that geomagnetic storms and other electromagnetic variations are associated with changes in the incidence of various diseases, myocardial infractions and strokes (Halberg et al., 2000). Some evidence has also been accumulated on the association between geomagnetic disturbances and increases in work and traffic accidents (Stoupel et al., 2004; Ptitsyna et al., 1996; Dorman, 2005). Others have also been reported on the association between geomagnetic disturbances and increases in number of road traffic and work (industrial) accidents (Stoupel et al., 2004; Srivastava and Saxena, 1980; Ptitsyna et al., 1996; Babayev et al., 2006). These studies were based on the hypothesis that a significant part of traffic accidents could be caused by the incorrect or retarded reaction of drivers to the traffic circumstances, the capability to react correctly being influenced by the changes in the environmental physical activity, particularly, sharp fluctuations of GMF.

Some studies revealed that the most significant effects of myocardial infarction, brain stroke and traffic accidents were observed on the days of GMF disturbances accompanied with Forbush decreases (Villoresi et al., 1995; Ptitsyna et al., 1998; Dorman, 2005). There are numerous indications that solar activity and solar activity variability-driven time variations of the geomagnetic field can be hazardous in relation to human health state and safety. Some evidence has been reported on the association between geomagnetic disturbances and increases in work and traffic accidents (Ptitsyna et al., 1998). These studies were based on the hypothesis that a significant part of traffic accidents could be caused by the incorrect or retarded reaction of drivers to the traffic circumstances, the capability to react correctly being influenced by the environmental magnetic and electric fields. Reiter (Stoupel, 2004) found that work and traffic accidents in Germany were associated with disturbances in atmospheric electricity and in the geomagnetic field (defined by sudden perturbations in radio-wave propagation).

On the basis of 25 reaction tests, it was found also that the human reaction time, during these disturbed periods, was considerably retarded. On the basis of huge statistical data on several millions of medical events in Moscow and in St. Petersburg, there were found sufficient influence of geomagnetic storms accompanied with Cosmic Ray (CR) Forbush-decreases on the frequency of myocardial infarctions, brain strokes and car accident road traumas (Villoresi et al., 1994).

The most remarkable and statistically significant effects have been observed during days of geomagnetic perturbations defined by the days of the declining phase of Forbush decreases in CR intensity. During these days the average numbers of traffic accidents increase by ($17.4 \pm 3.1\%$) (Dorman, 2005). Some studies show correlations between high SA and GMA and increased number of traffic accidents (Stoupel et al., 2004).

In the present investigation, an attempt has been made to get possible relationship between solar, geomagnetic and cosmic ray activities with rate of traffic accident death in India. For this study, rate of traffic accident death, solar, geomagnetic and cosmic ray activity parameters for the period of 1989-2010 has been used.

EXPERIMENTAL DATA

In this study, the rate of male and female death by traffic accident in India, yearly average of solar activity parameters sunspot number, solar flare index, coronal index and cosmic ray intensity, geomagnetic activity parameter Kp, Ap and Dst indices for the period 1989-2011, has been taken into consideration. The data of traffic accident has been taken from the National Crime Records Burea (NCRB) 2010 records, ministry of home affairs of India. The data of solar activity parameters, sunspot number, solar flare index and coronal index are taken from the Solar-Terrestrial Physics (STP) solar data (www.ngdc.noaa.gov/stp/solardataservices). The data of cosmic ray intensity yearly average count rates of Oulu super neutron monitor has been used. The data of geomagnetic activity parameters has been taken from Omni web data system (Table 1).

Analysis and results method of analysis and results

In this study, statistical method of correlation has been used. The correlation is one of the most commonly used statistics. A correlation is a single number that describes the degree of relationship between two variables. Correlation coefficient, symbolized as r, is a numerical summary of a bivariate relationship and can range from -1.00 to $+1.00$. Any r that is positive indicates a direct or positive relationship between two measured variables. Negative r indicates indirect or inverse relationship. The formula for the correlation is:

$$r = \frac{N \sum XY - (\sum X)(\sum Y)}{\sqrt{\left[N \sum X^2 - (\sum X)^2 \right] \left[N \sum Y^2 - (\sum Y)^2 \right]}}$$

Where: N=number of pairs of scores, $\sum XY$ =sum of the products of

Table 1. Yearly average of geomagnetic, solar activity parameter, cosmic ray intensity and rate of traffic accident during period of 1989-2010.

Year	Kp index	Dst index	Sunspot numbers	Solar flare index (FI)	Coronal index (CI)	Yearly average of cosmic ray intensity	Rate of death
1989	843	−905.42	157.6	17.39	17.52	5480	20.8
1990	775.42	−638	142.6	12.2	15.99	5416	21.1
1991	918.5	−932.83	145.7	15.16	14.74	5432	22.1
1992	787.5	−617.75	94.3	7.74	10.97	5922	22.5
1993	723.1	−501.42	54.6	4.23	6.88	6203	21.8
1994	822.67	−629.92	29.9	1.58	4.36	6280	21.2
1995	657.92	−511.67	17.5	0.86	2.86	6387	24.3
1996	578.1	−331.92	8.6	0.42	1.98	6503	23.6
1997	496.42	−440.42	21.5	1.01	3.08	6545	24.5
1998	613.67	−518.67	64.3	4	6.35	6399	26.6
1999	665.25	−398.17	93.3	6.39	8.93	6203	27.6
2000	718.42	−581.33	119.6	7.61	9.74	5784	25.5
2001	637.42	−539.75	110.9	6.8	10.53	5889	26.4
2002	686.33	−638.67	104.1	4.56	10.14	5806	24.8
2003	870.33	−527.25	63.56	3.46	6.21	5759	24.3
2004	663.33	−372.58	40.44	1.6	6.4	6093	25.5
2005	643.25	−480.17	29.78	1.91	4.64	6156	26.7
2006	490.58	−353.1	15.18	0.54	4.54	6478	28.1
2007	460.83	−250.1	7.5	0.47	2.11	6632	30
2008	444.1	−239.58	2.86	0.03	1.6	6662	29.7
2009	274.17	−337	3.09	0.02		6804	30.5
2010	382	−186.92	16.5	0.39		6623	32.4

paired scores, ΣX = sum of x scores, ΣY = sum of y scores, ΣX^2 = sum of squared scores, ΣY^2 = sum of squared score.

The scale of correlation coefficient is:

0.8 to 1.0 or -0.8 to -1.0 (very large relationship)
0.6 to 0.8 or -0.6 to-0.8 (large relationship)
0.4 to 0.6 or -0.4 to -0.6 (good medium relationship)
0.2 to 0.4 or 0.2 to -0.4(weak relationship)
0.0 to 0.2or 0.0 to -0.2 (weak or no relationship)

RESULTS

Large negative correlation with correlation coefficient, -0.84 between rate of death by accident and Kp indices.
Large negative correlation with correlation coefficient, -0.79 between rate of death by traffic accident and Ap indices.
Large positive correlation with correlation coefficient, 0.79 rate of death by traffic accident and Dst indices.
Good medium correlation with correlation coefficient, -0.56 was found between rate of death by traffic accident and yearly average of sunspot number.

Negative correlation with correlation coefficient, -0.60 has been found between rate of death by traffic accident and yearly average of solar flare index.
Negative correlation with correlation coefficient, -0.56 has been found between rate of death by traffic accident and yearly average of coronal index.
Positive correlation with correlation coefficient, 0.69 has been found between rate of death by traffic accident and yearly average of cosmic ray counts rate.

Conclusion

The long term study confirms results of number of previous observations on links between timing of human death and environmental physical factors (Gmitrov and Ohkubo, 2002, Gmitrov and Gmitrova, 2004; Stoupel, 2002, 1999; Stoupel et al., 2004, 2005, 2007). Results of this study also shows that there is strong relationship between death by traffic accident and geomagnetic activity, solar activity parameters (sunspot numbers, solar flare index, coronal index, Kp index, Ap index, Dst index) and cosmic ray activity. The significant correlation with correlation coefficient of -0.84, -0.79, and 0.79

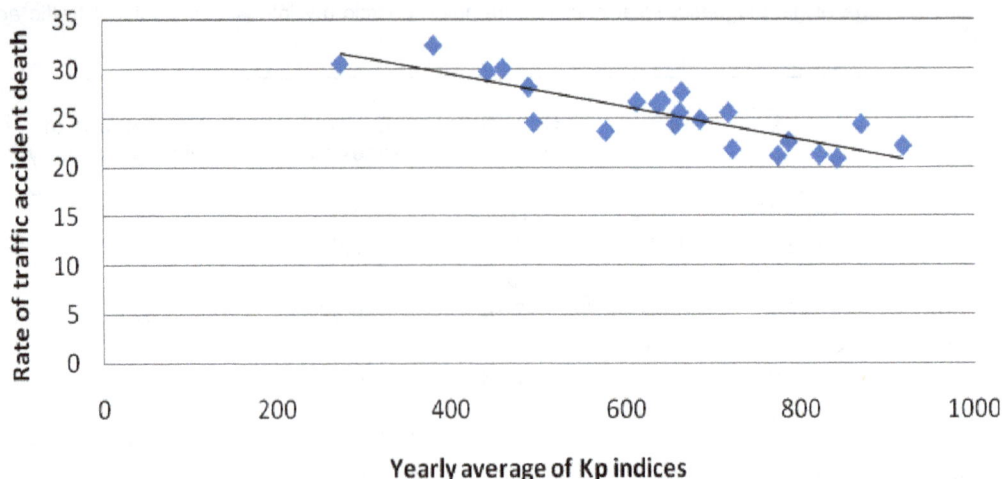

Figure 1. Scatter plot between yearly average of Kp indices and rate of death by traffic accident during period of 1989-2010 showing negative correlation with correlation coefficient -0.84.

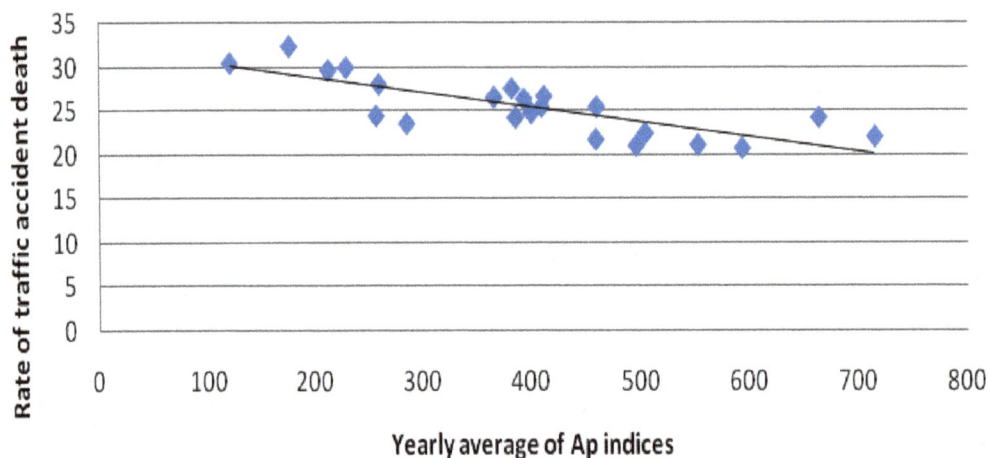

Figure 2. Scatter plot between yearly average of Ap indices and rate of death by traffic accident during period of 1989-2010 showing negative correlation with correlation coefficient -0.79.

between yearly average of Kp, Ap, Dst index and rate of death by traffic accident shown in Figures 1, 2, 3 and -0.56, -0.60, -0.56 between yearly average of sunspot number, solar flare index ,coronal index and death by traffic accident shown in Figures 4, 5 and 6 and 0.6, between yearly average of cosmic ray count rate and death by traffic accident shown in Figure 7; confirms the relationship of solar, geomagnetic activity parameter, cosmic ray activity and rate of death by traffic accident.

Although, there are several factors of traffic accident like medical-biological; social; sleeping while driving, etc; as well as terrestrial weather changes – meteorological conditions. Nevertheless, alongside with aforementioned medical-biological, meteorological, social and other affecting factors, disturbances and variations in space

weather (geophysical and cosmic ray activity) can also play a significant role in traffic accidents as a trigger factor, increasing the risk of traffic accidents.

Previous investigators have determined that human brain activity is influenced by geomagnetic activity (Babayev and Allahverdiyeva, 2007; Allahverdiyev et al., 2001). Disturbances of geomagnetic conditions causes negative influence, seriously disintegrate brain's functionality, activate braking processes and amplify the negative emotional background of an individual. It is also shown that geomagnetic disturbances affect mainly emotional and vegetative spheres of humans. A human being's physiological state and brain's bioelectrical activity, affected by geomagnetic activity negatively, may result in inadequate reaction and its retardation in relation

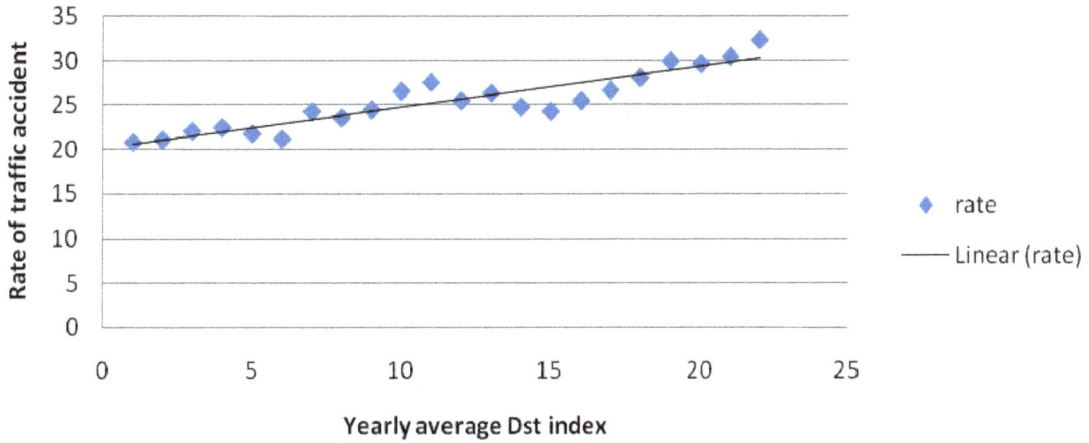

Figure 3. Scatter plot between yearly average of Dst indices and rate of death by traffic accident during period of 1989-2010 showing positive correlation with correlation coefficient .079.

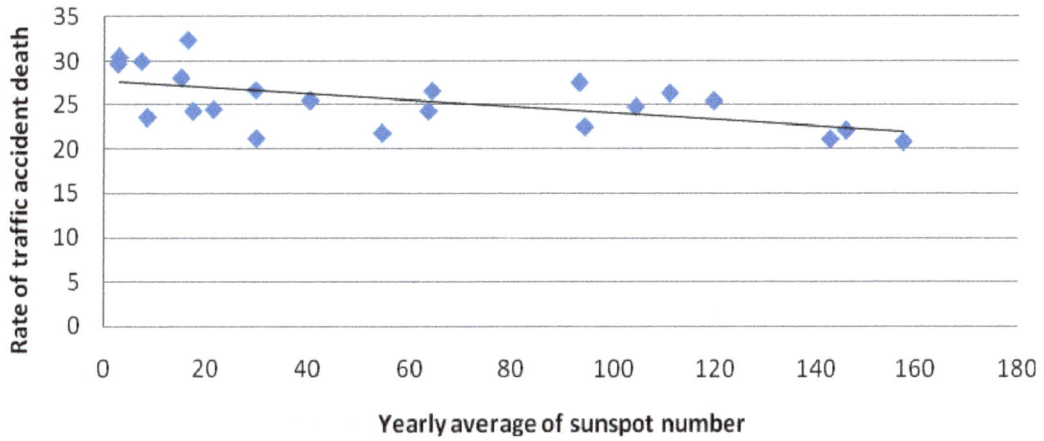

Figure 4. Scatter plot between yearly average of sun spot number and rate of death by traffic accident during period of 1989-2010 showing negative correlation with correlation coefficient -0.56.

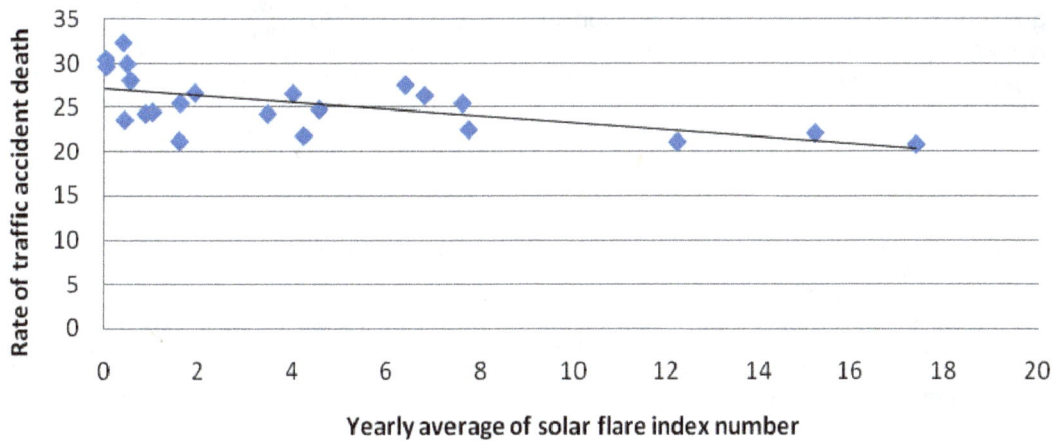

Figure 5. Scatter plot between yearly average of solar flare index and rate of death by traffic accident during period of 1989-2010 2010 showing negative correlation with correlation coefficient -0.60.

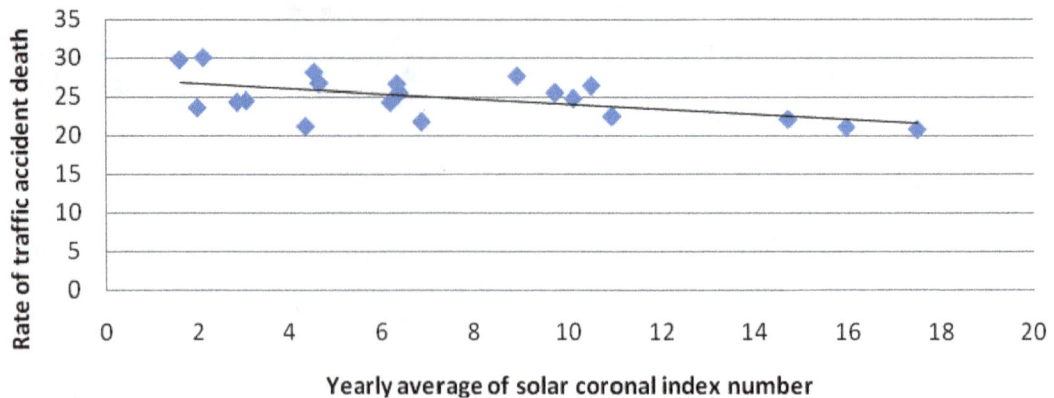

Figure 6. Scatter plot between yearly average of solar flare index and rate of death by traffic accident during period of 1989-2010 showing negative correlation with correlation coefficient -0.56.

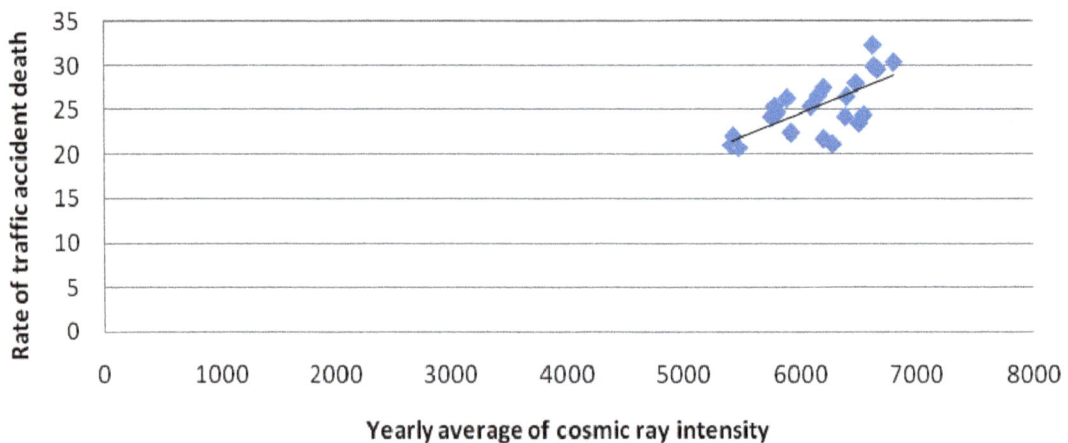

Figure 7. Scatter plot between yearly average of cosmic ray count rate and rate of death by traffic accident during period of 1989-2010 showing positive correlation with correlation coefficient 0.69.

to the situation leading to the relatively increased number of traffic accidents. In this study, the phenomena with solar, geomagnetic and cosmic ray activity in relation to India and obtained significant correlation. For general picture of the phenomena, further study is needed for different latitude and longitude.

REFERENCES

Allahverdiyev AA, Babayev ES, Khalilov EN, Gahramanova NN (2001). Possible space weather influence on the functional state of the human brain. Proceedings of the ESA Space Weather Workshop "Looking Towards a European Space Weather Programme". P. 17.

Babayev ES, Allahverdiyeva AA (2007). Effects of geomagnetic activity variations on the physiological and psychological state of functionally healthy humans: Some results of Azerbaijani studies. Adv. Space Res. 40:1941.

Babayev ES, Hashimov AM, Yusifbeyli NA, Rasulov ZG, Asgarov AB (2006). Geomagnetic storm risks to electric power distribution and supply systems at mid-latitude locations and their vulnerability from space weather. Technical and Physical Problems in Power Engineering" (Proc. 3rd Int. Conf. TPE-2006, Turkey, 29-31 May 2006), Gazi University, Ankara, 1097.

Cornelissen G, Halberg F, Breus Tk, Syutkina EV, Baevskii RM, Weydahl A, Watanabe Y, Otsuka K, Siegelova J, Fiser B, Bakken E (2002). Non-photic Solar Associations of Heart Rate Variability and Myocardial Infarction." J. Atm. Solar-Terr. Phys. 64:707.

Dimitrova S, Mustafa FR, Stoilova I, Babayev ES, Kazimov EA (2009). Possible influence of solar extreme events and related geomagnetic disturbances on human cardio-vascular state: results of collaborative Bulgarian-Azerbaijani studies. Adv. Space Res. 43:641.

Dimitrova S, Stoilova I, Cholakov I (2004). Influence of local geomagnetic storms on arterial blood pressure. Bioelectromagnetics 25:408.

Dorman LI (2005). Space Weather and Dangerous Phenomena on the Earth: Principles of Great Geomagnetic Storms Forecasting by Online Cosmic Ray Data. Ann. Geophys. 23:2997.

Dorman LI, Iucci N, Ptitsyna NG, Villoresi G (2001). Cosmic ray as indicator of space weather influence on frequency of infracts myocardial, brain strokes, car and train accidents, in: Proc. 27th ICRC (Hamburg). P. 3511.

Dzvonik O, Stetiarova J, Kudela K, Daxner P (2006). A monitoring of space weather effects on some parameters of mental performance and health in aviation personnel. Stud. Psychol. 48:273.

Gmitrov J, Gmitrova A (2004). Geomagnetic field effect on cardiovascular regulation. Bioelectromagnetics 25:92.

Gmitrov J, Ohkubo C (2002). Artificial static and geomagnetic field interrelated impact on cardiovascular regulation. Bioelectromagnetics 23:329.

Gurfinkel Iu I, Liubimov VV, Oraevskii VN, Parfenova LM, Iur'ev A (1995). The effect of geomagnetic disturbances in capillary blood flow in ischemic heart disease patients. Biofizika 40:793.

Halberg F, Cornelissen G, Otsuka K, Watanabe Y, Katinas GS, Burioka N (2000). Cross-spectrally coherent 10.5 and 21-year biological and physical cycles, magnetic storms and myocardial infarctions. Neuroendocrinol. Lett. 21:233.

Kay RW (1994). Geomagnetic storms: association with incidence of depression as measured by hospital admission. Br. J. Psychiatry 164(3):403.

Persinger MA, Richards PM (1995). Vestibular experiences of humans during brief periods of partial sensory deprivation are enhanced when daily geomagnetic activity exceeds 15–20 nT. Neurosci. Lett. 194(1-2):69.

Ptitsyna NG, Villoresi G, Dorman LI, Iucci N, Tyasto MI (1998). Natural and man-made low-frequency magnetic fields as a potential health hazard. UFN (Uspekhi Physicheskikh Nauk). 168(7):767.

Ptitsyna NG, Villoresi G, Kopytenko YA, Kudrin VA, Tyasto MI, Kopytenko EA, Iucci N, Voronov PM, Zaitsev DB (1996). Coronary heart diseases: an assessment of risk associated with workexposure to ultra low frequency magnetic fields." Bioelectromagnetics 17:436.

Srivastava BJ, Saxena S (1980). Geomagnetic –biological correlations –Some new results. Indian J. Radio Space Phys. 9:121.

Stoupel E (1999). Effect of geomagnetic activity on cardiovascular parameters. J. Clin. Basic Cardiol. 2:34.

Stoupel E (2002). The effect of geomagnetic activity on cardiovascular parameters. Biomed. Pharmacother. 56:247.

Stoupel E (2006). Cardiac arrhythmia and geomagnetic activity. Indian Pacing Electrophysiol. J. 6:49.

Stoupel E, Babayev ES, Mustafa FR, Abramson EP, Israelevich P (2007). Acute myocardial infarction (AMI) occurrence – environmental links. Baku 2003–2005 data. Medical science monitor. Int. Med. J. Exp. Clin. Res. P. 175.

Stoupel E, Domarkiene S, Radishauskas R, Israclcvich P, Abramson E, Sulkes J (2005). In women myocardial infarction occurrence is much stronger related to environmental physical activity than in men-a gender or an advanced age effect? J. Clin. Basic Cardiol. 8:59.

Stoupel E, Kalediene R, Petrauskiene J, Domarkiene S, Radishauskas R, Abramson E, Israelevich P, Sulkes J (2004). Three Kinds of Cosmophysical Activity: Links to Temporal Distribution of Deaths and Occurrence of Acute Myocardial Infarction. Med. Sci. Monit. 10:80.

Villoresi G, Breus TK, Dorman LI, Iucci N, Rapoport SI (1994). The influence of geophysical and social effects on the incidences of clinically important pathologies (Moscow 1979–1981). Phys. Med. 10(3):79-91.

Villoresi GL. Dorman I, Ptitsyna NG, Iucci N, Tiasto MI (1995). Forbush decreases as indicators of health - hazardous geomagnetic storms", Proc. 24th Intern. Cosmic Ray Conf. Rome. 4:1106-1109.

Watanabe Y, Hillman DC, Otsuka K, Bingham C, Breus TK, Cornelissen G, Halberg F (1994). Cross-spectral coherence between geomagnetic disturbance and human cardiovascular variables at non-societal frequencies. Chronobiologia 21(3-4):265.

Zhadin MN (2001). Review of Russian literature on biological action of DC and low frequency AC magnetic fields. Bioelectromagnetics 22:27.

8

Aftershock sequence analysis of 19 May, 2009 earthquake of Lunayyir lava flow, northwest Saudi Arabia

Al-Zahrani H.[1]* , Al-Amri A. M.[1, 3], Abdel-Rahman K.[1, 2] and Fnais M.[1]

[1]Department of Geology and Geophysics, King Saud University, Riyadh, Saudi Arabia.
[2]Department of Seismology, National Research Institute of Astronomy and Geophysics, Cairo, Egypt.
[3]Water Resources Exploration Chair In the Empty Quarter, King Saud University, Riyadh, Saudi Arabia.

Aftershock sequence of 19th May, 2009 Lunayyir earthquake (Mw 5.7) has been recorded by deploying seismic stations immediately after the occurrence of the main shock. Analysis of this sequence clarified that; the major part of cumulative seismic moment released during the first hours after the occurrence of the main shock; their orientation is north west (NW) parallel to the Red Sea axial trend along Najd faulting system; it is clustered at two depths; from 5-10 and 15-25 km beneath Lunayyir area; illustrated successive periods of maxima and minima; decayed rapidly following the relation of n(t)=37.28 t-0.6. Fault plane solutions for five large events indicate normal faulting is the major mechanism. These findings are in agreement with the north east (NE) (transform fault) trend runs crossing the Red Sea into Shield area where the activity has been initiated and travelled through the Lunayyir area. While the NW Najd faulting system runs parallel to the Red Sea axial trend and intersected with NE fault trend underlying Lunayyir area. Accordingly, it can be concluded that, the recent earthquake activities at Lunayyir lava flow area are related to the present-day Red Sea tectonics. Monitoring earthquake activities through installing permanent seismic network around the area is highly recommended.

Key words: aftershock sequence, fault plane solution, stress, and Lunayyir.

INTRODUCTION

Lunayyir lava flow area lies between Latitudes 25°.10 - 25°.17 N and Longitudes 37°.45 - 37°.75 E in the northwestern part of Saudi Arabia (Figure 1). It is oriented, generally, northwest-southeast (Laurent and Chevrel, 1980) parallel to the Red Sea axial trend. It has been experienced in October 2007 an earthquake swarm with maximum magnitude of 3.2, while in 2009 another earthquake swarm has been initiated on 19th April and increased gradually till its maximum on 19th of May where the main shock (M$_w$ 5.7) was occurred. Great number of aftershocks with maximum magnitude of 4.8 was recorded. This earthquake activity was close to some scattered urban communities (for example, Al Ays town, 40 km southeast of the epicenter). Accordingly, Saudi government evacuated more than 40,000 people

(Pallister et al., 2010).

Field survey has been done for the epicentral area immediately after the mainshock occurrence (Fnais, 2010; Jónsson et al., 2010; Al-Zahrani, 2010). Depending on these investigations, there are crakes with different directions while the total length of surface rupture was extended up to 8 km and 90 cm of offset with the main shock (Fukao and Kikushi, 1987). It is known that the stress drop on the fault plane due to the occurrence of an earthquake produces an increase of effective shear stress around the rupture area (Chinnery, 1963). The transferring in the static stress may explain the location of some aftershocks away from the fault.

GEOLOGICAL AND TECTONIC SETTING

Lunayyir lave flow area imposed in the Shield area that composed of hard rocks and surficial soft sediments

*Corresponding author. E-mail: alzaharani.h@hotmail.com.

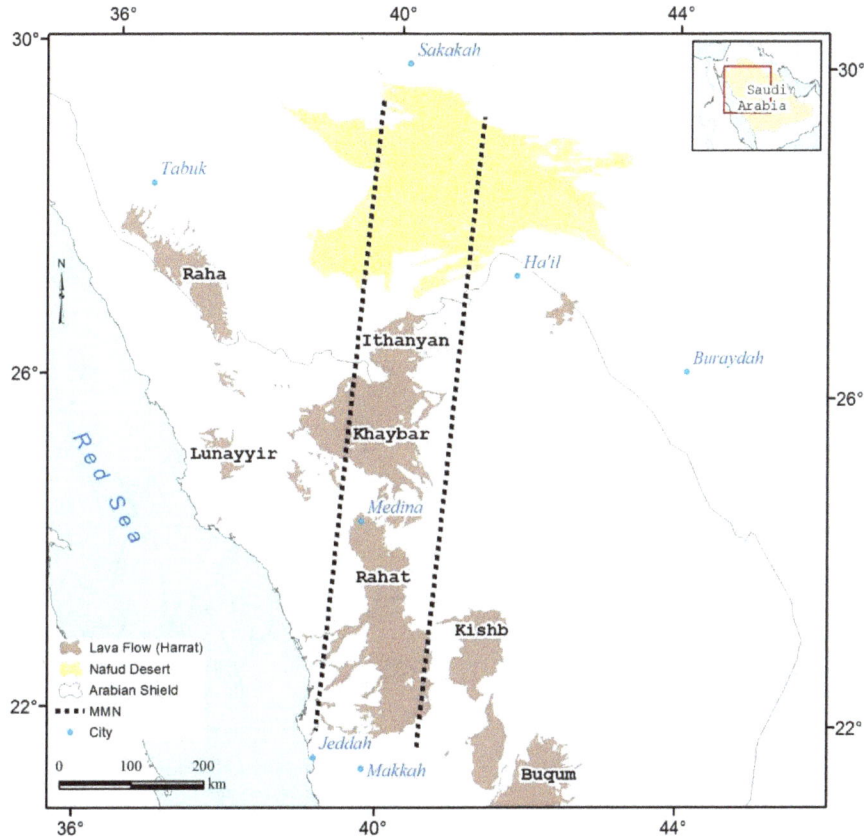

Figure 1. Location of Lunayyir Lava flow.

ranges in age from precambrian to recent. Precambrian rocks represented by gneisses and schists that had been subjected to strong folding and faulting. Tertiary to Holocene basalts composed of alkaline olivine basalt, whereas quaternary sediments extend along some Wades in the northern part of the area.

Tectonically, Lunayyir area was subjected to two episodes of tectonic movements synchronized with the Red Sea floor spreading through pre-early Miocene rifting period (Girdler, 1969; Anon, 1972; Rose et al., 1973). It has been controlled-to great extent-by the regional stress regime of the western Arabian plate associated with the Cenozoic development of the Red Sea. Harrat Lunayyir probably faulted during Cenozoic rift time where the up arching period was parallel to the Red Sea coast. During Late Miocene-Pliocene, the alkali basalt invaded into the Harrat Lunayyir. According to the above-mentioned, there are different fault trends prevailing the area and oriented NE-SW, NNW-SSE, and NW-SE (Figure 2).

RECORDING OF AFTERSHOCK SEQUENCE

King AbdulAziz City for Science and Technology (KACST) and King Saud University (KSU) installed two sub-networks of seismic stations at the epicentral area on 20th May for recording the aftershock sequence (Figures 3 and 4, and Table 1).

KACST sub-network consists of 15 short-period seismic stations and one of broadband, while KSU sub-network composed of 6 stations of short period with the same configuration of KACST stations. Each of the short period stations is equipped by short period SS-1 seismometer (Kinemetrics Inc.); Quantira digitizer as seismic data logger (Q_{330}) and Baler (20GB hard disk) for storing of recorded earthquakes; and 12-volt battery and solar panel for continuous charging, while, broadband station include broadband Streckeisen STS-2 seismometer (Kinemetrics Inc.). The data has been collected with sampling rate of 300 SPS for the broadband and short period as well.

First arrival times of P and S-waves were picked to compute the hypocentral parameters using modified HYPO71PC software version (Lee and Valdes, 1985). Furthermore, some of recorded data was analyzed using HYPOINVERSE software (Klein, 1987) which calculates the standard errors of hypocentral parameters as well. Two of 1-D velocity models (Makris et al., 1978; Al-Amri et al., 2008) have been used through data processing in this study. Residual values from crustal models were compared with no significant differences have been

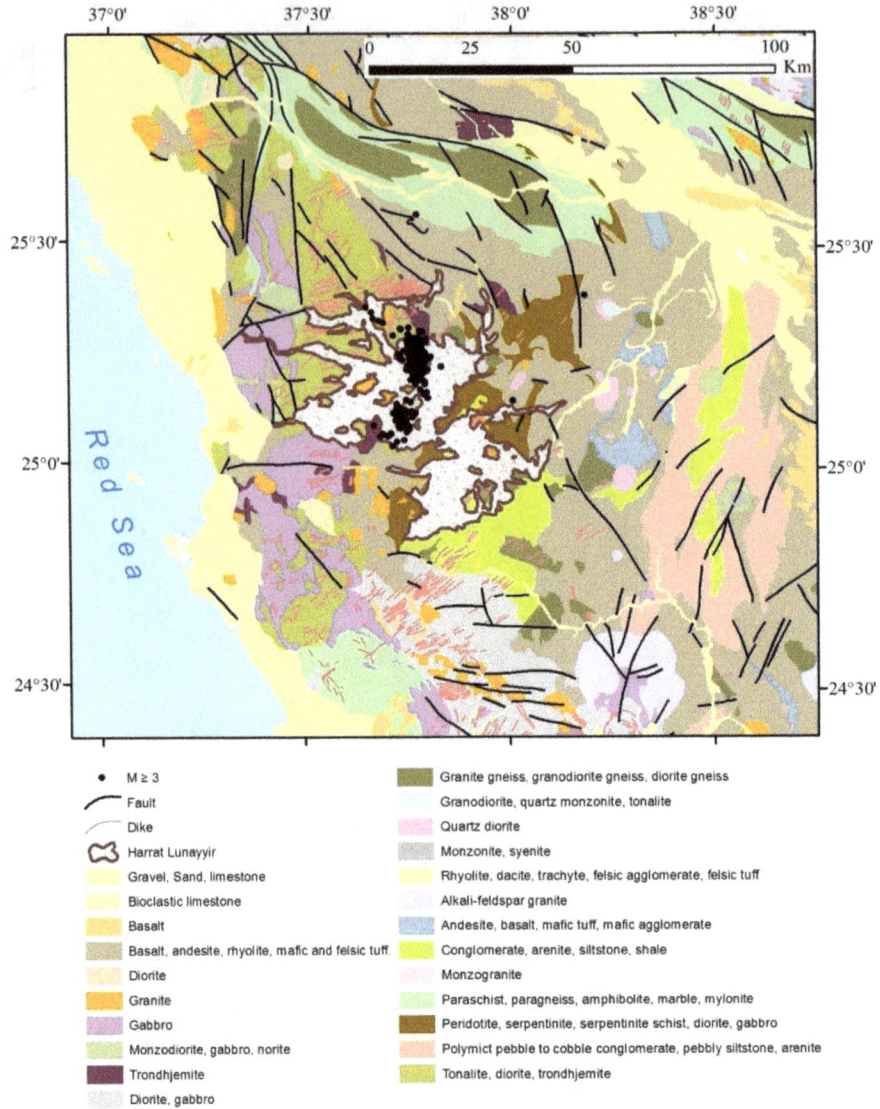

Figure 2. Geological setting around Lunayyir area.

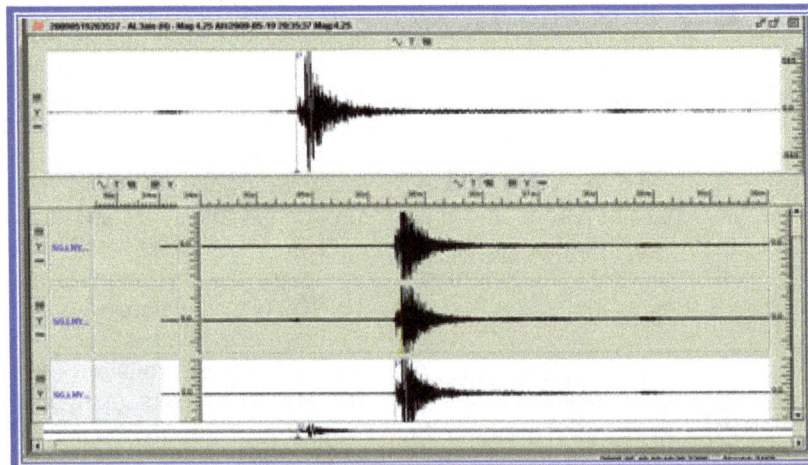

Figure 3. Waveform of 2009-05-19 aftershock (20:35:37 and M = 4.25).

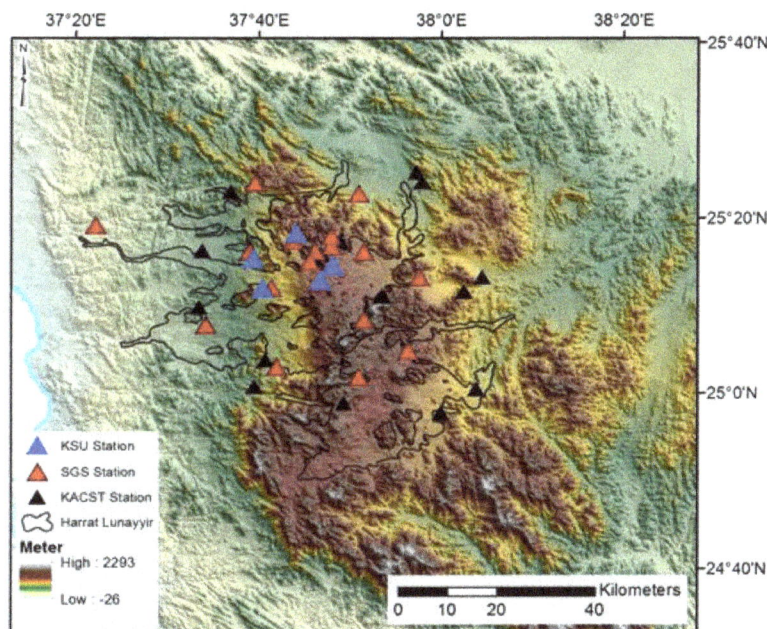

Figure 4. Map of two seismic deployments (KSU and KACST).

Table 1. Parameters of KACST and KSU sub-networks at Lunnayyir area.

Station code	Latitude (N)	Longitude (E)	Elevation (m)	Seismometer	Data logger
STN01	25.2554	37.7698	1010	STS-2	Q_{330}
STN02	25.0618	37.6792	559	SS-1	Q_{330}
STN03	25.1851	37.8929	946	SS-1	Q_{330}
STN04	25.2640	37.7805	970	SS-1	Q_{330}
STN05	25.1637	37.5574	393	SS-1	Q_{330}
STN06	25.1934	38.0424	600	SS-1	Q_{330}
STN07	25.4015	37.9651	600	SS-1	Q_{330}
STN08	25.3823	37.6161	600	SS-1	Q_{330}
STN09	24.9631	37.9994	600	SS-1	Q_{330}
STN10	25.0120	37.6592	600	SS-1	Q_{330}
STN11	25.2877	37.8155	956	SS-1	Q_{330}
STN12	25.2708	37.5637	600	SS-1	Q_{330}
STN13	25.0072	38.0650	600	SS-1	Q_{330}
STN14	24.9802	37.8231	600	SS-1	Q_{330}
STN15	25.2204	38.0755	600	SS-1	Q_{330}
STN16	25.4202	37.9542	600	SS-1	Q_{330}
KSU01	25.2134	37.7794	1011	SS-1	Q_{330}
KSU02	25.2423	37.8026	957	SS-1	Q_{330}
KSU03	25.3030	37.7355	962	SS-1	Q_{330}
KSU04	25.2555	37.6571	551	SS-1	Q_{330}
KSU05	25.1989	37.6743	551	SS-1	Q_{330}

noticed between HYPO71 and HYOINVERSE programs, where the epicentral difference was less than 0.5 km. whereas the depths, which more sensitive to the velocity model, differed by about 1 km on average. Consequently, the hypocentral parameters given by HYPO71 were quite realistic.

Accordingly, 1050 of well-located aftershocks during the period of observation (20th May till 20th June) have

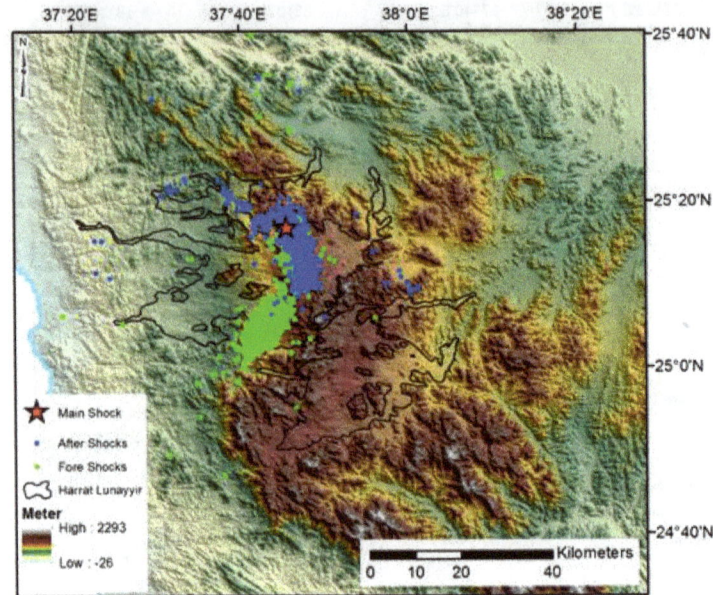

Figure 5. Aftershock sequence from 20 May - 19 June.

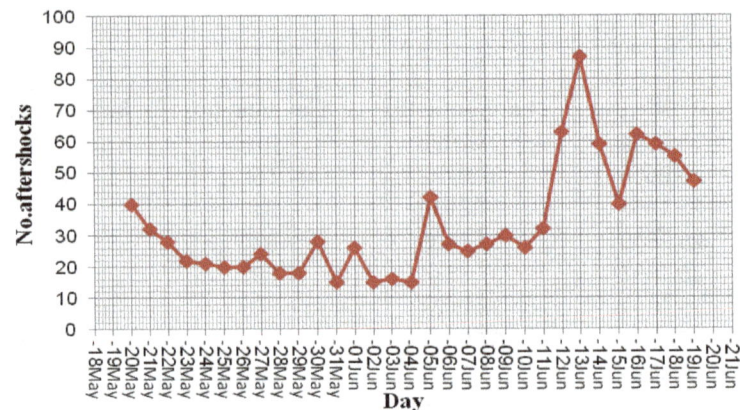

Figure 6. Relationship between numbers of aftershocks per day.

been analyzed through this study (Figure 5). Depending on Figure 5 aftershocks are extend northeast at the beginning and changed later into northwest. Figure 6 illustrates three episodes of maxima as follows; from 29th May - 2nd June; from 4th - 6th June; and from 11th - 13th June. These episodes were overlapped by three episodes of minima. These aftershocks are of crustal origin where it is clustered at two depth intervals between 5 to 10 km and from 15 to 25 km (Figure 7).

RATE OF DECAY FOR AFTERSHOCKS SEQUENCE

It is indicated that, the occurrence rate of aftershocks, N(t) obeys the modified Omori relation (Kisslinger, 1997;

Gheitanchi, 2003; Benito et al., 2004). The empirical relation for the rate of decay estimation has been suggested by Omori (Utsu, 1961). This relation combine the frequency of aftershocks N(t) per unit time t, following the main shock and represented by:

$$N(t) = k/t^{-c}$$

Where K and c are constants and should be estimated for each region.

More than 400 well-located aftershocks (M ≥ 2.0) have been used to estimate the rate of decay for Lunayyir earthquake swarm. The cyclic activation of aftershocks was indicated by their heterogeneities. Hence, first month of aftershock observation could be differentiated into

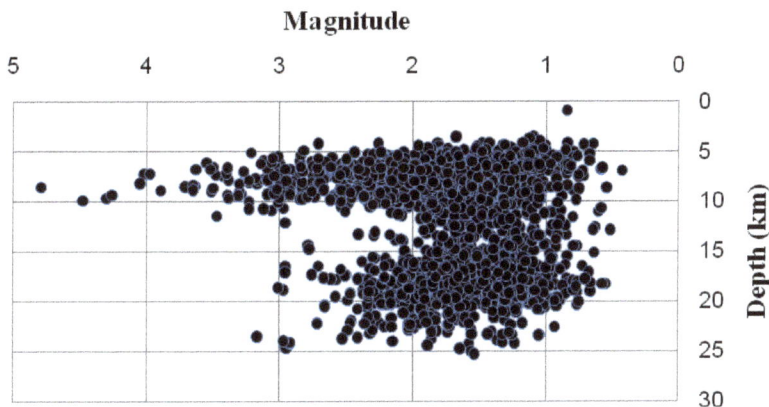

Figure 7. Magnitude – Depth relationship.

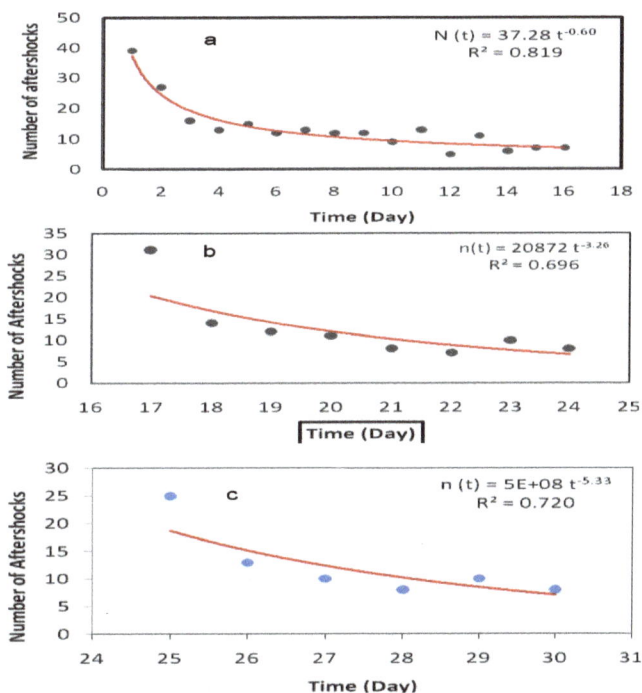

Figure 8. Rate of decay for the aftershock sequence (M ≥ 2).

three intervals; the first interval from 20th May till 4th June; the second from 5th-12th June, while the third from 13th - 19th June (Figure 8) as follows:

$N(t) = 37.28\ t^{-0.60}$, with $R^2 = 0.82$ (for the first two weeks),
$N(t) = 20872\ t^{-3.26}$, with $R^2 = 0.7$ (for the third week), and
$N(t) = 5E+08\ t^{-5.33}$, with $R^2 = 0.72$ (for the last week)

Where N (t) being the number of events by day, t is the time in days after the main shock, and R^2 is the correlation coefficient. Figure 8 indicates that, there are three different rates of decay through the first month of observation for the aftershocks of M ≥ 2.0. Another noticeable feature for Lunayyir aftershock sequence is

the releasing of major part of the cumulative seismic moment during the first hours after the main shock. Excluding the first day (20th May, 2009), no event with local magnitude above 4.0 was recorded during the period of recording.

FOCAL MECHANISM

The first motion polarities of P-wave have been picked for the large earthquakes (M ≥ 4.0) to estimate the fault plane solution for these aftershocks using the PMAN program (Suetsugu, 2003). The input parameters of the azimuths and take-off angles were calculated based on

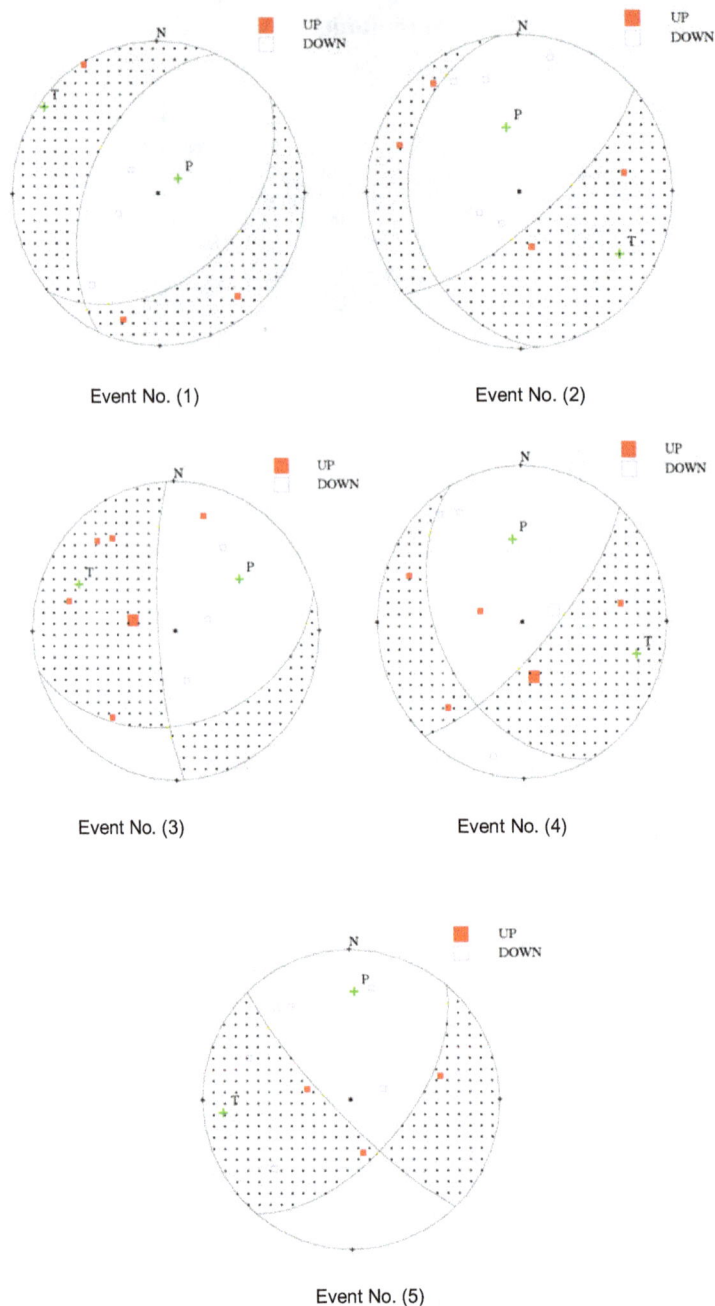

Event No. (1)   Event No. (2)

Event No. (3)   Event No. (4)

Event No. (5)

Figure 9. Fault plane solution for large aftershocks.

1-D velocity model of Al-Amri et al. (2008). Figure 9 shows fault plane mechanisms of large aftershocks by equal area projections of the lower focal hemisphere while their parameters are listed in Tables 2 and 3.

DISCUSSION

The recorded aftershocks are oriented NW parallel to the Red Sea axial trend. Furthermore, the focal mechanism of the large aftershocks illustrated NW trend is the main

faulting trend of normal mechanism. The extensional stress (*T*-axis) is directed NE, while NW trend represents the compressional stress (*P*-axis). These results are in agreement with the field measurements for the ground cracks and fissures accompanied with the main shock (Jónsson et al., 2010; Fnais, 2010). These aftershocks are of crustal origin where it is clustered at two depths of 5 to 10 km and 15 to 25 km beneath Lunayyir area. By considering the depth of the main shock was 9 km, it indicates that the rupture process was initiated at depth of 9 km and propagated with bilateral behavior of

Table 2. Source parameters for aftershocks used for focal plane solutions.

Event No.	Nodal Plane 1			Nodal Plane 1			P-axis		T-axis	
	Strike (°)	Dip (°)	Rake (°)	Strike (°)	Dip (°)	Rake (°)	Pl. (°)	Az. (°)	Pl. (°)	Az. (°)
1	204	51	-107	50	42	-70	76	54	4	-54
2	173	28	-143	49	73	-67	56	349	25	121
3	177	81	-127	75	38	-16	43	52	26	-65
4	43	74	-50	151	42	-156	46	353	19	104
5	136	81	-149	40	59	-11	28	3	14	-95

Table 3. Fault plane mechanism for large aftershocks.

Event No.	Date			Origin time			Location		Depth (Km)	Mag.
	y	m	d	H	M	S	Lat. (°N)	Long. (°E)		
1	2009	05	19	18	20	00	25.2820	37.7665	9.62	4.29
2	2009	05	19	19	21	27	25.2885	37.7622	7.20	4.01
3	2009	05	19	19	26	58	25.2733	37.7653	8.23	4.04
4	2009	05	19	20	35	37	25.3093	37.7303	9.30	4.25
5	2009	05	20	19	57	16	25.2975	37.7505	8.49	4.79

directivity. Aftershock sequence was represented by overlapping cycles of maxima and minima with different numbers and magnitudes. This may reflect the non-homogeneity in stress levels owing to the different episodes of magmatic dyke intrusions (Al-Amri and Fnais, 2009; Zahran et al., 2009). Most of the accumulated energy was released shortly after the main shock and decreased rapidly with different rates of decay which may be explains the presence of small scale ground cracks and fissures intersected with the major ruptures at the epicentral area (Mukhopadhyay et al., 2012; Al-Zahrani et al., 2012).

These results clarified the affecting of Lunayyir area with the present-day active tectonics related to the Red Sea floor spreading. Accordingly, the continuous monitoring for the volcano-tectonic activities and earthquake hazards at Lunayyir lava flow area are highly recommended.

ACKNOWLEDGEMENT

This project was supported by King Saud University, Deanship of Scientific Research, College of Science Research Center.

REFERENCES

Al-Amri A, Fnais M (2009). Seismo-volcanic investigation of 2009 earthquake swarms at Harrat Lunayyir (Ash Shaqah), Western Saudi Arabia. Int. J. Earth Sci. Eng. India (in press).

Al-Amri A, Rodgers A, Al-khalifah T (2008). Improving the Level of Seismic Hazard Parameter in Saudi Arabia Using Earthquake Location. Arabian J. Geosci. 1:1-15 DOI 10.1007/s12517-008-0001-5.

Al-Zahrani H (2010). An Investigation of Seismo-volcanic Sources of Harrat Lunayyir, NW Al-Madinah Al- Munawwarah. M.Sc. Thesis, Fac. Sci., King Saud Univ., 102 P.

Al-Zahrani H, Fnais M, Al-Amri A, Abdel-Rahman K (2012). Tectonic framework of Lunayyir area, northwest Saudi Arabia through aftershock sequence analysis of 19 May, 2009 earthquake and aeromagnetic data. Int. J. Phys. Sci. 7(44):5821-5833.

Anon (1972). The Sea that is really an ocean. New Scientist 34:414-445.

Benito B, Cepeda J, Martinez Diaz J (2004). Analysis of the spatial and temporal distribution of the 2001 earthquakes in El Savador, in Rose, W.I., Bommer, J.J., Lopez, D.L., Carr, M.J., and Major, J.J, eds., Natural hazards in El Salvador: Boulder, Colorado. Geol. Soc. Am. special paper 375:1-18.

Chinnery M (1963). The stress changes that accompany strike-slip faulting. Bull. Seismol. Soc. Am. 53:921-932.

Fnais M (2010). Coseismic Ruptures caused by 19 May 2009 Earthquake (Mw 5.7), Western Saudi Arabia. Bull. Egypt Geophys. Soc. (in press).

Fukao Y, Kikuchi M (1987). Source retrieval for mantle earthquakes by iterative deconvolution of long-period P-waves. Tectonophysics 144:249-269.

Gheitanch M (2003). Analysis of the 1990 Fork (Darab), southern Iran, earthquake sequence. J. Earth Space Phys. 29(1):13-19.

Girdler R (1969). The Red Sea-a geophysical background. In: Degens ET, Ross D. A. (eds): Hot brines and recent heavy metal deposits in the Red Sea. Springer, New York pp. 59-70.

Jónsson S, Pallister J, McCausland W, El-Hadidy S (2010). Dyke Intrusion and Arrest in Harrat Lunayyir, western Saudi Arabia, in April - July 2009. Geophys. Res. 12, EGU2010-7704.

Kisslinger C (1997). Aftershocks and fault zone properties. Adv. Geophys. 38:1-35.

Klein RW (1987). Hypocenter location program. HYPOINVERSE, part 1: user guide, open file report. U.S .Geological Survey, Menlo Park, California 113 pp.

Laurent D, Chevrel S (1980). Prospecting for Pozzolan on Harrat Lunayyir. Ministry of Petrol and Mineral Resources.

Lee W, Valdes C (1985). HYPO71PC: A personal computer version of the HYPO71 earthquake location program. U.S .Geological Survey Open File Report 85-749, 43 pp.

Makris J, Allam A, Mokhtar T, Basahel A, Dehghani, Bazari (1978). Crustal structure of northeast region of Saudi Arabia and its transition to the Red Sea. Internal Report, National Research Institute of Astronomy and Geophysics, Egypt.

Mukhopadhyay, B, Mogren S, Mukhopadhyay M, Dasgupta S (2012). Incipient status of dyke intrusion in top crust – evidences from the Al-Ays 2009 earthquake swarm, Harrat Lunayyir, SW Saudi Arabia. Geo. Nat. Hazards Risk. 1:1-19.

Pallister J, McCausland W, Jónsson S, Lu Z, Zahran H, El-Hadidy S, Aburukbah A, Stewart I, Lundgern P, White R, Moufti M (2010). Broad accommodation of rift-related extension recorded by dyke intrusion in Saudi Arabia. Nat. Geosci. 8 pp.

Rose D, Whitmarsh R, Ali S, Boudreaux J, Coleman R, Fleisher R, Girdler R, Manheim F, Matter A, Nigrini C, Stoffers P, Supko P (1973). Red Sea, Drilling. Science. 179:377-380.

Suetsugu D (2003). PMAN The program for focal mechanism diagram with p-wave polarity data using the equal-area projection. IISEE Lecture Note, Tsukuba, Japan pp. 44-58.

Utsu T (1961). A statistical study on the occurrence of aftershocks. Geophys. Mag. 30:521-605.

Zahran H, McCausland W, Pallister J, Lu Z, El-Hadidy S, Aburukba A, Schawali J, Kadi K, Youssef A, Ewert J, White R, Lundgren P, Mufti M, Stewart I (2009). Stalled eruption or dike intrusion at Harrat Lunayyir, Saudi Arabia?. Am. Geophys. Union, abstract, V13E-2072.

9

Excessive correlated shifts in pH within distal solutions sharing phase-uncoupled angular accelerating magnetic fields: Macro-entanglement and information transfer

Blake T. Dotta, Nirosha J. Murugan, Lukasz M. Karbowski and Michael A. Persinger

Biophysics Section, Biomolecular Sciences Program, Laurentian University Sudbury, Ontario, Canada.

Entanglement or "excess correlation" between physical chemical reactions separated by significant distances has both theoretical and practical implications. In 24 experiments, the inverse shifts in pH were noted in two quantities of spring water separated by 10 m that shared rotating magnetic fields (0.5 µT) with changing angular velocities when one solution was injected with proton donors (weak acetic acid). The values of increased pH in the "entangled" (non-injected) beakers were 0.01, 0.03, and 0.07 for water volumes of 100, 50, and 25 cc, respectively. The associated fixed amount of energy of ~10^{-21} J per molecule from the coordinated fields in the two loci was related to the change in numbers of H^+ within these volumes and predicted the time required to produce the maximum shift in pH. These results suggest that macroentanglement as a potentially inexpensive method of transfer of information over long distances may have practical application.

Key words: Convergent loci, entanglement, weak magnetic fields, pH, communications.

INTRODUCTION

Demonstration of macroentanglement for discrete reactions over non-traditional distances that do require conventional electromagnetic transmission has significant practical importance for future, inexpensive and private modes of communication. Although there have been elegant demonstrations of entanglement involving electron spins and gases (Ahn et al., 2000; Fickler et al., 2012; Hoffman et al., 2012; Julsgaard et al., 2001), the equipment is expensive and limited in availability. Dotta and Persinger (2012) reported that two photochemical reactions separated by 10 m but that shared identical circular rotating magnetic fields whose phase and group velocities were uncoupled, to satisfy the conditions calculated by Tu et al. (2005), responded as if the two separate spaces were the same locus. Simultaneous injection of single reactants in the solutions in the separate spaces produced a doubling of photon emissions as measured by photomultiplier tubes. The effect, which was visually conspicuous, was equivalent to injecting twice the amount into one reaction. It involved energies in the order of 10^{-11} to 10^{-12} J and was evident at distances of 3 km. Separate studies suggested that similar effects are demonstrable with pairs of cell cultures or human brains (Dotta et al., 2011). Recent source localization from quantitative electroencephalography revealed remarkable excess correlation between brain activities for pairs of individuals separated by about 300 km but who shared the same changing angular velocity circular magnetic fields (Burke et al., 2013).

In the pursuit of understanding the mechanism, we

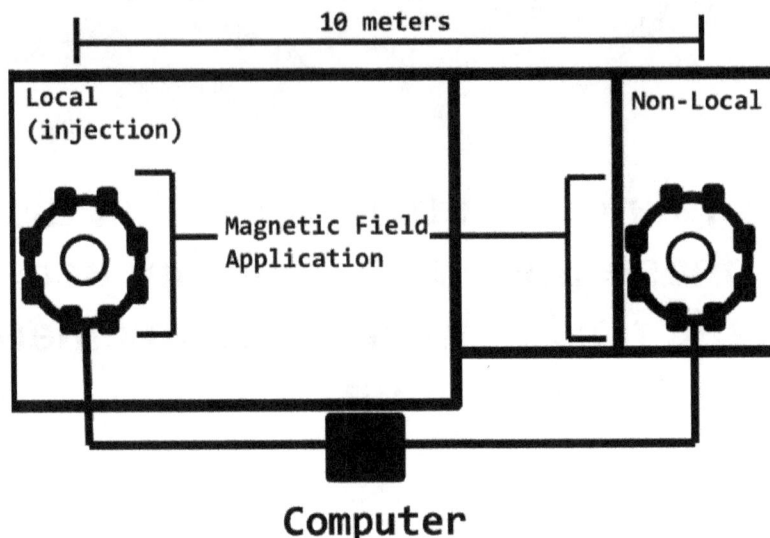

Figure 1. Diagram of the experimental design.

explored less expensive procedures to demonstrate this potent effect. Both theory and calculations indicate that when two spaces share the same configuration of dissociation of phase and group velocity of circularly accelerating or decelerating magnetic fields, a discrete amount of energy is shared. Several of our experiments involving photon transmission through tissue indicated that the hydrogen ion (DeCoursey, 2003), particularly the hydronium atom (H_3O^+), is a primary candidate. In the present experiment, the pH of two solutions separated by 10 m but sharing the same configuration of magnetic fields that was associated with the photon doubling effect were monitored over 20 min after a small amount of proton donor (a weak acid) was injected into one solution. Here we present clear evidence that the injection of a proton donor into one solution produced the predictable decrease in pH that was accompanied by a reliable minute quantitative increase in pH in the distal solution when both shared the same field configuration. This conspicuous effect meets the criteria of excess correlation for macroentanglement and the transfer of a discrete amount of energy over non-traditional distances.

MATERIALS AND METHODS

There were a total of 24 different experiments completed on separate days (one per week). Beakers containing the same volume of spring water, either 25, 50, or 100 cc were placed in the center of each of two circular arrays of 8 coupled solenoids as described by Dotta and Persinger (2012). The equipment is also described in U.S. Patent 6,312,376: b1: November 6, 2001; Canadian Patent No. 2214296. Each pair of solenoids that were reed switches or relays (250Ω) were arranged within small plastic (film) canisters and connected such that they were pairs of north and south poles. The circumference of the equally spaced solenoids that were separated by 45° from each other was ~60 cm.

The flasks containing the water were placed in the middle of the circular arrays (Figure 1) where a power meter indicated the average field strength of the applied fields was 0.5 to 1 μT as measured by a power meter.

This experiment involved exposing the samples of spring water (4 mM of HCO_3; 1.77 mM Ca; 76 μM of Cl; 1.3 mM; of Mg, 41.9; μM of NO_3; 61 μM SO_4 ; 17.9 μM K; 43.5 μM Na) in 125 cc flasks for 18 min to a counterclockwise rotating computer generated magnetic field whose wave pattern (phase) was decelerating while the group velocity (rotation of the wave around the array) was accelerating at 20+2 ms. This means that the duration of the field presentation at each solenoid decreased by 2 ms such that the duration for the first solenoid was 20 ms and 4 ms for the 8th solenoid. After the optimal time of 6 min of this exposure (part 1), the field pattern was changed to an accelerating phase pattern and a decelerating group velocity of 20-2 ms (part 2). This means that 2 ms was added to the duration of the field after the 20 ms at the first solenoid. The second field was applied an additional 12 min. The pH values for both beakers in each experiment were recorded separately once per second by Dr. Daq systems (Pico Technology, United Kingdom) which are sensitive to the .01 pH unit.

There were two separate circular arrangements of 8 solenoids within which the flasks of spring water (circles) were placed. A computer generated the changing angular velocity and phase modulated fields within the two arrays. The proton donor (acetic acid) was injected into the local flask. pH levels were measured every second from the local and the non-local flasks. The latter never received any injections.

After 4 min of exposure to the first field, 50 μl of 0.83 M acetic acid (proton source) was injected into the active beaker. Immediately after the onset of the second field configuration (8 min from the beginning of the experiment), 50 μL was injected into the active beaker once every min until 16th minute of the experiment (9 injections). Nothing was ever injected into the "entangled" beaker. To ensure the specificity of the effect, triplicates of experiments were completed where pH values were monitored in the same manner while the same sequences of injections of protons were injected into the active beaker but no experimental magnetic fields were present. Another triplicate of experiments involved measuring the pH values in both beakers over the same duration when no protons were injected and no fields were present to control for any

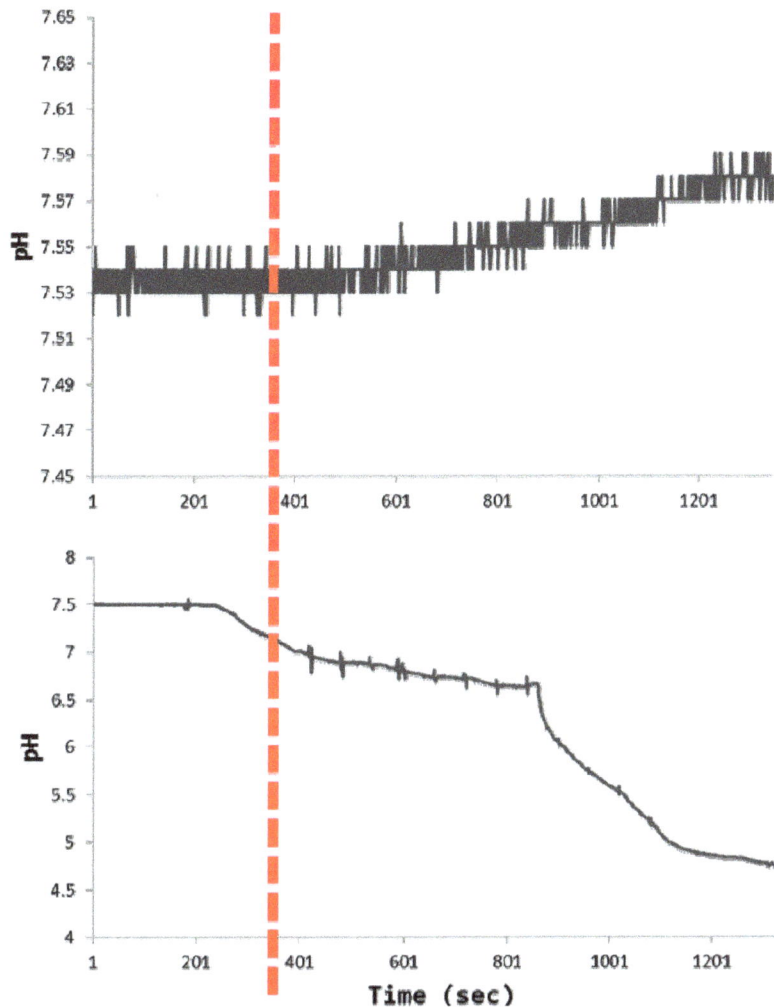

Figure 2. Upper panel: Example of the shift towards basic pH over time (in seconds) in the non-local flask after the initiation of the second component of the magnetic field configuration that was generated at both local and non local spaces. Lower panel: the response to sequential addition of 50 μL of acetic acid (indicated by small spikes) to the local flask during the same time period. Vertical line indicates activation of the "entanglement" field.

serial drift. There was no change in the pH values (flat line) from baseline for any of the pairs of beakers over the 18 min period for these conditions. The net shifts in pH for volume (25, 50, 100 cc of spring water, plus, control) were analyzed by SPSS 16 PC software.

RESULTS

An example of the shifts in pH over time in the non-injected beaker containing 50 cc of spring water that shared the same configuration of circularly rotating magnetic fields as well as the active beaker of spring water that was serially injected with 50 μL of acetic acid 10 m away in a second room is shown in Figure 2. The progressive increases in acidity in the injected beaker is obvious as well as the opposite drift towards basic pH in

the non-injected beaker. The vertical line indicates the activation of the second field whose onset was also associated with the entanglement effect reported by Dotta and Persinger (2012) for photon emissions.

Figure 3 shows the average net change in the pH over time within the non-injected beakers within the different volumes of spring water. Within the 50 cc volumes, the mean total shift (increase) in pH was 0.03 (SEM=0.006). The mean total shift for the 25 cc volumes was pH=0.07 (SEM=0.009). The mean total shift for 100 cc of water (M=0.01, SEM=0.008) did not differ from the control condition (M=0.01, SEM=0.003) when no fields were applied. The difference between the four experimental conditions was statistically significant [$F(3,20)=23.94$, p <.001]; the amount of variance explained, the effect size (eta^2 estimate), was 82%.

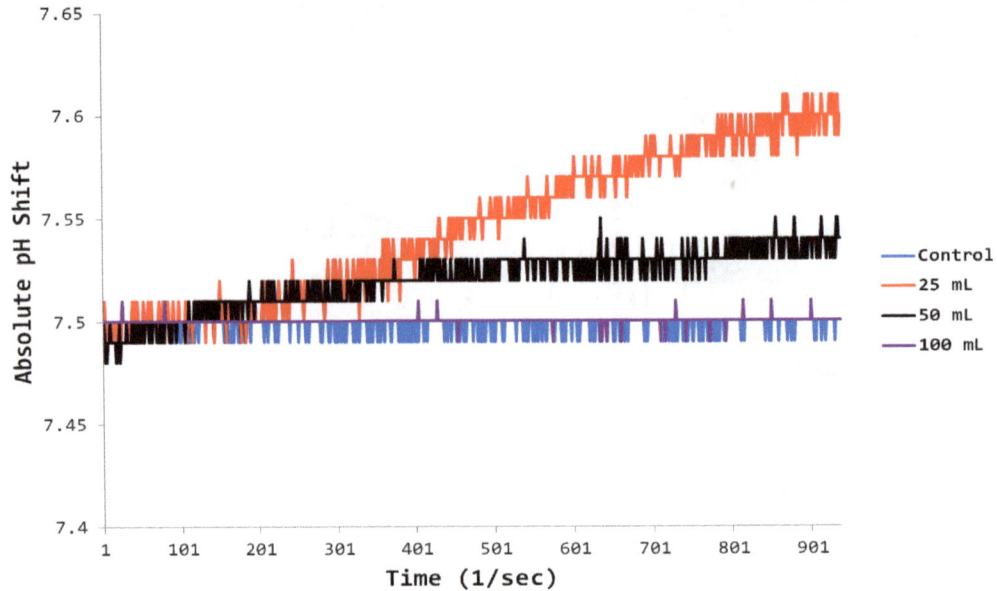

Figure 3. The absolute shift in pH within the 25 (red), 50 (black) and 100 (purple) cc volumes of spring water within the non-local flasks as a function of time from the onset of the entanglement phase. The blue line refers to control conditions (no shared field but acetic acid injected into the local flasks).

DISCUSSION

The results of this study indicate that strong excess correlation or entanglement between two separate spaces can occur when they are both exposed to circular (rotating) magnetic fields with appropriate changes in angular velocity. The effect was robust and consistent with the two separate spaces behaving as if they were the same locus. The acidification of the injected water (local flasks) was associated with the shift towards base in the "non-local" flasks. According to the traditional formula $[H^+]=10^{-pH}$, a shift of 0.01, 0.03, and 0.07 pH towards basic from a starting pH of 7.50 would be associated with a decrease of 1, 2, and $4 \cdot 10^{-9}$ M of H^+. When each of these concentrations is multiplied by the moles of water in each volume of 100, 50, and 25 cc, or 5.55 M, 2.78 M, and 1.39 M (1 mole of water=18 cc), the conserved molarity would be $5.55 \cdot 10^{-9}$ M. With $6.023 \cdot 10^{23}$ units per mole, there would be $3.35 \cdot 10^{15}$ H^+ involved with the phenomenon.

In comparison if we assume the functional median of the increased acidity in the local (injected) flasks was to about pH=5.6, the equivalent is $2.5 \cdot 10^{-6}$ M. The acetic acid was applied in 0.05 cc quantities to 25, 50, and 100 cc of water. The ratio of these volumes multiplied by the molarity of water in each volume is a constant of $2.77 \cdot 10^{-3}$ which results in a net concentration of $6.9 \cdot 10^{-9}$ M which is well within range of measurement error for the shift towards basic in the non local flasks.

If this is a physical process, the energy available from this special configuration of magnetic fields should converge with the shift in quantities of H^+ involved with the "superimposition" of the two spaces. From our perspective, the energy available from both magnetic fields (1 μT) is the product of the strength of the field ($kg \cdot A^{-1} \cdot s^{-2}$), unit charge ($A \cdot s$) and a measure of diffusivity ($m^2 \cdot s^{-1}$). Assuming $9 \cdot 10^{16}$ $m^2 \cdot s^{-1}$ (diffusion of light) and a unit charge ($1.6 \ 10^{-19}$ $A \cdot s$), the energy available for entanglement would be $\sim 1.4 \cdot 10^{-8}$ J and if applied over the time for the pH shift to occur (~ 1000 s), would be $14.4 \cdot 10^{-6}$ J for the cumulative energy. When divided by the number of H^+ involved with the shifts in pH ($3.35 \cdot 10^{15}$), the energy per proton is $4.3 \cdot 10^{-21}$ J per H^+. For comparison, the kT (where k is the Boltzmann constant and T is temperature in Kelvin) threshold that defines the thermal "noise" component for 25°C (298 K) is $\sim 4.1 \cdot 10^{-21}$ J. This is also within an order of magnitude of the $\sim 10^{-20}$ J quantum unit (Persinger, 2010) that may be involved as a universal quantity (Persinger et al., 2008).

ACKNOWLEDGEMENT

Thanks to Dr. W. E. Bosarge Jr, Chairman, Quantlab LLC for his support.

REFERENCES

Ahn J, Weinacht TC, Bucksbaum PH (2000). Information storage and retrieval through quantum phase. Science 287:462-466.
Burke RC, Gauthier MY, Rouleau N, Persinger MA (2013). Experimental demonstration of potential entanglement of brain

activity over 300 km for pairs of subjects sharing the same circular rotating, angularly accelerating magnetic fields: verification by s_LORETA, QEEG measurements. J. Consc. Explor. Res. 4: 35-44.

Decoursey TE (2003). Voltage-gated proton channels and other proton transfer pathways. Physiol. Rev. 83:476-579.

Dotta BT, Buckner CA, Lafrenie RM, Persinger MA (2011). Photon emissions from human brain and cell culture exposed to distally rotating magnetic fields shared by separate light-stimulated brains and cells. Brain Res. 388:77-88.

Dotta BT, Persinger MA (2012). A "doubling" of local photon emissions when two simultaneous, spatially separated, chemiluminescent reactions share the same magnetic field configurations. J. Biophys. Chem. 3:72-80.

Fickler R, Lapkiewicz R, Plick WN, Krenn M, Schaeff C, Ramelow S, Zeilinger A (2012). Quantum entanglement of high angular momenta. Science. 338:640-644.

Hoffman J, Krug M, Ortegel N, Gerard L, Weber M, Rosenfield W, Weinfurter H (2012). Heralded entanglement between widely separated atoms. Science. 337:72-75.

Julsgaard B, Kozhekin A, Polzik ES (2001). Experimental long-lived entanglement of two macroscopic objects. Nature. 413: 400-403.

Persinger MA (2010). 10^{-20} Joules as a neuromolecular quantum in medicinal chemistry: an alternative approach to myriad molecular pathways. Cur. Med. Chem. 17:3094-3098.

Persinger MA, Koren SA, Lafreniere GF (2008). A neuroquantological approach to how human thought might affect the universe. Neuroquantol. 6:262-271.

Tu LC, Luo J, Gilles GT (2005). The mass of the photon. Rept. Prog. Phys. 68:77-130.

10

Correction of topological errors in geospatial databases

Siejka M., Ślusarski M. and Zygmunt M.

Department of Land Surveying, University of Agriculture in Krakow, Poland.

The problems related to geometrical correctness of maps gathered and stored by the geographic information system (GIS) computer systems are still present nowadays. Firstly, digitization of traditional paper maps is still in progress all over the world (the maps are being converted to a vector form and stored in relational databases). Secondly, the existing digital maps are being constantly updated with use of manual and partially automated methods. The following article presents a system, developed in order to detect and correct topological errors of surface objects. The method of operating on virtual objects in computer aided design (CAD) software environment is suggested. The main advantage of this solution is its applicability in verification of topological errors in spatial databases that contain large numbers of objects.

Key words: Cadastre, geographic information system (GIS), land parcel identification system (LPIS), topology, computer aided design (CAD), software, spatial data.

INTRODUCTION

Spatial relations of cadastral objects are nowadays presented in two basic forms: digital and analog. The numerical form is illustrated by digital data. The analog form includes maps or plans composed of lines, points, symbols, and text (Gaździcki, 1995; Klajnšek and Žalik, 2005). In this situation, two basic models of spatial data can be distinguished: raster model and vector model.

The simple vector model has two basic disadvantages. One of them is the redundancy of data in situations when the given point belongs to two or more objects. In such cases, the point's coordinates must be saved in each of these objects. The other significant disadvantage is that the relations between objects can be detected using the methods of analytical geometry with complex and time-consuming calculations. Application of the vector model to real data bases, were a great number of objects are processed, causes significant increase of work time. In order to eliminate the disadvantages of simple vector model, topological vector model should be used. The spatial data in topological vector model consists of data describing topological relations and geometrical data

which are the coordinates of points. In topological data there is a redundancy that allows verification. The geometrical data on the other hand, have no redundancy (Gaździcki, 1995; Longley et al., 2005).

Using topological vector models, has become very popular nowadays in various geographic information system (GIS) software systems. An example of topological vector model is georelational model. In this model the geometrical properties of elements and the topological information related to them are stored in computer files. The values of the attributes on the other hand are stored in the tables of the relational database management system (RDBMS) (Nedas et al., 2007).

In all kinds of models we are dealing with the problem of topological inconsistency of objects. The problem is very common and of practical importance. This is why many software packages offer different solutions. It is obvious that the most desirable solution would be elimination of the errors and creation of consistent topology (Hope and Kealy, 2008). Another possible solution is to develop a GIS system that would use

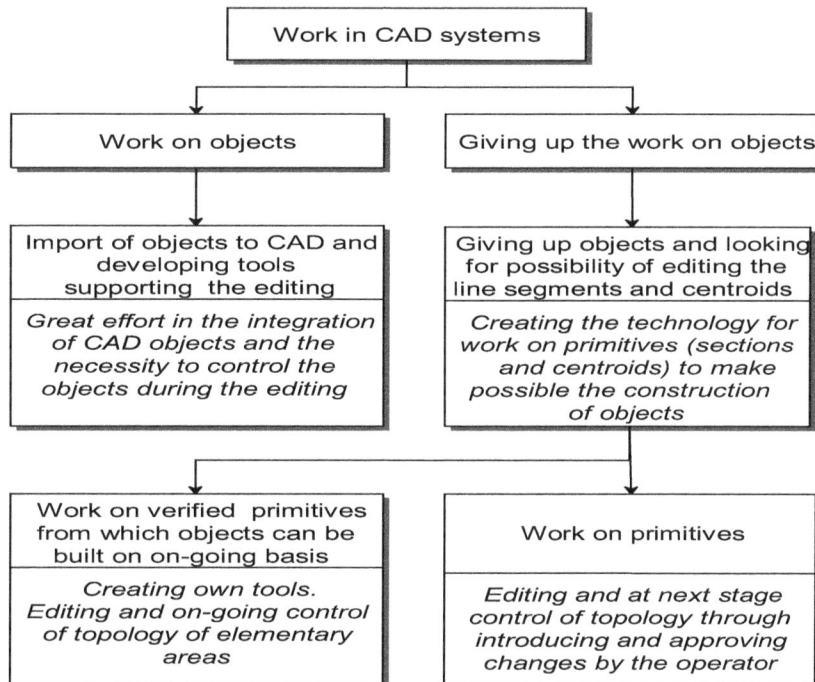

Figure 1. Scheme of alternative technologies of work in CAD.

probabilistic approach to handling uncertain borders (Lagacherie et al., 1996). However, such a solution requires a lot of money and it is time consuming (Klajnšek and Žalik, 2005).

During the processing of large amounts of data for regional or global studies, errors occur (Van Oosterom et al., 2006). It is necessary to eliminate these errors, and the method is dependent mainly on the kind of data processed (Zadravec and Žalik, 2009). In case when demanded positional accuracy of objects is not the most important, we can use highly automated technologies of topological error elimination. An example of such solution was presented in a study describing the methodology of topological error correction in GIS vector data (Maraş et al., 2010). For databases storing official data, such as cadastre, topological error elimination methods cannot decrease quality of the processed data. That is why in this case it is necessary to use a technology that would minimize the number of editing operations and allow to control in detail the correction of topological errors.

In proposed solution of topological errors verification which occurs in official data bases we concentrate on indication of potential errors. Correction of these errors is depends on the operator decision. When the operator was wrong, the error will be detected by the program in the next iteration.

METHODOLOGY

The problem of updating digital databases is related to employing

proper technologies. In the process of developing optimal technology it is crucial to firstly answer the following questions: whether or not to work on the objects using the present-day GIS software? Or look for another solution? Professional GIS systems, aside from unquestionable advantages, also have disadvantages. The main drawback of GIS systems when it comes to updating large amounts of data, is the low efficiency of editing separate elementary areas (plots, management field) existing in the form of objects. An alternative for GIS systems can be developing a technology based on computer aided design (CAD) software. There are two possible ways of working in CAD. The first one is working on objects. Importing objects to CAD, possibility of creating new objects and tools that support editing the objects. This method is burdened with considerable efforts related to the integration of CAD objects and the need to verify them while editing.

The second way is giving up on working on objects and looking for a possibility to edit segments and centroids. It is connected with developing a technology to work on primitives (segments, centroids) in order to allow building objects. Such a technology can be created in two ways. One of the ways is to work on verified primitives, from which objects can be built. In this case edition and verification of topology of elementary areas is performed on an ongoing basis. The other way is to work only on primitives. In this solution the first step is edition of objects. The next step is verification of topology by entering and confirming adjustments by the operator. A block diagram of alternative technologies of working in CAD is shown on Figure 1.

DETECTION AND ELIMINATION OF TOPOLOGICAL ERRORS

The proposed methodology of work is based on virtual objects – topo on fly (Figure 2). Virtual objects are properly created sets of primitives (segments as representations of borders of surface objects) and descriptions (as identifiers of closed areas). To allow

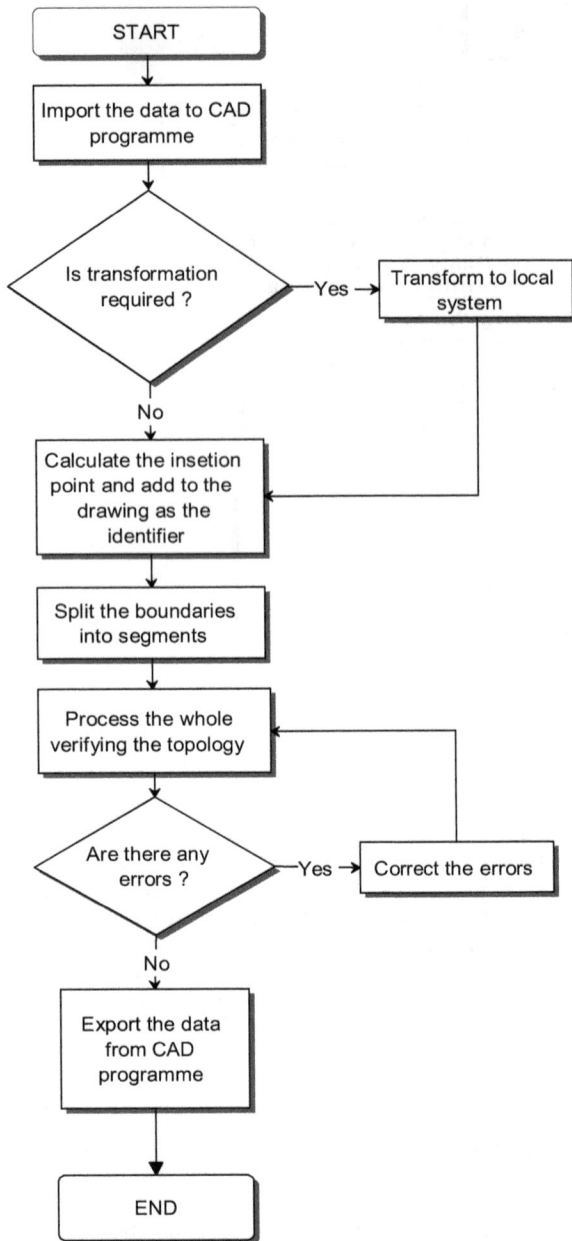

Figure 2. Algorithm of data processing.

(2) Verification and elimination of category I errors – errors in lines of surface objects that make it impossible to construct correct surface objects,

(3) Verification and elimination of category II errors – errors in identifiers of surface objects that make it impossible to unequivocally identify them,

(4) Verification and elimination of category III errors – other errors.

(5) Export of data from CAD software – possibility to save the data in any format of GIS software.

The program was developed in the language of Microsoft Visual Basic for Application (MVBA) for the platform of Bentley System software - PowerMap. MVBA offers algorithms and data structures widening fundamental powers of programming in visual basic. This environment supplies among others, algorithms for data processing, used in range searching, point location, checking of intersections, creating polygons, and also calculation of distances. Launch of the application verifying and eliminating topological errors takes place directly in the Bentley PowerMap work environment, through reading-in and starting the program created by MVBA environment (Figure 2).

Verification and elimination of Category I errors

The group of surface object line errors, that make it impossible to build correct objects, includes (Figure 3 to 5):

(a) Dangle errors,

(b) Intersection of borders of surface objects in points which do not belong to the borders,

(c) Near element error – special case.

The most commonly occuring category I error is dangle. It is a segment whose one end is not connected with any node. Such errors make it impossible to automatically generate outlines of surface objects. This error occurs when the line has not been drawn all the way to the node, or after an existing node has been removed. Algorithm checking this kind of error tests if for the beginning point (B) of checked linear object, exists at least one node overlaping investigated object. In the same way the end of the segment (E) is checked. This task is executed by the following algorithm:

```
For i=1 to n; n – all linear objects
  Is_dangleB=true
  Is_dangleE=true
  For j=1 to m; m – all object assigned to be checked
    If distance|B[i]-B[j]|=0 or distance|B[i]-E[j]|=0 then
      Is_dangleB=False
    End If
    If distance|E[i]-B[j]|=0 or distance|E[i]-E[j]|=0 then
      Is_dangleE=False
    End If
    If Is_dangleB=False and Is_dangleE=False then exit for
  Next j
  If Is_dangleB=true then add B[i] to the list of errors
  If Is_dangleE=true then add E[i] to the list of errors
Next i
```

Elimination of this error boils down to drawing the line to an existing node or adding a node in the place of intersection with another line. Intersection error occurs when one line divides another line into two separate parts in a place where a node does not exist. Algorithm checking this kind of error verifies if two segments intersect. If the answer is yes, it calculates the node, and adds it to the list of errors.

proper preparation of vector data for the purpose of working in the described technology (on virtual objects), aside from borders of objects one should also insert an identifier in the form of text. The description must be inside the representation of an object. This implies the necessity to check whether this condition is fulfilled, because calculation of center of gravity of an object does not always give positive result. The problem was solved basing on the algorithm described by O'Rourke (1998). Only the data prepared in such a way can be checked for topological correctness. Verification of topological errors in the presented solution is conducted in the following steps:

(1) Importing the data to CAD software – possibility to read the data that comes from many sources, organizing it and moving to one data format,

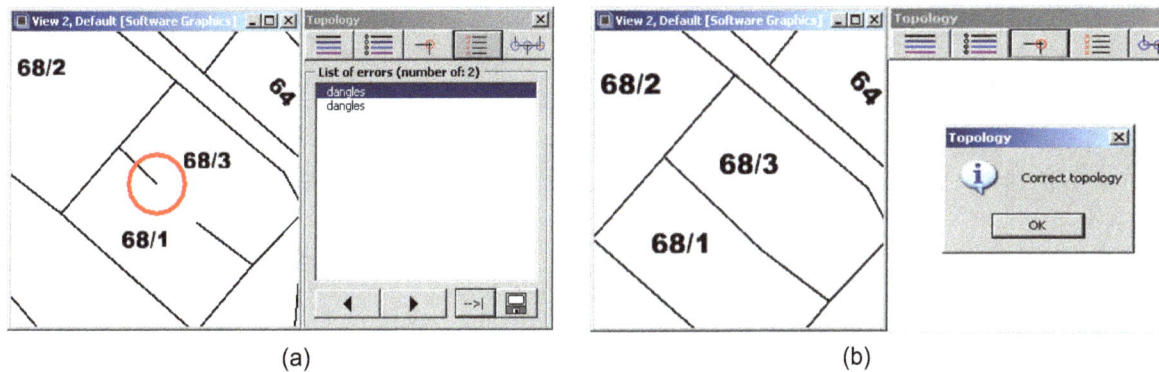

Figure 3. Indication of dangle error, a) before correction, b) after correction.

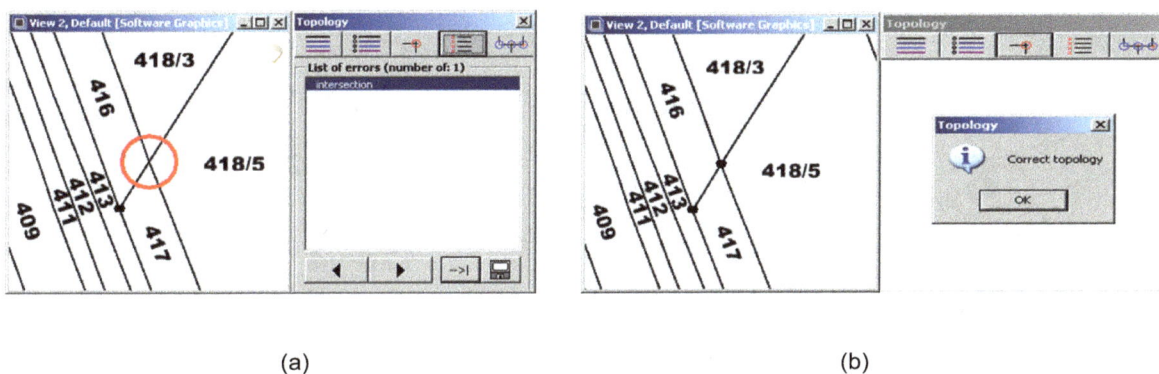

Figure 4. Indication of intersection error, a) before correction, b) after correction

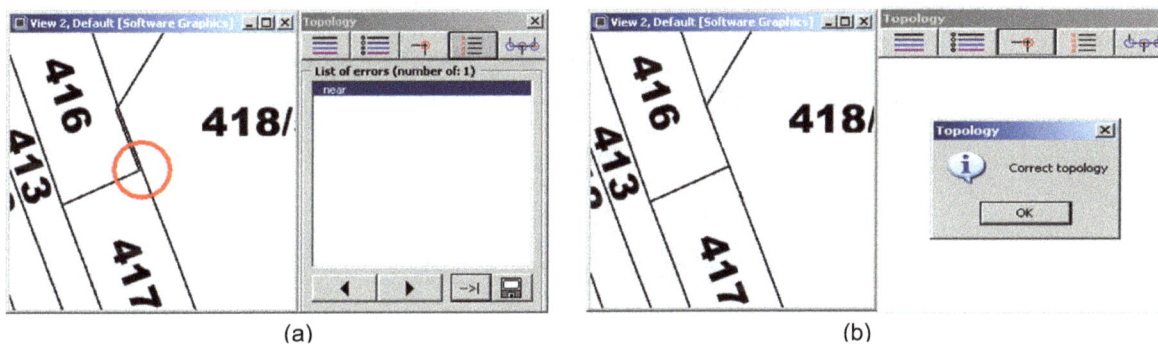

Figure 5. Indication of near element error – special case, a) before correction, b) after correction.

The checking algorithm makes use of intersections verification function implemented in MVBA environment.

```
For i=1 to n; n – all linear objects
  For j=1 to m; m – all objects assigned to be checked
    If Is_intersection [i]-[j] = True  then
      add an intersection point to the list of errors
    End If
  Next j
Next i
```

This kind of error can be automatically corrected by designating and assuming the point of intersection. However, one should remember that in case of official databases the processed information is a legal document, there should not be any automatic changes without authorization. According to the assumed classification of errors, near element belongs to category I only if the border lines partially overlap (Figure 5). Algorithm checking this kind of error verifies multicollinearity of points. If checked points B[j], E[j] lie on the line segment [i], and are not its nodes – appropriate information is added to the list of errors.

(a)   (b)

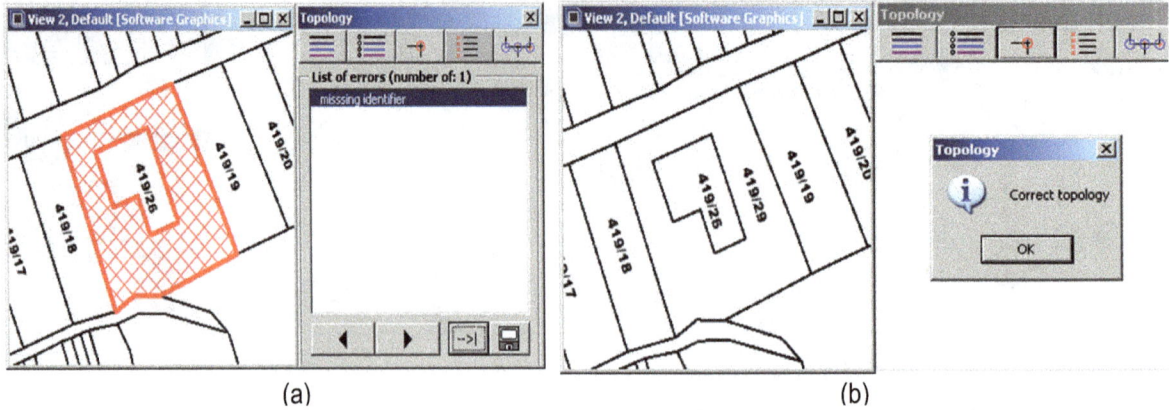

Figure 6. Missing surface object identifier, (a) before correction, (b) after correction.

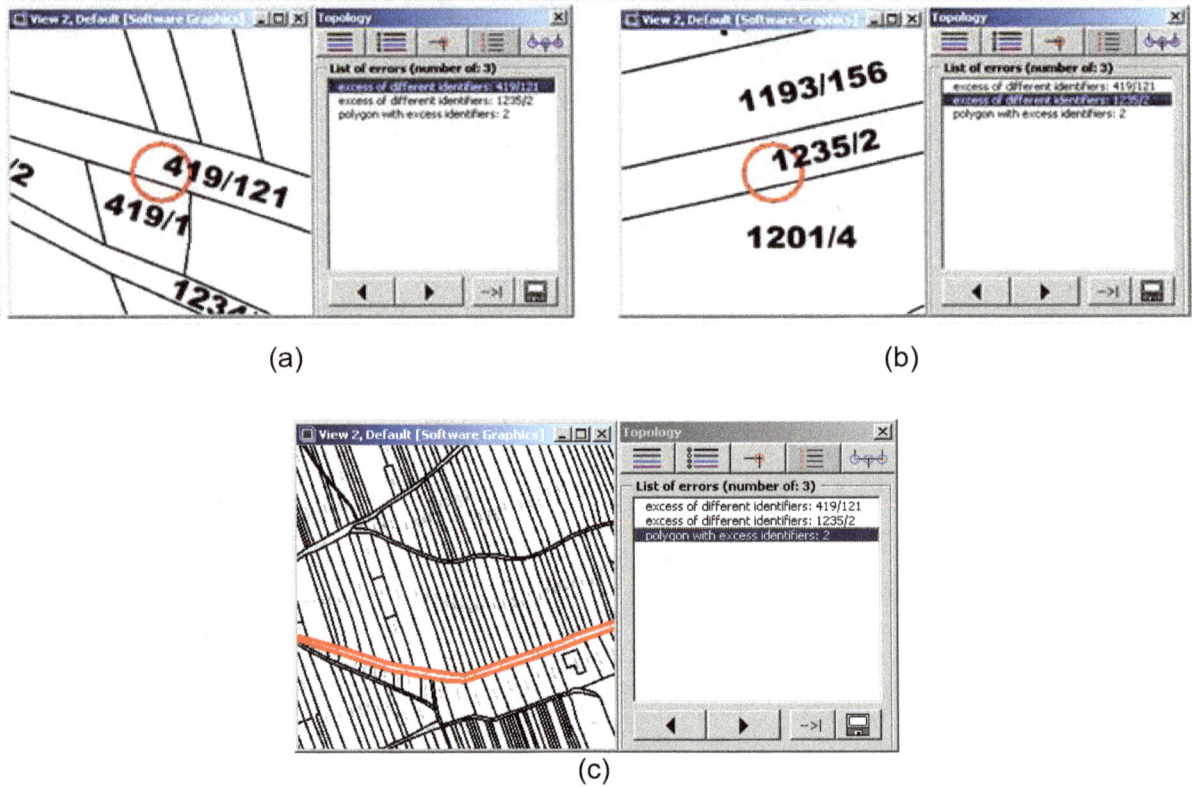

(a)   (b)

(c)

Figure 7. Surface object described by more than one identifier – before correction; a) identifier 419/121, b) identifier 1235/2, c) closed surface object described by two different identifiers.

```
For i=1 to n; n – all linear objects
   For j=1 to m; m – all objects assigned to be checked
      If distance perpendicular between B[j] and Line [i] < ε then
         add B[j] to the list of errors; ε=0.000001
      End If
      If distance perpendicular between E[j] and Line [i] < ε then
         add E[j] to the list of errors; ε=0.000001
      End If
   Next j
Next i
```

This type of error can be automatically corrected.

Correction and elimination of Category II errors

The group of surface object identifier errors, that make it impossible to unequivocally identify objects, includes (Figures 6 to 9):

(a) Missing surface object identifier,

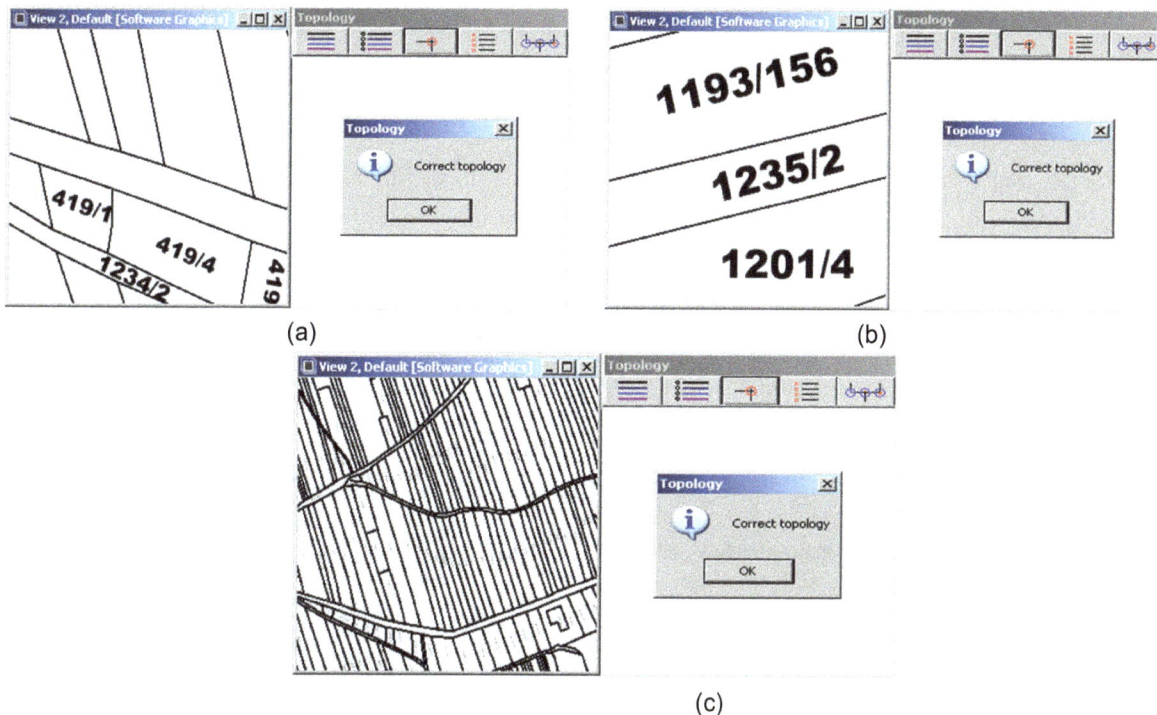

Figure 8. Surface object described by more than one identifier – after correction; a) elimination of identifier 419/121, b) leaving a correct identifier 1235/2, c) investigated polygon, after correction does not prove any error.

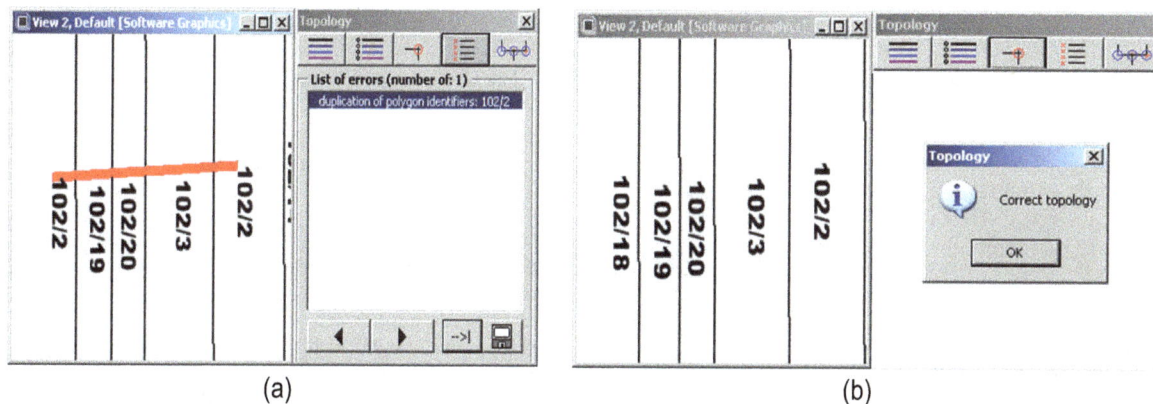

Figure 9. Surface object identifier duplication , a) before correction, b) after correction.

(b) Too many identifiers within one surface object,
(c) Duplication of surface object identifiers.

Missing surface object identifier (for example plot number, name of management field) is one of the most common topological errors. Such an error is especially difficult to detect in case of long objects (roads, rivers, etc.). According to general assumptions, identifier should be located within the outline of an object. The control algorithm of topology indicates the missing identifier error by thickening the edges and hatching. Checking algorithm verifies if identifiers belong to each polygon. If it can assign one identifier – the check is correct. If no identifiers are assigned, appropriate information is added to the list of errors.

For i=1 to n; n – all polygons
  Is_in_polgon=False
  For j=1 to m; m – all identifiers selected for checking
    If Is_identifier [j] inside polygon [i] then Is_in_polygon=True
  Next j
  If Is_in_polygon=False then add polygon [i] to the list of errors
Next i

The error is eliminated by inserting the missing identifier. Inside a surface object there can be only one identifier. If the software detects more identifiers, it indicates the error by thickening the edges. Fixing this error boils down to deleting the excess descriptive elements. This kind of error can be eliminated by a

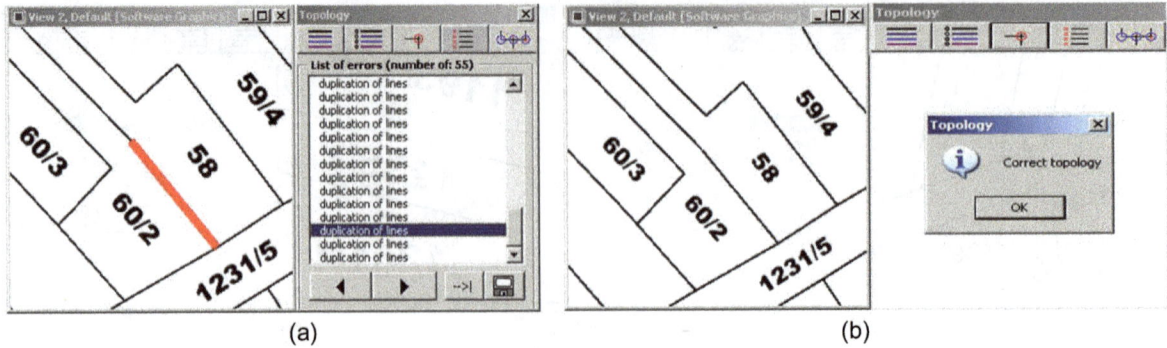

Figure 10. Indication of "duplication" error a) before correction b) after correction.

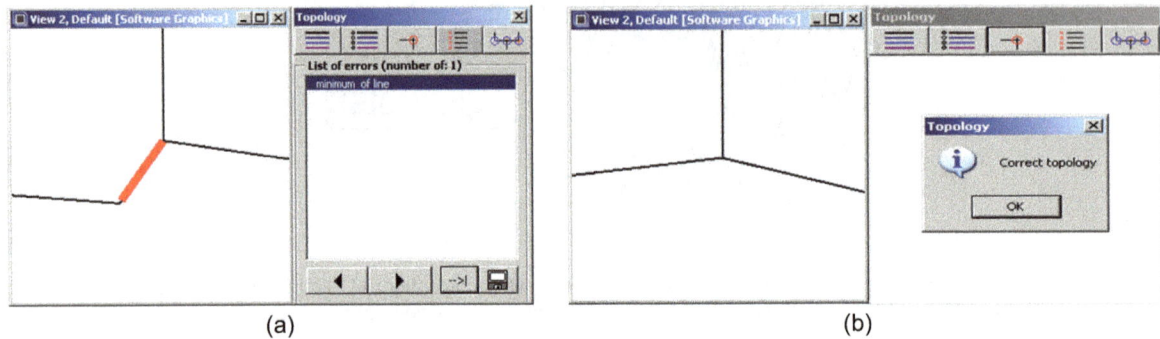

Figure 11. Indication of "minimum" error a) before correction b) after correction.

topology control algorithm only if the identifiers are identical. Checking algorithm verifies if identifiers belong to each polygon on the investigated area. If it can assign one identifier to one polygon – the check is correct. For more than one identifier assigned to one polygon appropriate information is added to the list of errors.

```
For i=1 to n; n – all polygons
  Is_in_polygon=0
  For j=1 to m; m – all identifiers selected for checking
    If Is_identifier [j] inside polygon [i] then
      Is_in_polygon= Is_in_polygon+1
    End If
  Next j
  If Is_in_polygon>1 then add polygon [i] to the list of errors
Next i
```

Identifiers of polygons in the framework of one elaboration must ensure unequivocality of identification. Error of duplication of polygons consists in assigning the same identifier to more than one polygon. Detection of the error is performed by identifiers sorting and comparison of neighboring elements of the table. In the case of existing repetitions appropriate information is added to the list of errors.

```
Sort n
For i=1 to n-1; n – number of all identifiers
  If n[i]=n[i+1]  then add n[i] to the list of errors
Next i
```

In case of official data that contain legal information, the correction must be supervised.

Verification and elimination of Category III errors

Category III errors are the errors, whose correction improves the database structure (Figure 10 to 15):

(a) "Duplication" error,
(b) "Minimum" error,
(c) Presence of an identifier outside of elaborated area,
(d) Area below the minimum,
(e) Near element error,
(f) Identifier beyond dictionary.

"Duplication" error occurs when there are lines drawn repeatedly basing on the same pair of points. Both topology control algorithms and applied software can ignore such errors. Omitting this error is a necessary condition while working on closed objects, such as SHP. Algorithm looking for errors sorts lines according to coordinates (x,y) of the beginning and end. Neighboring elements of table containing lines are compared.

```
Sort lines
For i=1 to n-1
  If Bx[i]=Bx[i+1] and By[i]=By[i+1] and Ex[i]=Ex[i+1] and
    Ey[i]=Ey[i+1]  then add line[i] to the list of errors
  End If
Next i
```

According to the authors, redundancy of boundary lines is the cause of errors in the export of information to other software. This error is eliminated by removing excess elements, which can be done automatically.

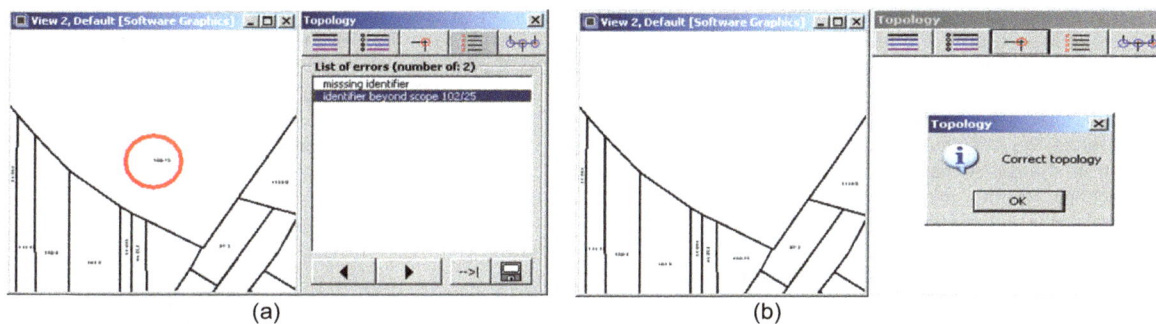

Figure 12. Identifier outside of elaborated area a) before correction, b) after correction.

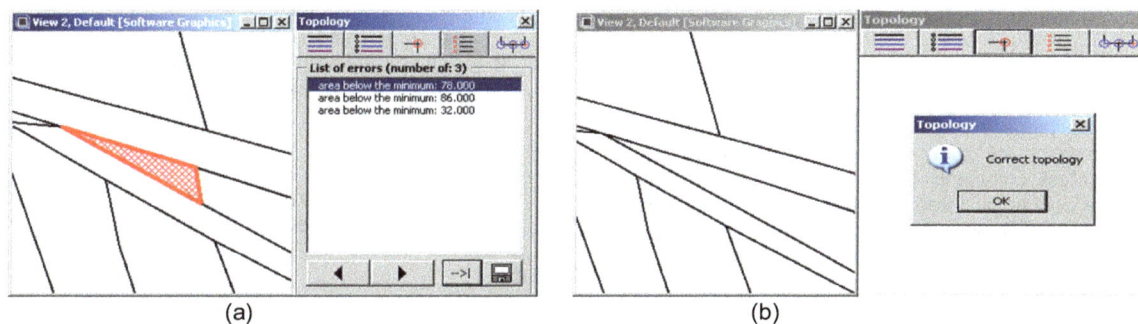

Figure 13. Area below the minimum, a) before correction, b) after correction.

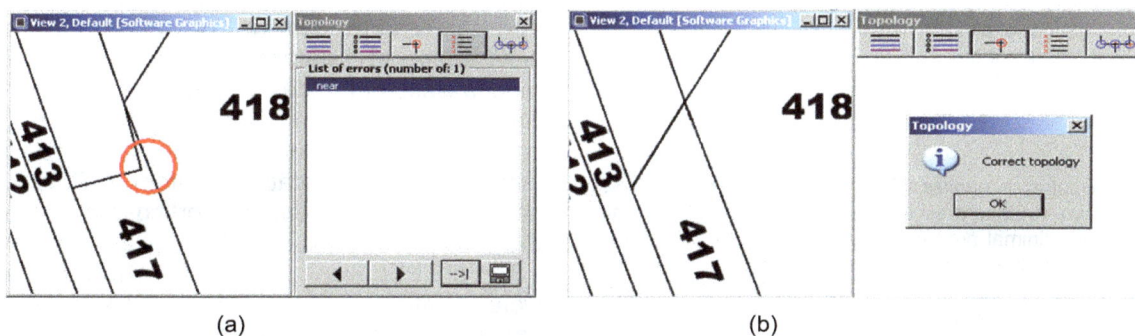

Figure 14. Indication of near element error, a) before correction, b) after correction.

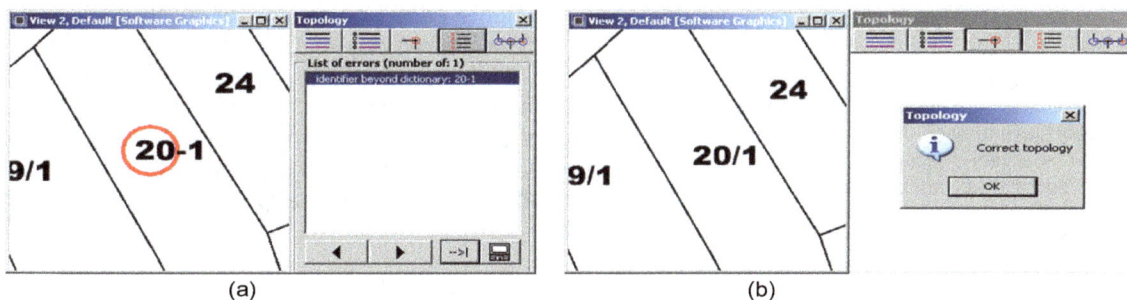

Figure 15. Identifier beyond dictionary, a) before correction, b) after correction

"Minimum" type lines are not visible on a map and it may seem that they do not have influence on the quality of working with a digital map. However, elements such as lines of zero length are "littering" the map, making it difficult to precisely connect elements or verify the coordinates. In case when there are many such elements, they significantly influence the effectiveness of software that processes the information from the digital maps. Such zero length lines can be deleted automatically. Another minimum type error are lines longer than zero, but shorter than the minimum length defined by the user. Deleting such elements is not necessary, but it influences the quality of the map drawing and the speed of information processing. Algorithm looking for errors calculates the length of each line and compares it with given criterion.

```
For i=1 to n; n – number of all lines
    If distance[i]< min and distance[i]<>0 then
        add line[i] to the list of errors (min)
    End If
    If distance[i] = 0 then add line[i] to the list of errors (zero)
Next i
```

This error can be eliminated by deleting the minimum object as long as topological correctness is sustained. Topology control algorithm indicates the presence of surface object identifiers outside of described area. Most commonly the source of such errors are corrections of the course of borders, narrow elements located on the outskirts of the described area. In this situation the correction can be done in two ways. One way is to automatically delete the identifier outside of the described area and then filling the blank area with a proper identifier. The other way is to move the identifier inside the edited surface object.

During the check-up the algorithm verifies assignment of particular identifiers to the polygons. If there are left on the list identifiers which were not assigned to a polygon, they form set of identifiers which are outside the range of investigated polygons.

```
For i=1 to n; n – number of all identifiers
    If identifier[i].use=False then
        add identifier[i] to the list of errors
    End If
Next i
```

The error of minimal area applies to objects whose area is below the defined criterion. In case presented on Figure 13, 100 m^2. The criterion set for the minimal area depends on the scale of the map and it can be a result of an incorrect vectorization. Algorithm processes all ranges comparing their area with the given criterion. When ranges of area smaller than given criterion are detected, appropriate information is added to the list of errors.

```
For i=1 to n; n – number of all polygons
    If area(polygon[i]) < criterion then
        add polygon[i] to the list of errors
    End If
Next i
```

Correction of this error depends on the user's decision, especially in case of official databases that include legal information. Elements lying or running very close to each other or unnaturally acute angles between the lines are a very special kind of errors that result from the process of vectorization of analog maps. For every line drawn on the map the verification procedure searches the buffer area defined by the user. Algorithm checking this error calculates the distance of a point to a line. If the distance of checked points B[j], E[j] to a line [i] is smaller from given criterion and the points are not the nodes of the line [i] appropriate information is added to the list of errors.

```
For i=1 to n; n – all linear objects
    For j=1 to m; m – all linear objects assigned to chech-up
        If distance perpendicular between B[j] and Line [i] < min then
            add B[j] to the list of errors
        End If
        If distance perpendicular between E[j] and Line [i] < min then
            add E[j] to the list of errors
        End If
    Next j
Next i
```

User's decision is necessary to fix this error. Using an identifier other than the ones defined in thematic dictionary is indicated as an error. Checking algorithm verifies in the used identifier on the list of permitted values. In the case of lack of information about error is added.

```
For i=1 to n; n – all identifiers
    Is_identifier_permitted=False
    For j=1 to m; m – all permitted identifiers
        If identifier[i]=identifier_permitted[j] then
            Is_identifier_permitted=True
            Exit For
        End if
    Next j
    If Is_identifier_permitted=False then
        add identifier[i] to the list of errors
    End If
Next i
```

Correction of such an error requires user's decision and an analysis of the source materials.

RESULTS

The presented system of topological correctness verification of surface objects was employed in modernization of cadastre in selected regions of Poland. The next application was verification of topological correctness of plots and management areas as a part of land parcel identification system (LPIS). LPIS is a geoinformation system supporting the management of the common agricultural policy in the European Union. As a part of LPIS over 1.5 million parcels in southern Poland were verified. At present in Poland in the works connected with managing LPIS commonly the solutions based on the GIS software are used. It is in the first place connected with the format of data in which spatial data are written.

Editorial effectivity of polygons in GIS systems and objects based on vector model in CAD systems, shows clear supremacy of CAD systems. Specific character of data bases in LPIS system needs processing of large numbers of data in short period of time. Hence, execution of a task consisting in updating LPIS data base including 1.5 million parcels, was possible to accomplish in required short time, thanks to application of suggested method.

DISCUSSION

Topological correctness of objects is a problem that

requires solving in all areas related to creating spatial databases. This applies both to GIS and cadastre (Gröger and Plümer, 2011; Laurini and Milleret-Raffort, 1994). The presented system of verification of topological correction of surface objects can be freely used in databases containing a large number of objects. This includes both GIS databases and official databases such as cadastre or LPIS. The system was developed as an application for the products of Bentley Systems family and not as independent software. Thanks to this, it can use all the tools in the base program.

An additional advantage of the proposed system is the possibility to read the data originating from many sources and saved in different formats, such as extensible markup language/geography markup language (XML/GML), shapefile (SHP), and comma separated values (CSV). The elaborated technology does not perform the check-up of done graphical operations all the time, but only on demand of the operator. This situation may cause creation of errors as a result of operator actions, which will be detected after successive activation of the application. Further works should be aimed on improvements of the program. Particular attention must be drawn on making possible topological correctness check-up immediately during editorial works of polygons.

Conclusion

In official databases such as cadastre or LPIS, where the data is treated as legal document, the possibilities of fixing the errors automatically are very limited. The future elaborations should include further research and tests of the methodology of geometrical correctness automatization.

REFERENCES

Gröger G, Plümer L (2011). Topology of surfaces modeling bridges and tunnels in 3D-GIS. Comput. Environ. Urban Syst. 35:208-216.

Gaździcki J (1995). Cadastral systems. PPKW. Warsaw. P. 123.

Hope S, Kealy A (2008). Using topological relationships to inform a data integration process. Trans. GIS. 12(2):267-283.

Klajnšek G, Žalik B (2005). Merging polygons with uncertain boundaries. Comput. Geosci. 31:353-359.

Lagacherie P, Andrieux P, Bouzigues R (1996). Fuzziness and uncertainty of soil boundaries: from reality to coding in GIS. In: Burrough, P.A., Frank, A.U. (Eds.). Geographic Objects with Indeterminate Boundaries. GISDATA2. Taylor and Francis. London. pp. 275–286.

Laurini R, Milleret-Raffort F (1994). Topological reorganization of inconsistent geographical databases: A step towards their certification. Comput. Graphics 18(6):803-813.

Longley PA, Goodchild MF, Maquire DJ, Rhind DW (2005). Geographic Information Systems and Science. John Wiley & Sons Ltd. Chichester. P. 519.

Maraş SS, Maraş HH, Aktuğ B, Maraş EE, Yildiz F (2010). Topological error correction of GIS vector data. Int. J. Phys. Sci. 5(5):476-483.

Nedas KA, Egenhofer MJ, Wilmsen D (2007). Metric details of topological line-line relations. Int. J. Geogr. Infor. Sci. 21(1):21-48.

O`Rourke J (1998). Computational Geometry in C. Cambridge University Press. P. 392.

Van Oosterom P, Lemmen Ch, Ingvarsson T, van der Molen P, Ploeger H, Quak W, Stoter J, Zevenbergen J (2006). The core cadastral domain model. Comput. Environ. Urban Syst. 30:627-660.

Zadravec M, Žalik B (2009). A geometric and topological system for supporting agricultural subsidies in Slovenia. Comput. Elect. Agric. 69:92-99.

11

Integrated geophysical investigation for post-construction studies of buildings around School of Science area, Federal University of Technology, Akure, Southwestern, Nigeria

A. O. Adelusi, A. A. Akinlalu and A. I. Nwachukwu

Department of Applied Geophysics, Federal University of Technology, Akure (FUTA), Ondo State, Nigeria.

An integrated geophysical investigation involving ground magnetic, very low frequency (VLF-EM), and electrical resistivity methods using dipole–dipole array and schlumberger vertical electrical sounding (VES) techniques were conducted around School of Science Area Obanla, Federal University of Technology, Akure, for post construction studies in assessing building foundation integrity. Two traverses were established in approximately E-W direction of length 170 to 200 m and station interval of 10 m, along which VLF-EM, ground magnetic and dipole–dipole measurements were carried out. Sixteen VES stations were occupied within the study area. The VLF- EM data were interpreted using the Karous Hjelt (KH) package and inverted into its 2D Pseudosection. The VES data were quantitatively interpreted using the partial curve matching technique and 1-D forward modelling with WinResist 1.0 version software. The dipole-dipole data were inverted into 2-D resistivity images using the DIPPRO™ 4.0 inversion software. The VLF-EM result mapped three near surface conductive zones suspected to be fractures/faults which are inimical to foundation integrity. The magnetic results delineated series of bedrock ridges and depression. The VES result delineated four major Geo-electric layers within the study area. The topsoil, weathered layer, fractured bedrock and fresh bedrock. The top soil (resistivity varies from 47 to 490 Ωm and thickness ranges from 0.7 to 3.9 m); weathered layer (resistivity varies from 13 to 207 Ωm and thickness ranges from 1.9 to 22.1 m), fractured bedrock (resistivity varies from 489.3 to 878.8 Ωm and thickness ranges from 2.4 to 19.6 m) and bedrock with resistivity 1094 to 96583 Ωm and depth to bedrock 2.6 to 24.8 m). The dipole-dipole results also mapped linear features (fracture) at distance 60 to 100 m and 100 to 120 m respectively along the two traverses. Then from the geophysical investigation, three major causes of potential failure in the area were identified, these are; failure due to lateral inhomogeneity of the subsurface layers, failure precipitated by differential settlement and failure initiated by geologic features such as fractures and faults.

Key words: Foundation integrity, lateral inhomogeneity, electromagnetic, resistivity, dipole-dipole.

INTRODUCTION

The rate of failed structures in Nigeria have increased in recent times (Oyedele et al., 2011). These structural failures are often times associated with the problem of poor quality of building materials, old age of buildings and improper foundation. In recent times, the land expanse in the Federal University of Technology, Akure have been opened to rapid development (Olayanju, 2011). Despite this rapid growth and development in the area, the impact of subsurface geologic structures in the area on the durability and easy maintenance of the erected structures have been seldom talked about. Vertical and near vertical cracks or discontinuities have been noticed in the walls of both old and recent buildings in the school (Bayode et al., 2012). This assertion can be attributed to the minimal attention towards the use of geophysics in foundation studies. In Engineering Geophysics and site investigation, structural information and physical properties of a site are sought (Sharma, 1997). This is so because the durability and safety of the engineering structural setting depend on the competence of the material, nature of the sub-surface lithology and the mechanical properties of the overburden materials.

Foundations are affected not only by design errors but also by foundation inadequacies such as sitting them on incompetent earth layers. When the foundation of a building is erected on less competent layers, it poses serious threat to the building which can also lead to its collapse. Therefore, there is need to evaluate the foundation integrity of the buildings around School of Science Area at the Federal University of Technology, Akure in terms of the subsurface structures and nature of the soil. This research is therefore targeted at revealing the use of Geophysical techniques as a reliable means of undertaking studies of construction sites as related to the Geologic nature of the environment thereby saving a lot of time and cost. Also, with the Science and Art of Geophysics, the basic problems of structures that have emerged problematic can be investigated and remediation actions can be taken.

Description of the study area

The area is located at Obanla, School of Science Area of the Federal University of Technology, Akure (FUTA), Ondo State, Nigeria. It is accessible through tarred road from Futa North gate. The site occupies an area of about 0.2 km. It lies between latitudes 7°18'21.3''N and 7°18'47.1''N (808000 and 808800) N in the Universal Transverse Mercator (UTM) scale, and longitudes 50°7'27.0'' and 50°8'19.2'' (734600 and 736200) E in the Universal Transverse Mercator scale (Figure 1).

Geology of the area

The campus is underlain by crystalline rock of the Precambrian basement complex of the Southwestern Nigeria (Rahaman, 1976; 1988). The lithologic units include migmatite gneiss complex, granitic gneiss and charnokites (Figure 2). Outcrops of biotite gneiss and granitic gneiss occur in some locations around the western part of the study area. Likewise some other

boulders of granite and charnokites occur at the western street of the study area. The fractured bedrock generally occur in a typical basement terrain (Odusanya and Amadi, 1989) in tropical and equatorial regions, weathering processes create superficial layers, with varying degree of porosity and permeability. These geologic events gave rise to such structures as folds, faults and fractures that are geologically associated with zones of weakness. Geophysical methods can however, map these geologic structures; hence their application is employed to study the subsurface geology of the area in order to ascertain if there are geologic structure that can affect foundations or cause building collapse.

METHODOLOGY

Two traverses of about 170 and 200 m, respectively, were established in an approximate E-W direction (Figure 3). Three geophysical methods involving the magnetic, very low frequency electromagnetic (VLF-EM) and the electrical resistivity methods were adopted for this survey. The electrical resistivity method utilized the dipole-dipole profiling and the vertical electrical sounding (VES) techniques. The dipole-dipole survey was used to determine the lateral and vertical variation in apparent resistivity of the subsurface beneath the two established traverses. The VES involved the use of Schlumberger array. Sixteen sounding stations were occupied along the two established traverses and the current electrode spacing (AB/2) was varied from 1 to 65 m. The electrical resistivity data was processed by plotting the apparent resistivity values against the electrode spread (AB/2). This was subsequently interpreted quantitatively using the partial curve matching method and computer assisted 1-D forward modeling with WinResist 1.0 version software (Vander Velpen, 2004). The dipole-dipole data were inverted into 2-D subsurface images using the DIPPRO™ 4.0 inversion software (Dippro, 2000). 2-D electrical imaging of the subsurface was obtained using dipole-dipole configuration.

The inter-electrode spacing of 10 m was adopted while inter-dipole expansion factor (n) was varied from 1 to 10. Resistivity values were obtained by taking readings using the ohmega resistivity meter. The ground magnetic survey involved measurements of total field component of the earth's magnetic field along the two traverses using GEM 8 proton precession magnetometer. The magnetic data were drift corrected and presented as profiles of relative magnetic intensity values against distance (Figure 4a, b). The automated Euler deconvolution software was used to estimate depth to basement along the traverses. The very low frequency electromagnetic data were processed by downloading the raw real and filtered real components from the Abem Wadi VLF-EM equipment. The data are presented as profiles (Figure 5a, b). The Abem Wadi measures the field strength and the phase displacement around the fracture zone. The EM data was interpreted and inverted into a 2-D section using the Karous-Hjelt filtering (Karous and Hjelt, 1983).

RESULTS AND DISCUSSION

Magnetic profiles

Along Traverse 1, the magnetic intensity contrast observed between -100nT and 700nT at distance between 20 and 30 m and between 500nT and about 1300nT at distance ranging between 40 and 60 m which

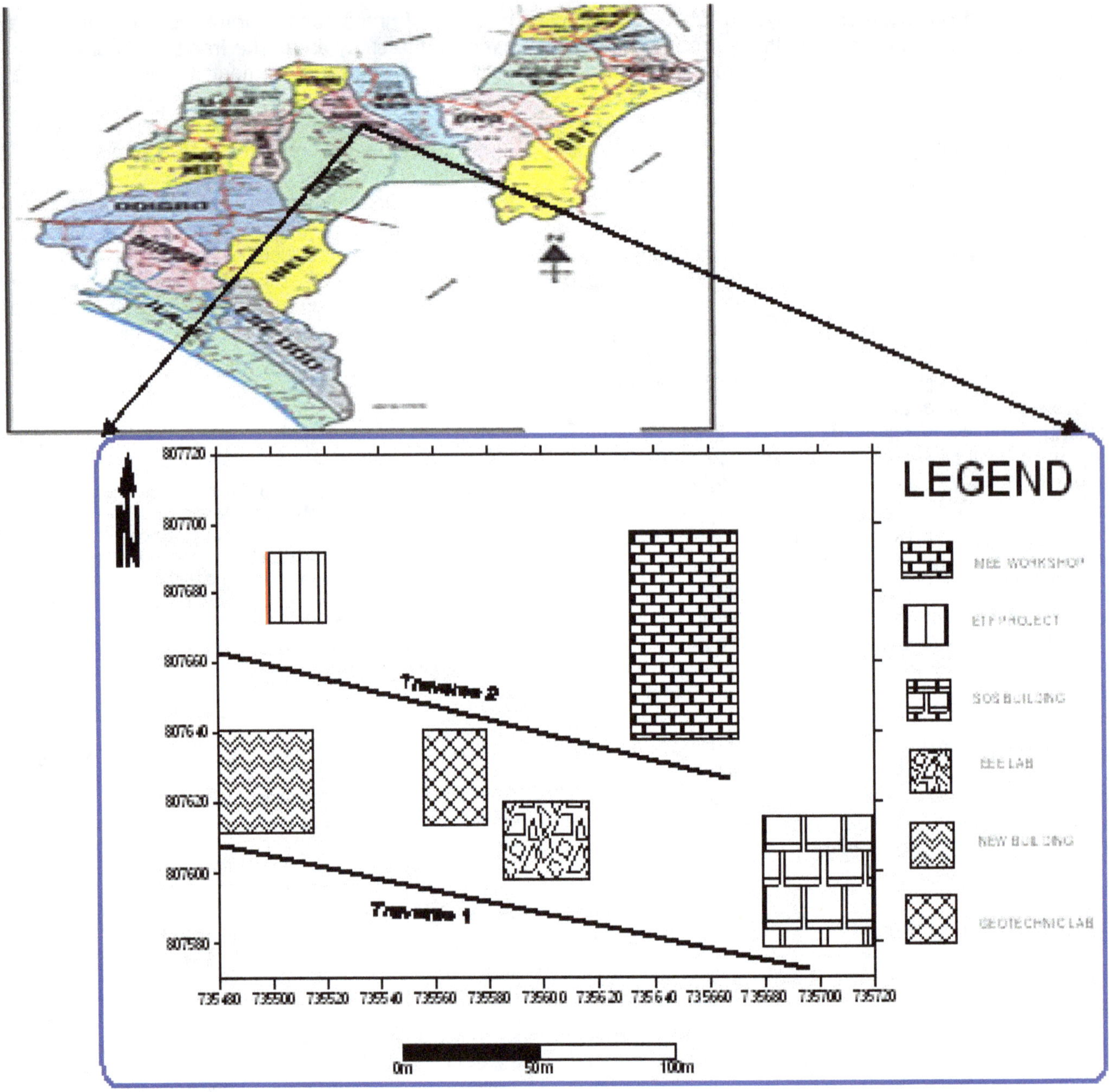

Figure 1. Location map of the study area.

are indicative of probable fracture zones. The automated Euler deconvolution software assist in the delineation of six probable fracture zones and estimated depth of between 5 m to about 20 m (Figure 4a). Continuous low magnetic intensity is observed at distance between 100 and 180 m which is typical of a fractured terrain although occurring at a greater depth. The magnetic intensity contrast along Traverse 2 observed between -2500 and -200nT at distance 30 to 50 m and between -50 and 500nT at distance 130 to 150 m are also indicative of probable fracture zones along the traverse. The Euler

technique helps in delineating four probable fracture zones at depth of between 5 and 10 m (Figure 4b).

VLF-EM profiles

On Traverse 1, the VLF-EM profile identified peak positive filtered real values which corresponds to probable fracture zones at distances 20, 80, 110, 130 and 160 m. These observations agrees with the conductive zones delineated by the KH section at

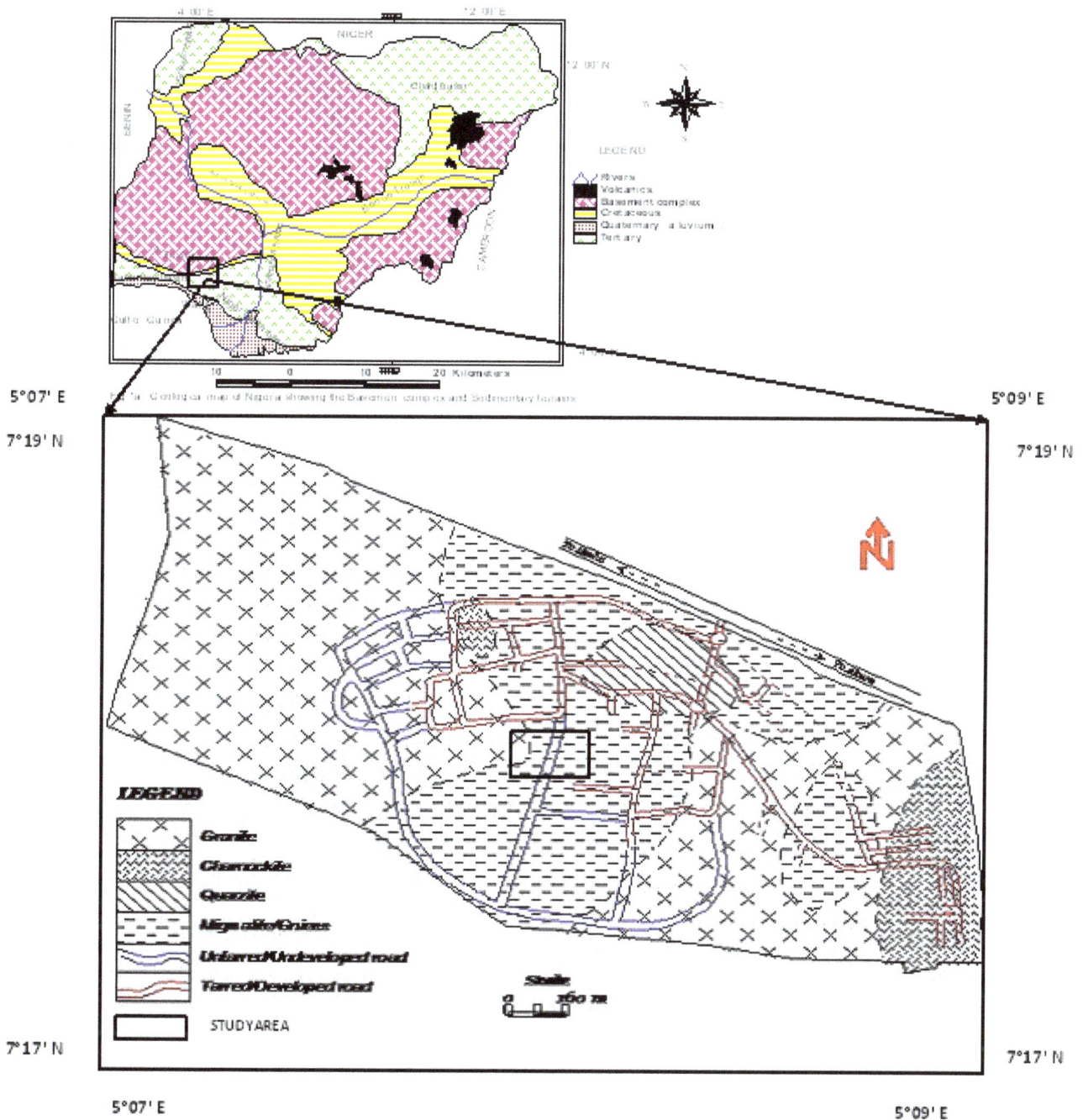

Figure 2. Geologic map of Nigeria top, geologic map of FUTA campus down (Modified after Kareem, 1995).

distances between 10 to 30 m, 60 to 80 m, 120 to 140 m and 160 to 180 m, respectively (Figure 5a). The conductive zone between 60 to 80 m is typical of a linear feature (fracture) because of its attitude that is dipping west. This agrees with the probable zone earlier delineated by the magnetic method at distance 40 to 60 m (Figure 5a). On Traverse 2, the VLF-EM profile identified peak positive filtered real values which correspond to probable fracture zones at distances 60

and 120 m. These coincides with the conductive zones delineated by the KH section at distances between 60 to 70 m at a very shallow depth and between 90 to 130 m which is typical of a linear feature and is dipping east (Figure 5b). The identified linear features have a significant depth extent. The identified linear feature coincides with a low magnetic intensity zone ranging between 60 and 120 m on the magnetic profile (Figure 4b).

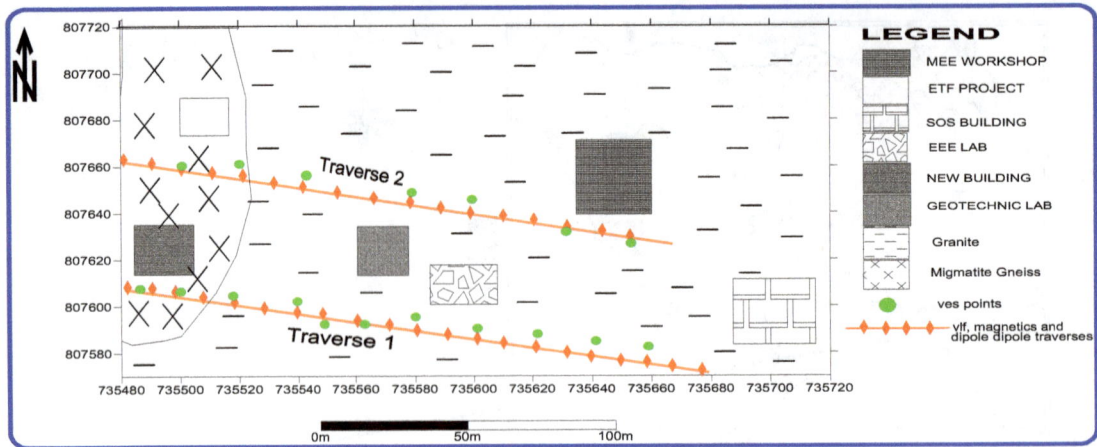

Figure 3. Data acquisition map of the study area.

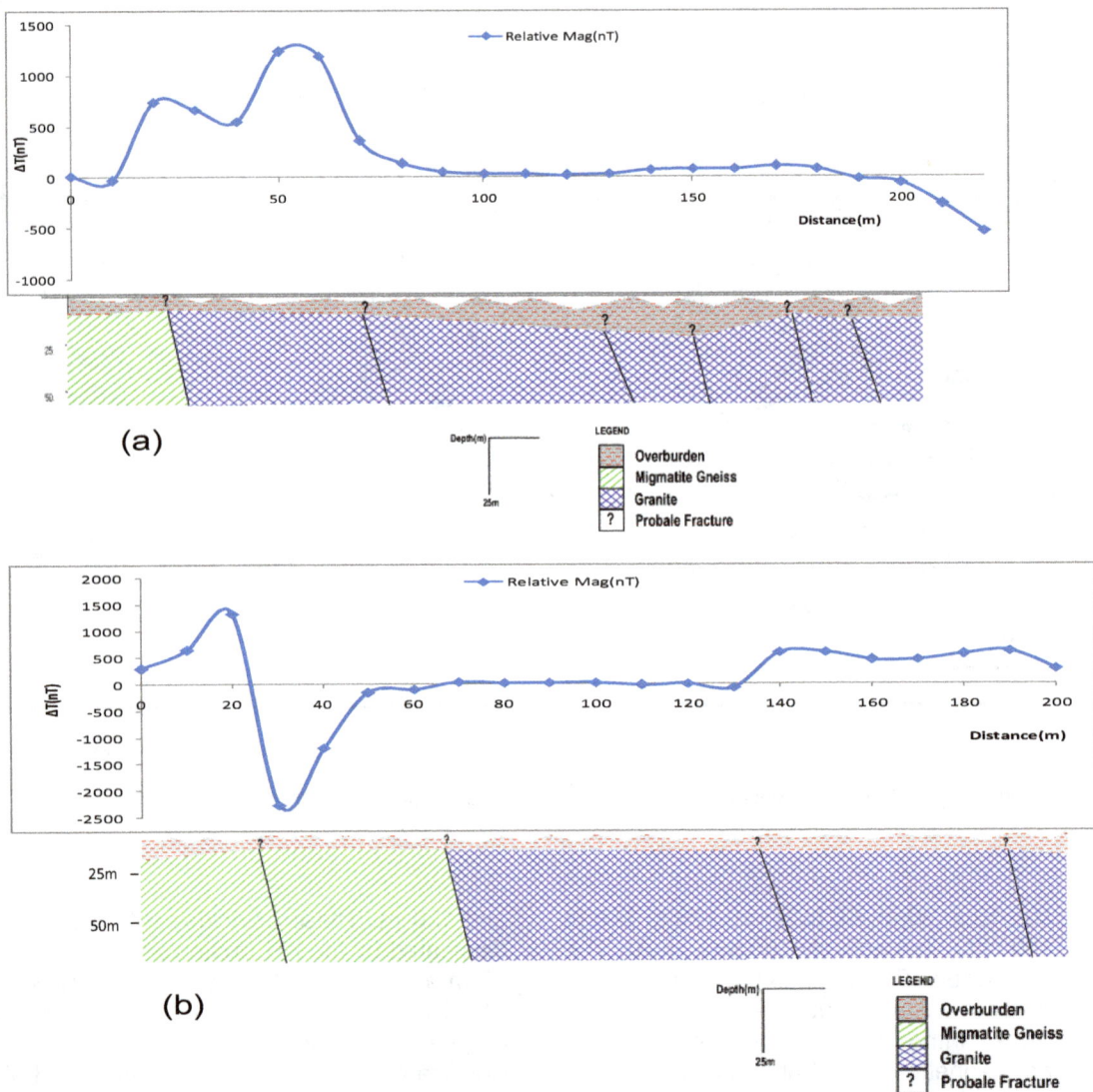

Figure 4. (a) Magnetic profile and its geomagnetic section along Traverse 1, (b) Magnetic profile and its geomagnetic section along Traverse 2.

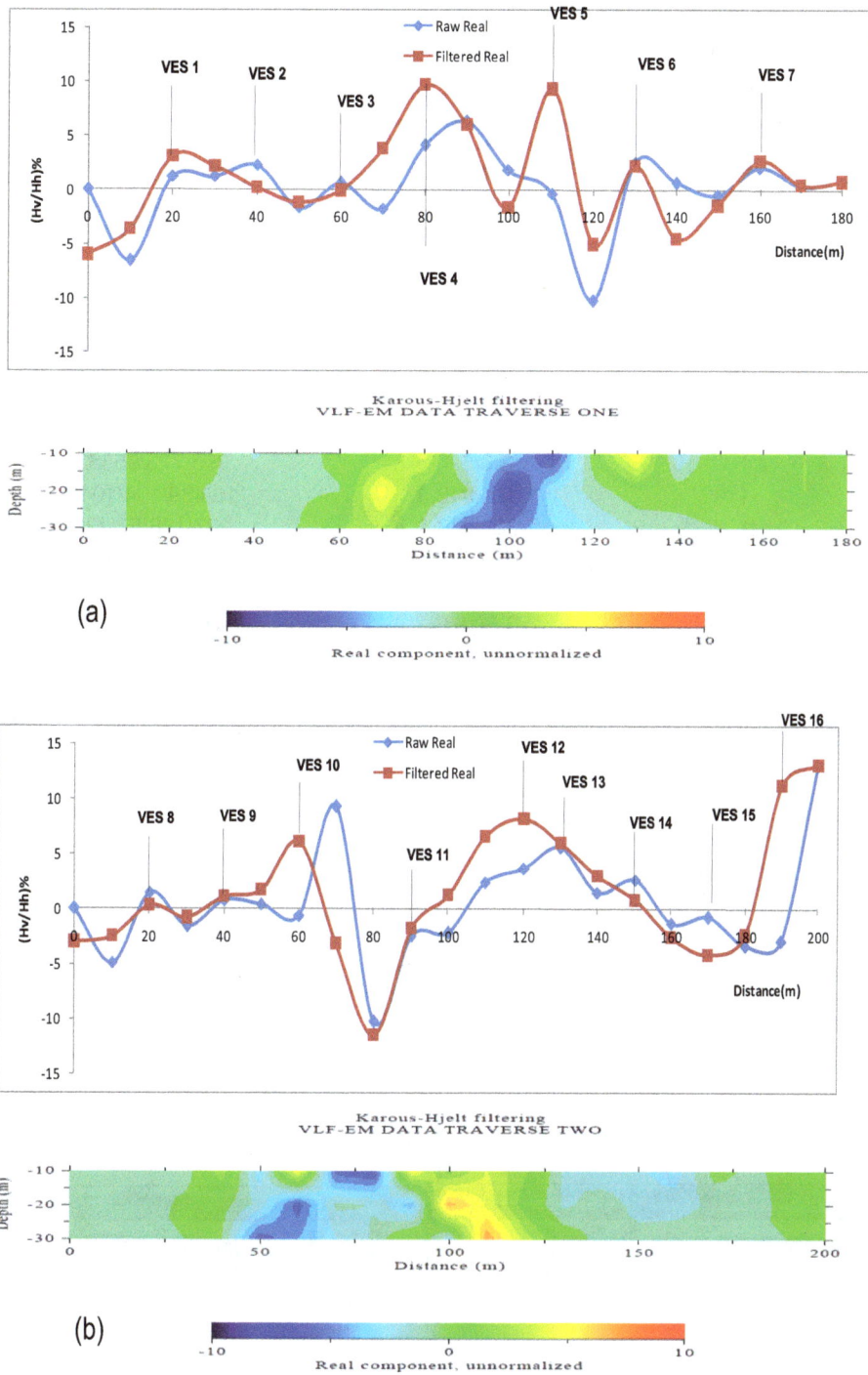

Figure 5. (a) VLF profile and KH Section along Traverse 1, (b) VLF profile and KH Section along Traverse 2.

RESISTIVITY SOUNDING CURVES AND GEOELECTRIC SECTIONS

Characteristics of VES curves

Table 1 gives a summary of the interpretation results of the VES curves at each of the studied localities. The number of layers varies between 3 and 5. Five curve types have been identified in all the locations. These include the A, H, HA, KH, and QH type. Typical VES curves were presented in (Figure 6) with the KH and A curve type dominating with 38 and 31%, respectively.

Table 1. Summary of geo-electric parameters.

VES stations	Traverse lines	Thickness (m) $h_1/h_2/h_3$---$/h_n$	Resistivity (ohm-m) $\rho_1/\rho_2/\rho_3/$---ρ_n	Type curves
1	2	0.6/1.7/7.7	193/101/207/547	HA
2	2	0.9/2.2/7.5	169/320/72/3287	KH
3	2	2.7/22.1	47/174/1673	A
4	2	1.4/3.2/10.6	47/178/72/5738	KH
5	2	2.1/7.2	152/34/1846	H
6	2	1.2/2.9/7.9	80/122/30/543	KH
7	2	2.3/8.7	80/101/216	KH
8	1	0.9/2.9/5.8	102/30/582/1388	HA
9	1	0.7/1.9	65/13/394	H
10	1	3.9/12.9	153/528/634	A
11	1	3.0/5.4	395/723/95863	A
12	1	1.1/2.4/10.0	110/489/104/10432	KH
13	1	1.3/5.9/17.4	227/198/57/222	QH
14	1	0.6/2.2/7.9	63/284/38/1094	KH
15	1	1.6/3.0	33/186/2549	A
16	1	0.6/19.6	209/879/1220	A

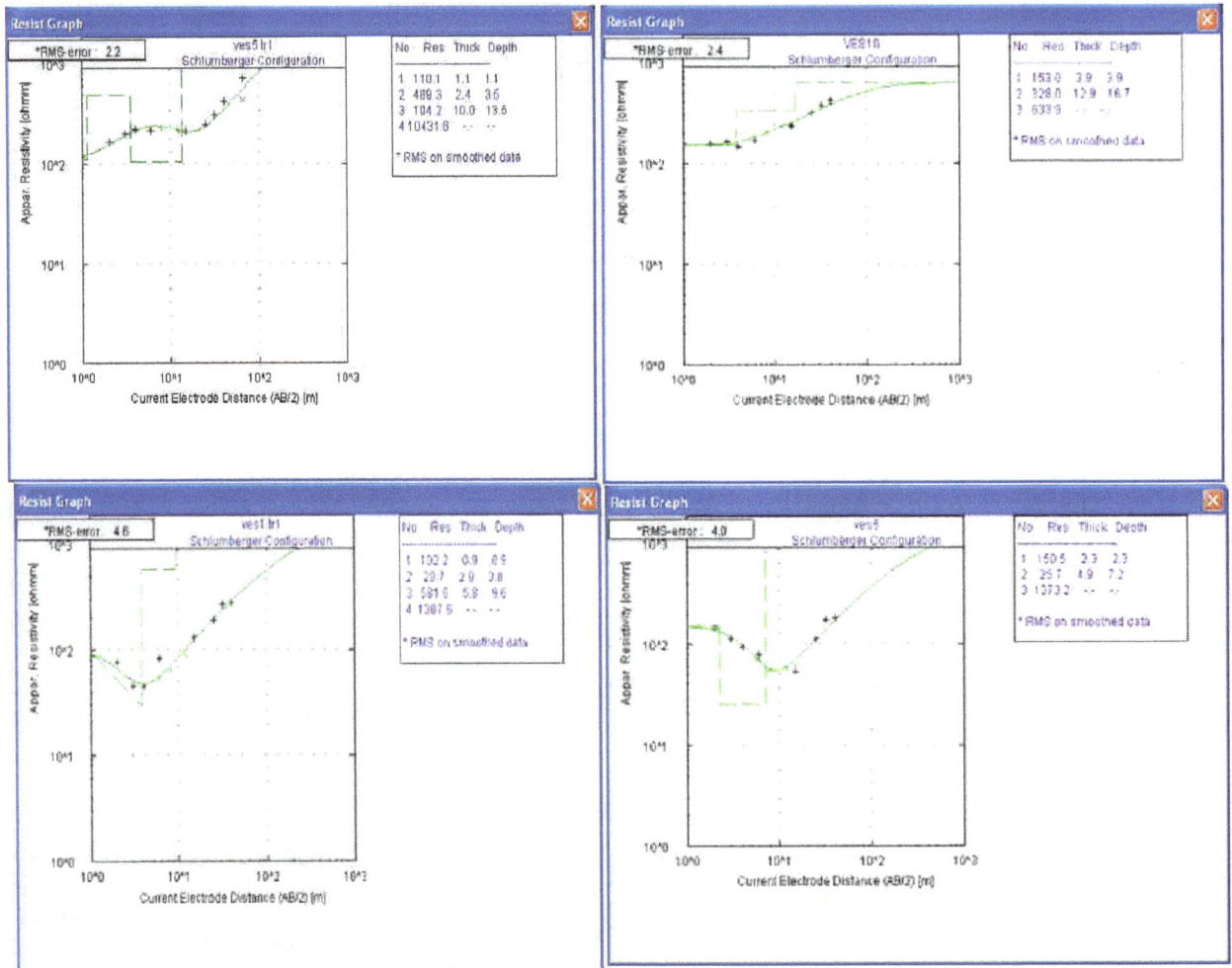

Figure 6. Typical VES curves from the study area (a=KH, b=A, c=HA, d=H).

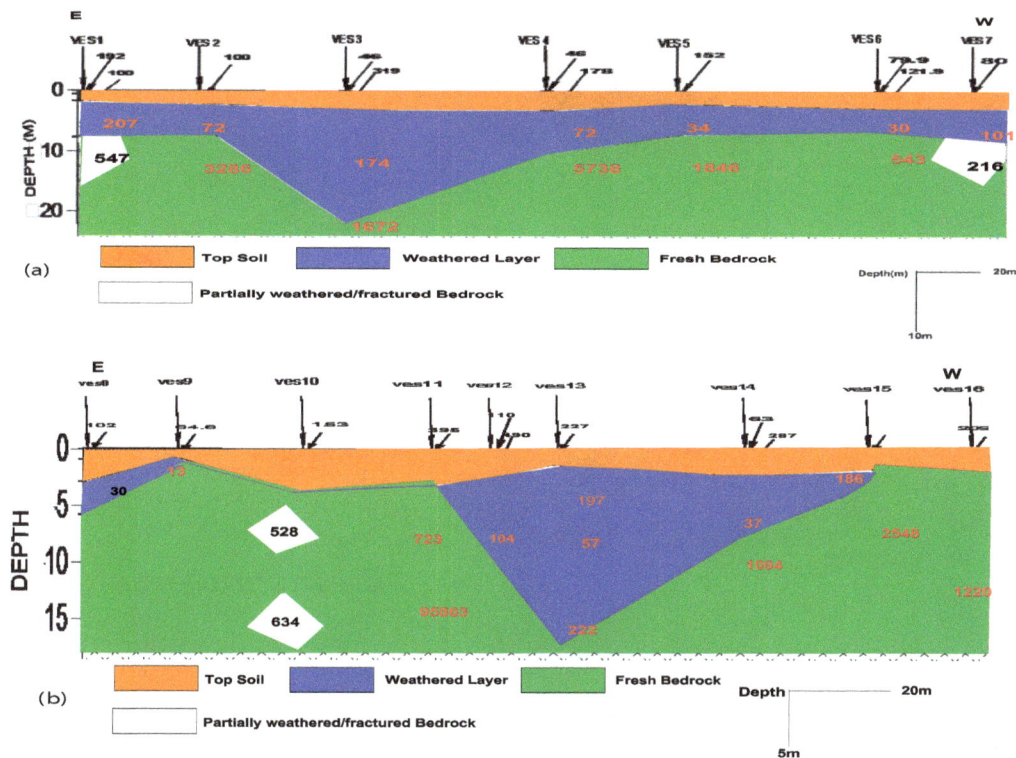

Figure 7. (a) Geo-electric section along Traverse 1, (b) Geo-electric Section along Traverse 2.

The HA and H curve type having 13% each and the QH curve type having 6%.

Geoelectric sections

The 2-D view of the geo-electric parameters (resistivity and depth) obtained from the inversion of the electrical resistivity sounding data are presented as geo-electric sections. The geo-electric section along Traverse 1 (Figure 7a) attempt to correlate the geo-electric sequence across the study area. Four subsurface geologic layers were delineated in the study area; the top soil, weathered layer, partly weathered/fractured basement and fresh bedrock. The topsoil (resistivity varies from 46.5 to 219 Ωm and thickness ranges from 0.6 to 2.7 m); weathered layer (resistivity varies from 72 to 207 Ωm and thickness ranges from 7.2 to 22.1 m), fractured bedrock (resistivity varies from 489.3 to 878.8 Ωm and thickness ranges from 2.4 to 19.6 m), bedrock resistivity (1139 to 5738 Ωm) and depth to bedrock (7.7 to 24.8 m).

On Traverse 2 (Figure 7b), four subsurface geologic layers were also delineated along this traverse. From the geo-electric section, the top soil, weathered layer, partly weathered/fractured basement and fresh bedrock were determined. The topsoil (resistivity varies from 63.5 to 490 Ωm and thickness ranges from 0.6 to 3.9 m); weathered layer (resistivity varies from 13 to 197 Ωm and

thickness ranges from 1.9 to 17.2 m, fractured bedrock (resistivity varies from 489.3 to 878.8 Ωm and thickness ranges from 2.4 to 19.6 m), bedrock resistivity (1094 to 96583 Ωm) depth to bedrock (2.6 to 19.5 m).

Dipole-dipole Pseudosections

The 2-D Pseudosection was produced from the dipole-dipole data taken along the two traverses (Figure 8a, b). It was set up to have a 2-Dimensional clear view of the subsurface because it shows an interpretation of unilateral data and its contours. This also gave similar information as the geo-electric section. It delineated topsoil, weathered/fractured layer (thickness 8 to 20 m) and the fresh bedrock. The resistive parts are seen at the lower part of the section which is the fresh bedrock while the green and blue coloured parts are the fractured part of the section. A suspected linear feature was delineated at distance 60 to 100 m (Figure 8a).

The 2-D pseudo-section was also produced from the dipole-dipole data taken along Traverse 2 (Figure 8b). This also gives similar information as the geo-electric section. It delineated topsoil, weathered/fractured layer and the fresh bedrock. The highly resistive parts are seen at the lower part of the section which is the fresh bedrock while the green and blue coloured parts are the weathered/fractured part of the section. A suspected

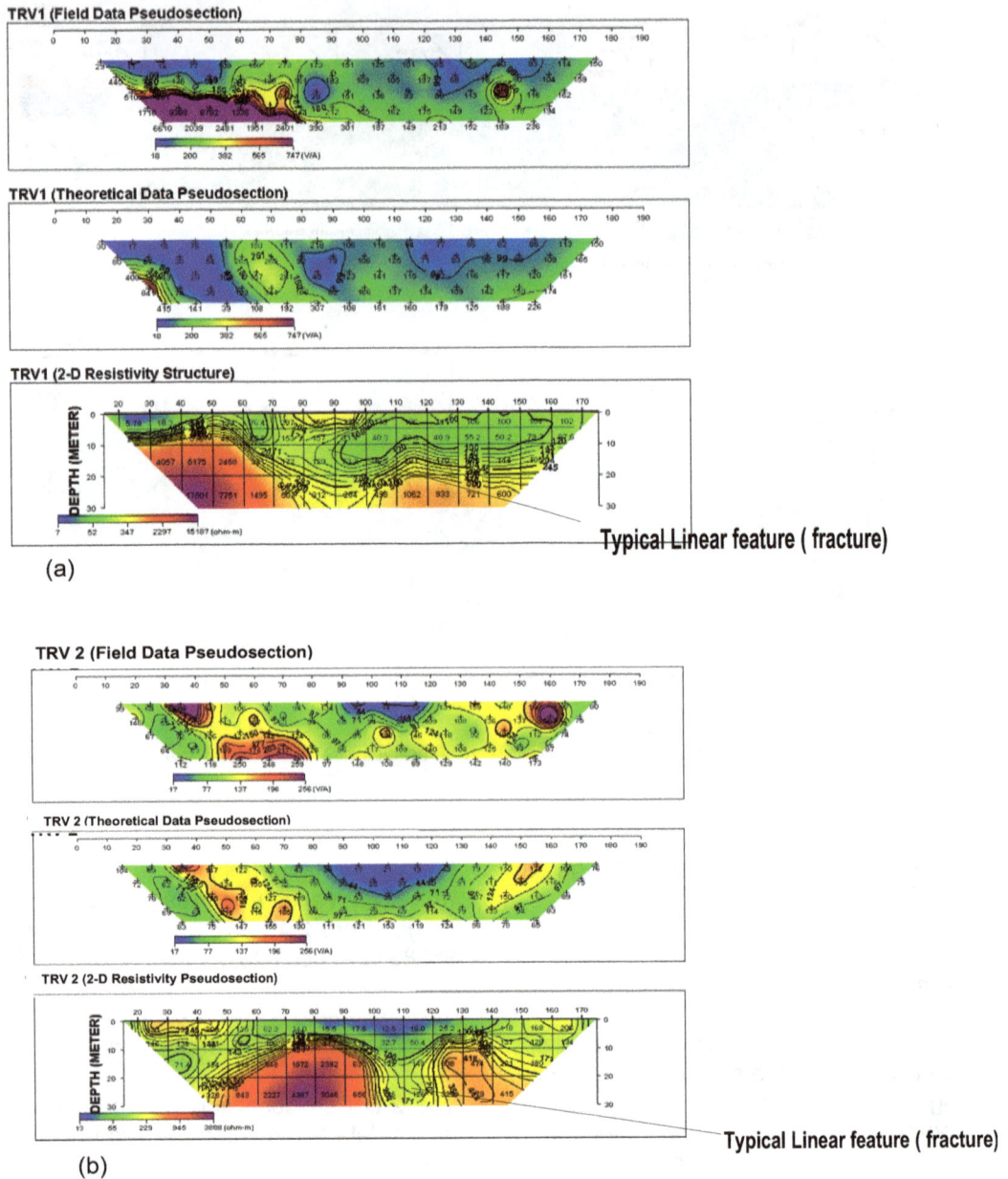

Figure 8. (a) Dipole -dipole pseudosection along Traverse 1, (b) Dipole -dipole pseudosection along Traverse 2.

linear feature was delineated at distance 100 to 120 m along Traverse 2 (Figure 8b).

ISORESISTIVITY AND ISOPACH MAPS

Isoresistivity map of topsoil

Figure 9a shows the isoresistivity map of the topsoil. The topsoil comprises of sandy clay/clayey sand formation with resistivity values ranging from 40 to 380 Ωm. The highest resistivity values were identified towards the

South western flank of the study area up to 380 Ωm which may be suitable for foundation works if the thickness is considerable. The lowest resistivity values were identified at the Northern, western, and southeastern region (40 Ωm) which is not suitable for foundation works.

Isopach map of topsoil

Figure 9b shows the Isopach map of the top soil. The map shows the largest thickness at the Southeastern

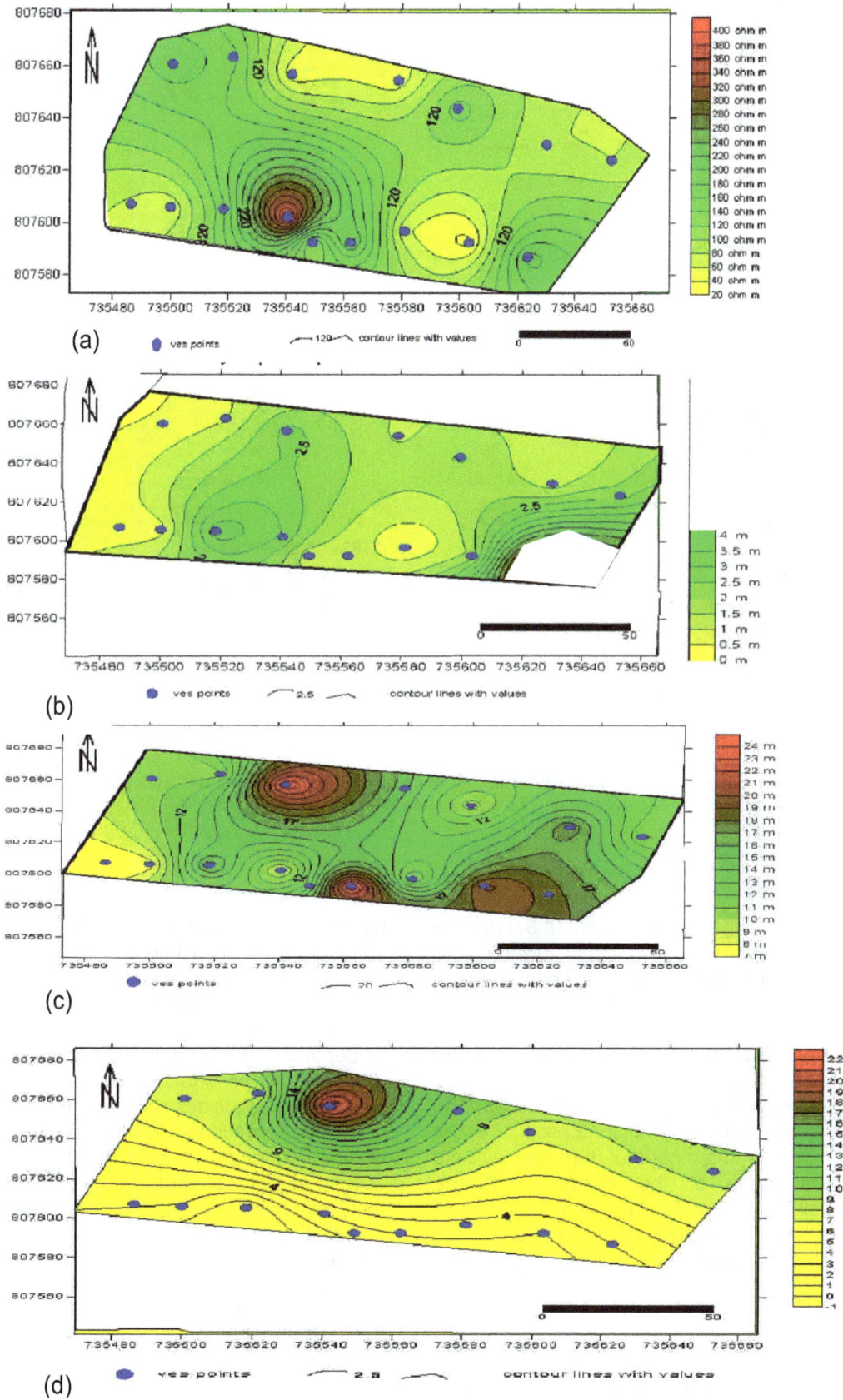

Figure 9. (a) Isoresictivity mapof top soil, (b) Isopach map of top soil, (c) Isopach map of overburden, (d) Isopach map of weathered layer.

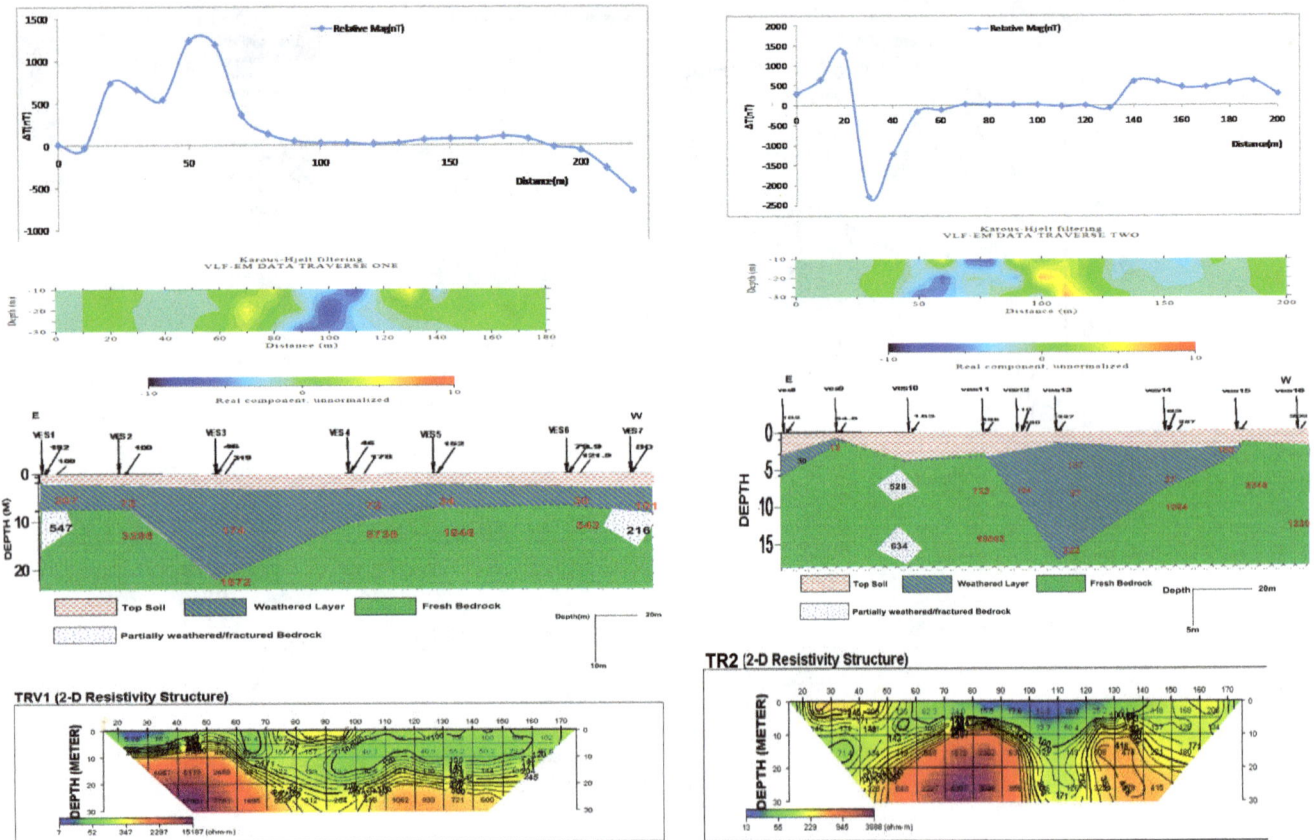

(a)   (b)

Figure 10. (a) Summary of results along Traverse 1 showing the magnetic profile, VLF-EM section, geo-electric section and dipole-dipole pseudosection, (b) Summary of results along Traverse 2 showing the magnetic profile, VLF-EM section, geo-electric section and dipole-dipole pseudosection.

flank of the study area with thickness of about 3.9 m. The lowest thickness values were identified towards the Northwestern flank of the study area up to 0.6 m.

Isopach map of overburden

Figure 9c shows the isopach map of the overburden. The overburden consists of two formation topsoil and the weathered layer. The southern parts of the area have the highest thickness value at the Northwestern and Southwestern flank of the study area with thickness of about 24 m. The lowest thickness values were identified towards the Western flank of the study area.

RESULTS

Figure (10a) shows the summary of profiles and sections obtained from various geophysical methods employed along Traverse 1. The magnetic intensity contrast

observed at distance 20 to 30 m coincides with the conductive zones delineated by the VLF-EM section at distance 10 to 30 m. This also agree with the low resistivity zone (fracture zone) observed on the dipole-dipole pseudo-section at distance between 20 and 40 m at depth 0 to 8 m. The linear feature delineated on the dipole-dipole pseudo-section at distance 60 to 100 m (Figure 8a) coincides with depression observed on the geo-electric section at distance between 80 to 140 m and is also delineated as conductive zones by the VLF-EM section at distance 60 to 80 m. These results reveal that the geophysical methods used for this study are complimentary.

Figure (10b) shows a magnetically quiet environment at distance ranging from 60 to 120 m. There might be existence of structures very close to the surface at this point since the VLF section was able to identify conductive zone at the distance which also agrees with low resistivity zone observed on the 2-D resistivity structure. The delineated linear feature (fracture) observed on the 2-D image occurring between

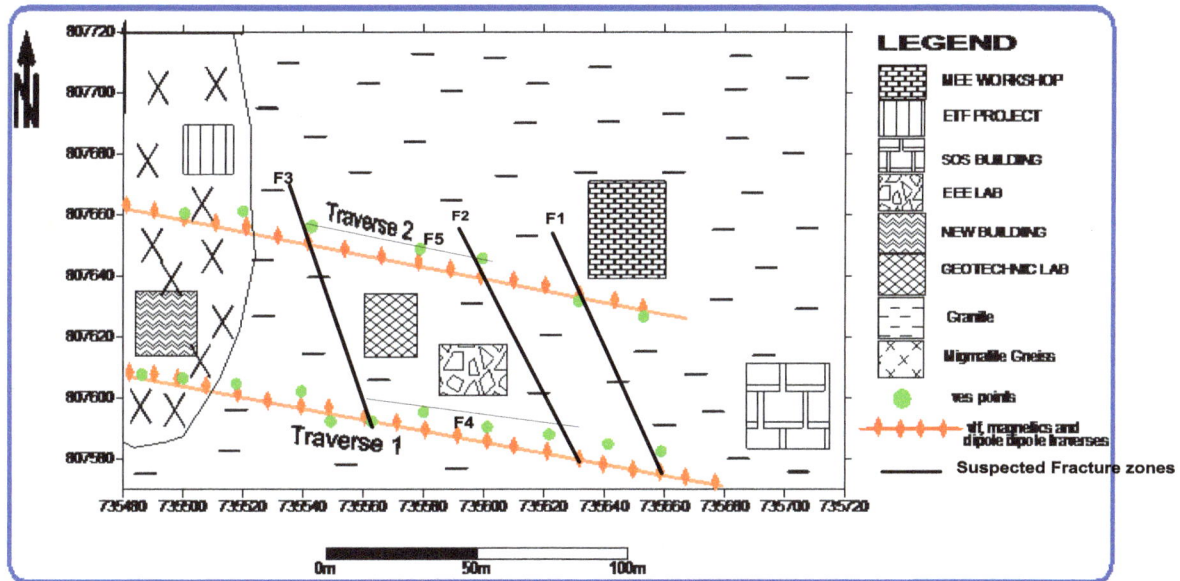

Figure 11. Structural map of the area.

100 and 120 m is also revealed on the VLF-EM section as a linear feature at distance 90 to 130 m.

Structural map of the area

The structural map developed from the VLF-EM, ground magnetic and dipole-dipole results are presented in Figure 11. Five fracture zones designated F_1 to F_5 were delineated. The ground magnetic combined with the VLF-EM method assist in delineating vertical discontinuities F_1, F_2 and F_3 while the 2-D resistivity image of the sub surface identified continuous near surface fracture zones at distance 60 to 170 m and 60 to 140 m on Traverse 1 and two, respectively. These fractures are designated F_4 and F_5 along Traverse 1 and 2, respectively. These structures suspected to be a major faulted zone underlies the study area especially the electrical electronics engineering (EEE) and Geotechnical lab. The presence of these structures poses danger to the continuous existence of the structures erected in this location.

Conclusion

In conclusion, the Geophysical methods were successful for post construction studies. The interpretation of VES, VLF–EM, magnetic and the dipole-dipole pseudo-section in the study area have allowed the delineation of incompetent zones in the study area. Five fracture zones designated as F_1 to F_5 believed to be a major faulted zone underlies the buildings in the study area with the 2-D image of the subsurface identifying near surface fractures that are inimical to engineering works. This

shows that subsurface geologic setting underlying the buildings is inhomogeneous and therefore can be said to be structurally deformed. The identified weak zones expose the buildings to future failure and eventual collapse. Based on the geophysical investigation, the competence of the subsurface of the study area can be generally classified as incompetent. Three major causes of potential failure in the area were also identified, these are; failure due to lateral inhomogeneity of the subsurface layers, failure precipitated by differential settlement and failure initiated by geologic features such as fractures and faults. Subsequent construction in the area should be founded on the fresh basement layer coupled with pile foundation to ensure the stability of the building.

ACKNOWLEDGEMENTS

Messrs. M. Bawallah and B. Osanyingbemi assisted during the data acquisition. Their participations are gratefully acknowledged

REFERENCES

Bayode S, Omosuyi GO, Abdullahi HI (2012). Post –foundation Engineering Geophysical investigation in Part of the Federal University of Technology, Akure, Southwestern Nigeria. J. Emerging Trends Eng. Appl. Sci. (JETEAS). 3(1):203-210.

Dippro for Windows (2000). DipproTM Version 4.0 Processing and Interpretation software for Dipole – Dipole electrical resistivity data. KIGAM, Daejon, South Korea.

Kareem WA (1995). Geological Map of Federal University of Technology, Akure. Unpulished M.Tech. Thesis, Dept of Applied Geology, Federal University of Tech. Akure. P. 109.

Karous M, Hjelt SE (1983). Linear Filter of VLF Dip-Angle Measurements. Geophys. Prospect. 31:782-794.

Odusanya BO, Amadi UMP (1989). An Empirical Resistivity Model for Predicting Shallow Groundwater Occurrence in the Basement Complex. Water Res. J. Nig. Asso. Hydrogeol. 2:77-87.

Olayanju GM (2011). Engineering Geophysical Investigation of a Flood Zone: A Case study of Alaba Layout, Akure, Southwestern Nigeria. J. Geol. Mining Res. 3(8):193-200.

Oyedele KF, Oladele S, Adedoyin O (2011). Application of Geophysical and Geotechnical Methods to Site Characterization for Construction Purposes at Ikoyi, Lagos, Nigeria. J. Earth Sci. Geotech. Eng. 1(1):87-100.

Rahaman MA (1976). A Review of the Basement Geology of Southwestern Nigeria. In Kogbe, C.A. (Editor), Geology of Nigeria. Elizabethan Publishing Co. pp. 41-48.

Rahaman MA (1988). Recent Advances in the Study of the Basement Complex of Nigeria. In Precambrian Geology of Nigeria, Geological Survey of Nigeria, Kaduna South. pp. 11-43.

Sharma VP (1997). Environmental and Enginering Geophysics, published by Cambridge University Press, United Kingdom. pp. 40-45.

Vander Velpen BPA (2004). WinRESIST Version 1.0 Resistivity Depth Sounding Interpretation Software. M. Sc Research Project, ITC, Delf Netherland.

12

Equivalence principle of light's momentum harmonizing observation from quantum theory to cosmology

Shinsuke Hamaji

Hyama Natural Science Research Institute, 403, Daiichi-Kiriya Building, 5-2, Chuo 2-chome, Nakano-ku, Tokyo 164-0011, Japan.

Unlike in Newtonian mechanics, in the theory of relativity, the coordinate axis is not fixed, and the time axis serves as the background for the observation. In other words, the theory of relativity is background independent (BI). The quantum theory created in the micro-world, in contrast, is background dependent (BD). Efforts to unify both by BI have not yielded much success. In addition, it is recognized that we cannot consider gravity in a quantum theory by BD. This paper focuses on neither the BI method based on the invariant speed of light and the inertial mass observed only in the uniform field nor the method to consider gravity in an existing quantum theory. By using the invariant mass and the speed of light in the frame of reference of the free space of the electromagnetism as well as the observed variable mass and wave speed, a method to link them to the equivalence of light's momentum was explained. In addition, some cases in which we can observe space from a micro-world by the equivalence principle were examined.

Key words: Background independent, gravity, free space, electromagnetism, invariant mass, variable mass, speed of light, wave speed.

INTRODUCTION

In this paper, given the established principles of electromagnetism, they were logically and rationally integrated with those of mechanics. The electromagnetic correlation between the speed and energy of light was defined and interrelated with the mechanics of the energy, momentum, mass, and speed of light. The question of whether gravity should be taken into consideration in the electromagnetic analysis of free space and vacuums is also addressed. Moreover, a deviation was made from the approach of the theory of relativity, which unifies electromagnetism and mechanics by disregarding gravity and assuming that the speed of light is constant. The two fields were unified by considering the interrelation of gravity and mass and

assuming a fixed speed of light in free space. This assumption is the most important aspect of the proposition of this paper.

Firstly, the known facts of classical physics were stated, including (1) the speed and energy of light waves, and (2) the mass and velocity of a material in free space. Some unique terms and symbols used in this paper were also defined, highlighting the difference between particle velocity (v) and wave speed (w) and went further to discuss the following issues about gravity, based on the assumptions already made:

i. Gravitational mass and the light wave speed under gravity.

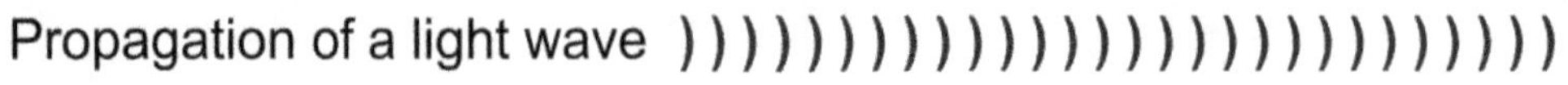

Propagation of a light wave)))))))))))))))))))))))))))))))

Figure 1. Propagation of a light wave.

ii. Mass and velocity of a material under gravity.

iii. Mass and speed of photons under gravity.

iv. The speed of light and the wave speed are different physical quantities.

v. Matter wave and uncertainty relation.

Finally, a conclusion from the abovementioned considerations was derived, and the correlation of speed, mass, momentum, energy, and quantum in the context of logical and rational integration of electromagnetism and mechanics were summarized. This integration is based on the interrelation of the physical quantities in terms of the equivalence principle by the momentum of light.

SPEED AND ENERGY OF LIGHT WAVES WITHOUT TAKING GRAVITY INTO CONSIDERATION

In electromagnetism, free space is a virtual space where no matter exists. It is defined by physical constants such as the speed of light in a vacuum (free space), magnetic permeability of a vacuum (free space), and permittivity of a vacuum (free space). Maxwell's equation of the speed of propagation of an electromagnetic field is as follows:

$$C = 1 \Big/ \sqrt{\varepsilon_0 \mu_0} \tag{1}$$

where C is the speed of light, and μ_0 and ε_0 are the permeability and permittivity of a vacuum (free space), respectively (Figure 1). The correlation between the electromagnetic wave energy (E), absolute value of the momentum of light (P), and speed of light (C) can be derived from the theory of electromagnetism alone:

$$E = PC \tag{2}$$

Furthermore, in the MKSA system or the International System of Units (SI), the absolute refractive index n_0 is obtained by dividing the speed of light in a vacuum (free space) (C) by the light wave speed in a medium (w). In other words, it is given by the phase speed.

$$n_0 = \frac{C}{w} = \sqrt{\frac{\varepsilon \mu}{\varepsilon_0 \mu_0}} \tag{3}$$

where μ, ε are the permeability and permittivity of the medium, respectively. It can be observed from the above that the light wave speed in a medium changes and is refracted relative to the speed of light in a vacuum (free space).

VELOCITY AND INERTIAL MASS OF MATTER WITHOUT TAKING GRAVITY INTO CONSIDERATION

According to the observations of Kaufmann's experiments on bending beta rays (Boorse and Motz, 1966), the inertial mass (m') varies when the velocity (v) of the center of gravity of an object's rest mass (m_0) changes (Figure 2).

$$\frac{m'}{m_0} = \frac{C}{\sqrt{C^2 - v^2}} = \frac{1}{\sqrt{1 - v^2/C^2}} \tag{4}$$

The momentum of light (P) is obtained from the correlation among the speed of light in free space (C), particle velocity (v), and wave speed (w):

$$P = m_0 C = m' \sqrt{C^2 - v^2} = m'w \tag{5}$$

The energy (E) is obtained by multiplying the momentum of light by the speed of light (C):

$$E = PC = m_0 C^2 = m_0 (v^2 + w^2) = m'wC \tag{6}$$

GRAVITATIONAL MASS AND LIGHT WAVE SPEED UNDER GRAVITY

If we consider free space as the standard, Earth's substances would exert a gravitational influence on the matter field of the surface. If the propagation speed of the light waves observed on Earth's surface is (w) and the gravitational potential on Earth's surface is given by 2ϕ = 2GM/r, then the sum (Equation 8) of the gravitational potential corresponds to the speed of light (C) in free space. It can thus be determined from the equation for the fixed speed of light (Equation 1) that the light wave speed decreases with increasing gravity (Figure 3).

$$w = f\lambda \tag{7}$$

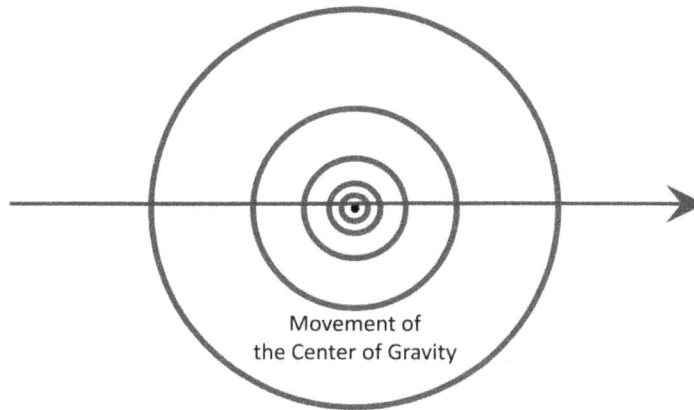

Figure 2. Movement of mass.

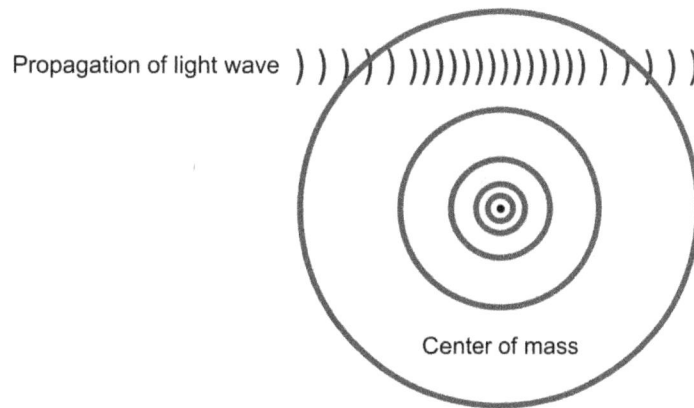

Figure 3. Perturbation between a light wave and a mass.

where f is the frequency of the wave and λ, the wavelength

$$C^2 = w^2 + 2\varphi \tag{8}$$

Gravitational mass (M) is the product of the material density ($C^2 - w^2$) and the radius (r) of the earth divided by the constant (2G). Therefore, there is no mass (M = 0) if there is no volume (r = 0).

$$M = (C^2 - w^2)r/2G = v_2{}^2\, r/2G \tag{9}$$

$$C^2 = w^2 + 2\varphi = v_2{}^2 + w^2 \tag{10}$$

This is equivalent to the potential ($v_2{}^2 = 2GM/r$) of the second cosmic velocity (v_2).

MATERIAL MASS AND VELOCITY UNDER GRAVITY

If the object's gravitational mass (M) is considered to be equal to the inertial mass (m) of its particle motion ($v^2, 2\varphi$), then the difference between the rest mass under gravity (m_0) and the inertial mass (m') can be obtained from the following:

$$C^2 = (v^2 + w^2) + 2\varphi \tag{11}$$

$$\frac{m'}{m_0} = \frac{\sqrt{C^2 - 2\varphi}}{\sqrt{C^2 - v^2 - 2\varphi}} = \frac{w_0}{w'} \tag{12}$$

$$n_0 = \frac{m}{M} = \frac{C}{\sqrt{C^2 - v^2 - 2\varphi}} = \frac{C}{w} \tag{13}$$

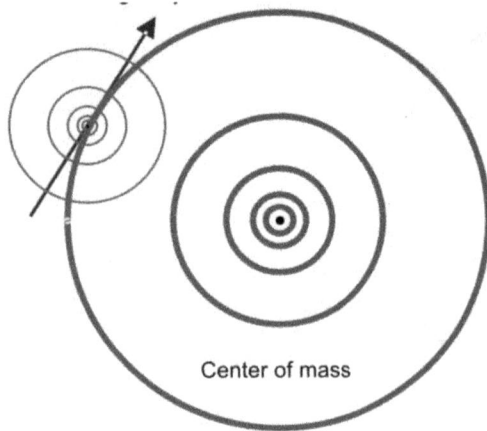

Figure 4. Perturbation in the movement of a mass.

Like the refraction rate n_0 of (Equation 3), n_0 of (Equation 13) also corresponds to the fixed speed of light in (Equation 1) (Figure 4). Furthermore, regarding the correlation between the momentum of light (P) and the energy (E), there is no equivalence between the gravitational mass and the inertial mass (M = m); however, an equivalent based on the momentum of light ($P = MC = mw$) can be obtained:

$$P = MC = m\sqrt{C^2 - v^2 - 2\varphi} = mw \qquad (14)$$

$$E = PC = MC^2 = M(v^2 + w^2 + 2\varphi) = mwC \qquad (15)$$

When stationary under gravity (v = 0), we can obtain an approximation of the kinetic energy ($mv^2/2$) on the basis of Newtonian mechanics:

$$E = M_2C^2 = M_2(w^2 + 2GM_1/r) \qquad (16)$$

$$GM_1M_2/r = M_2(C^2 - w^2)/2 = M_2v^2/2 = m_2v^2/2n_0 \qquad (17)$$

If we use the progress of an atomic clock installed in a GPS satellite as an example (Ashby, 2007),

Light wave speed on Earth's surface
$w_0 = 299,792,458 \text{ m/s}$
Geocentric gravitational constant
$GM = 3.986 \times 10^{14} \text{ m}^3/\text{s}^2$
Radius of Earth $r = 6,378,000$ m

Speed of light in free space $C = \sqrt{w_0^2 + 2GM/r} \qquad (18)$

Altitude of GPS satellite $h = 20,200,000$ m

Orbital velocity of GPS satellite $v = 3,874 \text{ m/s}$

Wave speed of GPS satellite $w' = \sqrt{C^2 - v^2 - 2GM/(r+h)}$
$$\qquad (19)$$

Progress of clock $w'/w_0 = 1 + (4.45 \times 10^{-10}) \qquad (20)$

SPEED AND MASS OF PHOTONS UNDER GRAVITY

The following can be derived from Einstein's photon hypothesis and the electromagnetic wave energy of (Equation 2):

$$E = PC = hf \qquad (21)$$

where h is Planck's constant and f, the frequency. Furthermore, the following can be derived from Equations 7, 15 and 21, although the frequency of the photon hypothesis is not given by $f = C/\lambda$ but by $f = w/\lambda$.

$$hf = mwC = mCf\lambda \qquad (22)$$

$$\lambda = \frac{h}{mC} \qquad (23)$$

This is also understandable from the Compton Effect (Greiner, 2001a). When a material is exposed to X-rays, some of the rays are scattered and become secondary X-rays. In the Compton Effect, the wavelengths of the secondary X-rays are larger than those of the incident ones.

$$\lambda_s - \lambda_i = \frac{h}{mC}(1 - \cos\theta) = \frac{w}{f_s} - \frac{w}{f_i} \qquad (24)$$

where λ_s is the wavelength of the secondary X-rays; λ_i, the wavelength of the incident X-rays; θ, the scattering angle; f_s, the frequency of the secondary X-rays; and f_i, the frequency of the incident X-rays. In this case, energy is lost under scattering, resulting in a reduced frequency ($f_s < f_i$). Moreover, because the light wave speed (w) remains unchanged, the wavelength increases ($\lambda_s > \lambda_i$) (Figure 5).

$$E = MC^2 = M(w_0^2 + 2\varphi_0) = m_0 w_0 C = m_0 Cf\lambda_0 = hf \qquad (25)$$

$$E = MC^2 = M(w'^2 + 2\varphi') = m'\dot{w}'C = m'Cf\lambda' = hf \qquad (26)$$

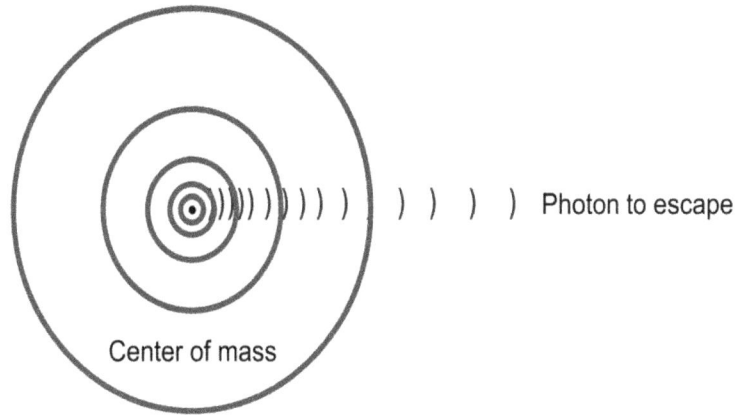

Figure 5. Photon escaping from a gravitational field.

However, with regard to the energy of the photons that escape from the gravitational field (that is, gravitational redshift), the frequency (f) remains unchanged and the gravitational potential ($\varphi_0 > \varphi'$) decreases, resulting in an increase in the light wave speed ($w_0 < w'$). Hence the wavelength increases ($\lambda_0 < \lambda'$) and the energy Equation 21 is maintained, even with a slight decrease in the inertial mass ($m_0 > m'$).

This is similar to Equation 20, where the progress of the atomic clock installed in the GPS satellite is faster than that of a clock on Earth's surface, owing to the weakened gravitational pull of the GPS satellite. This indicates that the propagated photon should be handled in the same manner as a material with mass.

SPEED OF LIGHT AND WAVE SPEED ARE DIFFERENT PHYSICAL QUANTITIES

The Shapiro Delay (Shapiro, 1964) is a combined effect of gravitational blueshift and redshift. In this phenomenon, the distance and frequency do not change, and the electromagnetic wave speed is only reduced by the expansion and contraction of the wavelength under changes in the gravitational field. This results in a delay in the arrival time (Hosokawa, 2004).

With regard to the relationships between the positions of several bodies, the Doppler Effect and gravitational blueshift and redshift are different phenomena. It has also been suspected that the speed of an electromagnetic wave does not change with frequency and energy (Abdo et al., 2009).

For gravitational blueshift,

$$E = MC^2 = M((w\downarrow)^2 + (2\varphi\uparrow)) = (m\uparrow)C(w\downarrow) = (m\uparrow)C(\lambda\downarrow)f = hf \qquad (27)$$

For gravitational redshift,

$$E = MC^2 = M((w\uparrow)^2 + (2\varphi\downarrow)) = (m\downarrow)C(w\uparrow) = (m\downarrow)C(\lambda\uparrow)f = hf \qquad (28)$$

For Doppler blueshift,

$$\uparrow E = (M\uparrow)C^2 = (M\uparrow)(w^2 + 2\varphi) = (m\uparrow)Cw = (m\uparrow)C(\lambda\downarrow)(f\uparrow) = h(f\uparrow) \quad (29)$$

For Doppler redshift,

$$\downarrow E = (M\downarrow)C^2 = (M\downarrow)(w^2 + 2\varphi) = (m\downarrow)Cw = (m\downarrow)C(\lambda\uparrow)(f\downarrow) = h(f\downarrow) \quad (30)$$

In the above equations, E is the energy; M, the gravitational mass; C, the speed of light; w, the electromagnetic wave speed; 2φ, the gravitational potential; h, Plank's constant; f, the frequency; m, the inertial mass; and λ, the wavelength.

The ratio of the speed of light to the wave speed is the same as the ratios of the progress of the clock. The ratios of the progress of the clock and the wave propagation speed are constant relative to stationary observer under given conditions, and the speed of light is always the same.

C = (C/w) fλ

However, the propagation speed of the electromagnetic wave changes with the position of the observer and with the conditions.

C/w = n: index of refraction

This is similar to how the propagation speed of light waves changes in different mediums.

C/n = w = fλ

In a sense, this can be used to determine when a

Table 1. Comparison according to the theory of lens and redshift.

Theory	Gravitation Phenomenon	Weak field→ Wave length	Frequency	Velocity	Energy	→Strong field→ wl	f	v	e	→Weak field (observer) wl	f	v	e
General theory of relativity	Lens	—	—	C	—	↓	↑	C	↑	—	—	C	—
	Redshift					—	—	C	—	↑	↓	C	↓
This paper	Lens	—	—	w	—	↓	—	↓	—	—	—	w	—
	Redshift					—	—	w	—	↑	—	↑	—

gravitational field is not a perfect vacuum (free space) and contains matter. If the binding energy is generated by the perturbation of objects separated by a distance, the potential energy would be restored. The equivalence principle of the momentum of light precisely relates all energy forms in space-time.

C = d/t

In the above equation, d is the distance and t, the time. A random change in the local energy can therefore be distinguished from a cosmology-like space-time transformation.

MATTER WAVE AND UNCERTAINTY RELATION

It is not incorrect for the relation E = hf to be complete in a matter wave (Greiner, 2001b). It is incorrect to ignore both the light wave and the matter wave as a Lorentz invariance without the equivalence principle and correlation expecting relations, namely, v = fλ or C = fλ. The speed of the quantum, which changes to a different type, is expressed as given below. The energy speed of each quantum is the speed of light.

$$C^2 = v^2 + \mathrm{w}_1^2 \rightarrow C = w_2 \qquad (31)$$

In the matter wave, a wavelength (speed) change and an inertial mass change have an inverse relation in a manner similar to the gravitational red (blue) shift of the electromagnetic wave.

$$E = MC^2 = M((\Delta w)^2 + (\Delta 2\varphi)) = (\Delta m)C(\Delta w) = (\Delta m)C(\Delta \lambda)f = n\mathrm{hf} \qquad (32)$$

The reference time that we set artificially does not fluctuate. Therefore, energy and time do not show uncertainty relations. An observable result of the energy fluctuates from the space-time of the established reference. Even if the wave speed of the quantum

fluctuates under gravity uncertainly, Planck's constant remains, and is the reason called the Hertzian oscillator.

The inverse relation of a wavelength (speed) change and an inertial mass change is similar to the group speed of the mixed wave and the relations of the phase velocity, and photons, matter waves, neutrino oscillations (Ahn et al., 2006), and so on can all be explained by the same mechanism. In the energy expression of the wave speed, speed lower than that of light in the mass-energy equivalence of mass have some meaning (Table 1).

CONCLUSION

The principles of electromagnetism state are that the speed of light is constant in free (virtual) space and that it is expressed by variable wave speeds in different mediums. Through this paper, it has been shown that the gravitational field could be expressed by the variable wave speed in different mediums. If the speed of light is assumed constant in the mass-energy equivalence, the invariant mass corresponds to the speed of light in free space, and the variable mass and variable wave speed in the matter field and the inertial system should be combined. The rest mass measured on the surface of Earth, which is not a free space, is not invariant. The progress of the clock by the ratio of the speed of light and the wave speed and the equivalence of the two masses must be directly linked to electromagnetism for the original unification of mechanics in only one supposition (C/w = m/M). The argument about whether a light wave is twisted or whether space-time is warped is not important. The inertia of light in the background has priority over the theory of relativity. Therefore, it is important how energy is observed from the view of free space. The equivalence principle of light's momentum harmonizes observations from quantum theory to cosmology.

Summary

The speed and velocity, mass, momentum, energy, and

quantum resulting from the interrelation of the physical quantities based on the equivalence principle by the momentum of light are defined as follows:

1. Speed and Velocity: The propagation speed of the light waves observed on Earth's surface is (w), and the gravitational potential (2Φ) on Earth's surface is given by Earth's mass plus the gravitational potential corresponding to the speed of light (C) in free space.
($C^2 = w^2 + 2\varphi$)

In the case of an object that has a center of gravity, the light wave speed (w) is decomposed into the particle velocity (v) and the wave speed (w).
($C^2 = v^2 + w^2 + 2\varphi$)

2. Mass: The invariable gravitational mass (M) is proportional to the square of the speed of light (C^2) and is the amount of matter that corresponds to the law of universal gravitation. The variable inertial mass (m) is proportional to the light wave speed (w) and is the amount of matter that corresponds to the law of conservation of momentum and Newtonian mechanics.

3. Momentum: The momentum of light (P), momentum (MC) obtained from the invariable gravitational mass (M) and the speed of light (C), and momentum (mw) obtained from the variable inertial mass (m) and the light wave speed (w) are all equivalent (that is, $P = MC = mw$) and correspond to the momentum of the material conforming to the conservation of momentum and the laws of motion of Newtonian mechanics.

4. Energy: For a substance that obeys the law of conservation of energy, the total energy (E) is calculated using the square of the speed of light (C^2) and the invariant mass (M), whereas the particle energy is calculated using the velocity (v) of the center of gravity, light wave speed (w), gravitational potential (2φ), speed of light (C), and momentum of light (P).

$$E = PC = MC^2 = M(v^2 + w^2 + 2\varphi) = mwC$$

5. Quantum: The digital measurement of energy (E) is satisfied with the law of conservation of energy using the product of Planck's constant (h), frequency ($f = w/\lambda$), and quantum number (n).

$$E = PC = MC^2 = mwC = nhf$$

ACKNOWLEDGEMENTS

Author would like to express his gratitude to all those who advised him on Internet forums and inquiry sites. Would also like to thank Editage for providing editorial assistance.

REFERENCES

Abdo AA, Ackermann M, Ajello M, Asano K, Atwood WB, Axelsson M, Baldini L, Ballet J, Barbiellini G, Baring MG, Bastieri D, Bechtol K, Bellazzini R, Berenji B, Bhat PN, Bissaldi E, Bloom ED, Bonamente E, Bonnell J, Borgland AW (2009). A limit on the variation of the speed of light arising from quantum gravity effects. Nature. 462(7271):331-334.

Ahn MH, Aliu E, Andringa S, Aoki S, Aoyama Y, Argyriades J, Asakura K, Ashie R, Berghaus F, Berns HG, Bhang H, Blondel A, Borghi S, Bouchez J, Boyd SC, Burguet-Castell J, Casper D, Catala J, Cavata C, Cervera A, Chen SM, Cho KO, Choi JH, Dore U, Echigo S, Espinal X, Fechner M, Fernandez E, Fujii K, Fujii Y, Fukuda S, Fukuda Y, Gomez-Cadenas J, Gran R, Hara T, Hasegawa M, Hasegawa T, Hayashi K, Hayato Y, Helmer RL, Higuchi I, Hill J, Hiraide K, Hirose E, Hosaka J, Ichikawa AK, Ieiri M, Iinuma M, Ikeda A, Inagaki T, Ishida T, Ishihara K, Ishii H, Ishii T, Ishino H, Ishitsuka M, Itow Y, Iwashita T, Jang HI, Jang JS, Jeon EJ, Jeong IS, Joo KK, Jover G, Jung CK, Kajita T, Kameda J, Kaneyuki K, Kang BH, Kato I, Kato Y, Kearns E, Kerr D, Kim CO, Khabibullin M, Khotjantsev A, Kielczewska D, Kim BJ, Kim HI, Kim JH, Kim JY, Kim SB, Kitamura M, Kitching P, Kobayashi K, Kobayashi T, Kohama M, Konaka A, Koshio Y, Kropp W, Kubota J, Kudenko Yu, Kume G, Kuno Y, Kurimoto Y, Kutter T, Learned J, Likhoded S, Lim IT, Lim SH, Loverre PF, Ludovici L, Maesaka H, Mallet J, Mariani C, Martens K, Maruyama T, Matsuno S, Matveev V, Mauger C, McConnel Mahn KB, McGrew C, Mikheyev S, Minakawa M, Minamino A, Mine S, Mineev O, Mitsuda C, Mitsuka G, Miura M, Moriguchi Y, Morita T, Moriyama S, Nakadaira T, Nakahata M, Nakamura K, Nakano I, Nakata F, Nakaya T, Nakayama S, Namba T, Nambu, R, Nawang S, Nishikawa K, Nishino H, Nishiyama S, Nitta K, Noda S, Noumi H, Nova F, Novella P, Obayashi Y, Okada A, Okumura K, Okumura M, Onchi M, Ooyabu T, Oser, SM, Otaki T, Oyama Y, Pac MY, Park H, Pierre F, Rodriguez A, Saji C, Sakai A, Sakuda M, Sakurai N, Sanchez F, Sarrat A, Sasaki T, Sato H, Sato K, Scholberg K, Schroeter R, Sekiguchi M, Seo E, Sharkey E, Shima A, Shiozawa M, Shiraishi K, Sitjes G, Smy M, So H, Sobel H, Sorel M, Stone J, Sulak L, Suga Y, Suzuki A, Suzuki Y, Suzuki Y, Tada M, Takahashi T, Takasaki M, Takatsuki M, Takenaga Y, Takenaka K, Takeuchi H, Takeuchi Y, Taki K, Takubo Y, Tamura N, Tanaka H, Tanaka K, Tanaka M, Tanaka, Y, Tashiro K, Terri R, T'Jampens S, Tornero-Lopez A, Toshito T, Totsuka Y, Ueda S, Vagins M, Whitehead L, Walter CW, Wang W, Wilkes RJ, Yamada S, Yamada Y, Yamamoto S, Yamanoi Y, Yanagisawa C, Yershov N, Yokoyama H, Yokoyama, M, Yoo J, Yoshida M, Zalipska J (2006). Measurement of neutrino oscillation by the K2K experiment. Phys. Rev. D 74:072003. hep-ex/0606032.

Ashby N (2007). Relativity in the global positioning system. Living Rev. Relativity 6.

Boorse HA, Motz L (1966). The World of the Atom. Basic Books, Inc., pp. 502-512.

Greiner W (2001a). The Quantization of Physical Quantities, Quantum Mechanics: An Introduction. Springer, pp. 1-8.

Greiner W (2001b). Wave Aspects of Matter, Quantum Mechanics: An Introduction. Springer, pp. 29-66.

Hosokawa M (2004). 2 Basics of time and frequency standard 2-3 Relativisitic effects in time and frequency standards. Review of the Communications Research Laboratory 49(1/2):15-23.

Shapiro II (1964). Fourth test of general relativity. Phys. Rev. Lett. 13(26):789-791.

13

Surface waves propagation in fibre-reinforced anisotropic elastic media subjected to gravity field

A. M. Abd-Alla[1] , T. A. Nofal[1], S. M. Abo-Dahab[1,2] and A. Al-Mullise[1]

[1]Mathematics Department, Faculty of Science, Taif University, Saudi Arabia.
[2]Mathematics Department, Faculty of Science, Qena, Egypt, Qena 83523, Egypt.

The objective of this paper is to investigate the surface waves in fibre-reinforced anisotropic elastic medium subjected to gravity field. The theory of generalized surface waves has firstly developed and then it has been employed to investigate particular cases of waves, viz., Stoneley waves, Rayleigh waves and Love waves. The analytical expressions for waves velocity and attenuation coefficient are obtained in the physical domain by using the harmonic vibrations. The wave velocity equations have been obtained in different cases. The numerical results are given and presented graphically. Comparison was made with the results obtained in the presence and absence of gravity and parameters for fibre-reinforced of the material medium. The results indicate that the effect of gravity and parameters of fibre-reinforced of the material medium are very pronounced.

Key words: Fibre-reinforced media, surface waves, Stoneley waves, Rayleigh waves, Love waves, gravity.

INTRODUCTION

The dynamical problem of propagation of surface waves in a homogeneous and non-homogeneous elastic and thermoplastic media are of considerable importance in earthquake, engineering and seismology on account of the occurrence of non-homogeneities in the earth's crust, as the earth is made up of different layers. Abd-Alla et al. (2011) investigated propagation of Rayleigh waves in generalized magneto-thermoelastic orthotropic material under initial stress and gravity field. A Stoneley and Rayleigh wave in a non-homogeneous orthotropic elastic medium under the influence of gravity has been investigated by Abd-Alla and Ahmed (2003). Abd-Alla (1999) studied propagation of Rayleigh waves in an elastic half-space of orthotropic material. Abd-Alla and Ahmed (1999) investigated propagation of Love waves in a non-homogeneous orthotropic elastic layer under initial stress overlying semi-infinite medium. Rayleigh waves in

a magnetoelastic half-space of orthotropic material under the influence of initial stress and gravity field investigated by Abd-Alla et al. (2004). Elnaggar and Abd-Alla (1989) studied Rayleigh waves in magneto-thermo-microelastic half-space under initial stress. Abd-Alla and Ahmed (1996) discussed Rayleigh waves in an orthotropic thermoelastic medium under gravity field and initial stress. Propagation of Rayleigh waves in a rotating orthotropic material elastic half-space under initial stress and gravity investigated by Abd-Alla et al. (2012). Wu and Chai (1994) studied propagation of surface waves in anisotropic solids: theoretical calculation and experiment Wu and Liu (1999) investigated the measurement of anisotropic elastic constants of fiber-reinforced composite plate using ultrasonic bulk wave and laser generated Lamb wave.

The group velocity variation of Lamb wave in fiber

reinforced composite plate studied by Sang-Ho et al. (2007). Fu and Zhang (2006) investigated the continuum-mechanical modelling of kink-band formation in fibre-reinforced composites. Espinosa et al. (2000) discussed the modeling impact induced delamination of woven fiber reinforced composites with contact/cohesive laws. Wave propagation in materials reinforced with bi-directional fibers presented by Weitsman and Benveniste (1974). Weitsman (1972) introduced the wave propagation and energy scattering in materials reinforced by inextensible fibers. Dai and Wang (2006) considered the stress wave propagation in piezoelectric fiber reinforced laminated composites subjected to thermal shock. Tadashi (2000) studied the propagation of Rayleigh waves along an obliquely cut surface in a directional fiber-reinforced composite. Rogerson (1992) investigated the Penetration of impact waves in a six-ply fibre composite laminate. Weitsrian (1992) studied the reflection of harmonic waves in fiber-reinforced materials. Huang et al. (1995) investigated the effect of fibre-matrix interphase on wave propagation along, and scattering from, multilayered fibres in composites.

Transfer matrix approach. Singh and Singh (2004) investigated the reflection of plane waves at the free surface of a fibre-reinforced elastic half-space. Sengupta and Nath (2001) studied the surface waves in fibre-reinforced anisotropic elastic media. Sapan and Ranjan (2011) studied the surface wave propagation in fiber-reinforced anisotropic elastic layer between liquid saturated porous half space and uniform liquid layer. Chattopadhyay et al. (2002) investigated the reflection of quasi-P and quasi-SV waves at the free and rigid boundaries of a fibre-reinforced medium. Baljeet (2007) discussed the wave propagation in an incompressible transversely isotropic fibre-reinforced elastic media. Baljeet (2005) studied the wave propagation in thermally conducting linear fibre-reinforced composite materials. Abd-Alla et al. (2000) studied the thermal stresses in a non-homogeneous orthotropic elastic multilayered cylinder.

Recently, Abd-Alla and Abo-Dahab (2012) investigated the rotation and initial stress effects on an infinite generalized magneto-thermoelastic diffusion body with a spherical cavity. Abouelregal and Abo-Dahab (2012) discusses the dual phase lag model on magneto-thermoelasticity infinite non-homogeneous solid having a spherical cavity.

The present investigation is to study the propagation of surface waves in fibre-reinforced anisotropic elastic medium subjected to gravity field leading to particular cases such as Rayleigh waves, Love waves and Stoneley waves. The waves velocity and attenuation coefficient are obtained in the physical domain by using the harmonic vibrations. The effects of the gravity, anisotropy and parameters for fibre-reinforced of the material medium on surface waves are studied simultaneously.

FORMULATION OF THE PROBLEM

Let M_1 and M_2 be two fires-reinforced elastic anisotropic semi-infinite solid media. They are perfectly welded in contact to prevent any relative motion or sliding before and after the disturbances and that the continuity of displacement, stress etc. hold good across the common boundary surface. Further, the mechanical properties of M1 are different from those of M_2. These media extend to an infinite great distance from the origin and are separated by a plane horizontal boundary and M_2 is to be taken above M_1.

Let Oxyz be a set of orthogonal Cartesian coordinates and let O be the any point of the plane boundary and Oz points vertically downward to the medium M_1. We consider the possibility of a type of wave travelling in the direction Ox, in such a manner that the disturbance is largely confined to the neighborhood of the boundary and at any instant, all particles in any line parallel to y-axis have equal displacements. These two assumptions conclude that the wave is a surface wave and all partial derivatives with respect to y are zero.

Further let us assume that u, w are the components of displacements at any point (x,y,z) at any time t. It is also assumed that gravitational field produces a hydrostatic initial stress is produced by a slow process of creep where the shearing stresses tend to become smaller or vanish after a long period of time. The equilibrium equation of the initial stress is in the form

$$\frac{\partial \tau}{\partial x} = 0, \quad \frac{\partial \tau}{\partial z} + \rho g = 0.$$

The dynamical equations of motion for three-dimensional elastic solid medium under the influence of initial stress and gravity (Sengupta and Nath, 2001) are

$$\frac{\partial \tau_{11}}{\partial x} + \frac{\partial \tau_{12}}{\partial y} + \frac{\partial \tau_{13}}{\partial z} + \rho g \frac{\partial w}{\partial x} = \rho \frac{\partial^2 u}{\partial t^2}, \tag{1a}$$

$$\frac{\partial \tau_{21}}{\partial x} + \frac{\partial \tau_{22}}{\partial y} + \frac{\partial \tau_{23}}{\partial z} + \rho g \frac{\partial w}{\partial y} = \rho \frac{\partial^2 v}{\partial t^2}. \tag{1b}$$

$$\frac{\partial \tau_{31}}{\partial x} + \frac{\partial \tau_{32}}{\partial y} + \frac{\partial \tau_{33}}{\partial z} + \rho g \frac{\partial w}{\partial z} = \rho \frac{\partial^2 w}{\partial t^2}. \tag{1c}$$

where, $\frac{\partial w}{\partial z} = -\left(\frac{\partial u}{\partial z} + \frac{\partial v}{\partial y}\right)$

The Equations (1) becomes

$$\frac{\partial \tau_{11}}{\partial x} + \frac{\partial \tau_{13}}{\partial z} + \rho g \frac{\partial w}{\partial x} = \rho \frac{\partial^2 u}{\partial t^2}, \tag{2a}$$

$$\frac{\partial \tau_{21}}{\partial x} + \frac{\partial \tau_{23}}{\partial z} = \rho \frac{\partial^2 v}{\partial t^2}, \tag{2b}$$

$$\frac{\partial \tau_{31}}{\partial x} + \frac{\partial \tau_{33}}{\partial z} - \rho g \frac{\partial u}{\partial z} = \rho \frac{\partial^2 w}{\partial t^2} \tag{2c}$$

where, ρ be the density of the material medium, g be the acceleration due to gravity and $\tau_{ij} = \tau_{ji} \; \forall \; (i, j = 1, 2, 3)$ are the stress components.

The constitutive equations for a fibre-reinforced linearly elastic anisotropic medium with respect to a preferred direction $\vec{a}$ are Sapan and Ranjan (2011):

$$\tau_{ij} = \lambda \, e_{kk} \, \delta_{ij} + 2 \, \mu_T \, e_{ij} + \alpha \left(a_k a_m e_{km} \delta_{ij} + e_{kk} \, a_i \, a_j \right)$$
$$+ 2(\mu_L - \mu_T)\left(a_i \, a_k \, e_{kj} + a_j \, a_k \, e_{ki} \right) + \beta(a_k a_m e_{km} \, a_i \, a_j) \qquad (3)$$

where, $e_{ij} = \frac{1}{2} \left(u_{i,j} + u_{j,i} \right)$ are components of strain, $\alpha, \; \beta, (\mu_L - \mu_T)$ are reinforced anisotropic elastic parameters, $\mu_L, \; \mu_T$ are elastic parameters, $\vec{a} = (a_1, a_2, a_3)$, $a_1{}^2 + a_2{}^2 + a_3{}^2 = 1$.

If $\vec{a}$ has components that are (1, 0, 0) so that the preferred direction is the x- axis, (2) simplifies, as follows

$$\tau_{11} = (\lambda + 2\alpha + 4\mu_L - 2\mu_T + \beta) \frac{\partial u}{\partial x} + (\lambda + \alpha) \frac{\partial w}{\partial z},$$
$$\tau_{22} = (\lambda + \alpha) \frac{\partial u}{\partial x} + \lambda \frac{\partial w}{\partial z},$$
$$\tau_{33} = (\lambda + \alpha) \frac{\partial u}{\partial x} + (\lambda + 2\mu_T) \frac{\partial w}{\partial z}, \qquad (4)$$
$$\tau_{12} = \mu_L \frac{\partial v}{\partial x},$$
$$\tau_{13} = \mu_L \left(\frac{\partial w}{\partial x} + \frac{\partial u}{\partial z} \right),$$
$$\tau_{23} = \mu_T \frac{\partial v}{\partial z}.$$

By substituting Equations (4) in (2a), (2b) and (2c) it becomes

$$(\lambda + 2\alpha + 4\mu_L - 2\mu_T + \beta) \frac{\partial^2 u}{\partial x^2} + (\lambda + \alpha + \mu_L) \frac{\partial^2 w}{\partial x \partial z} + \mu_L \frac{\partial^2 u}{\partial z^2} + \rho g \frac{\partial w}{\partial x} = \rho \frac{\partial^2 u}{\partial t^2}, \qquad (5)$$

$$\mu_L \frac{\partial^2 v}{\partial x^2} + \mu_T \frac{\partial^2 v}{\partial z^2} = \rho \frac{\partial^2 v}{\partial t^2}, \qquad (6)$$

$$\mu_L \frac{\partial^2 w}{\partial x^2} + (\lambda + \alpha + \mu_L) \frac{\partial^2 u}{\partial x \partial z} + (\lambda + 2\mu_T) \frac{\partial^2 w}{\partial z^2} - \rho g \frac{\partial u}{\partial x} = \rho \frac{\partial^2 w}{\partial t^2}. \qquad (7)$$

To examine dilatation and rotational disturbances, we introduce two displacement potentials φ and ψ by the Lame's potential method

$$u = \frac{\partial \varphi}{\partial x} - \frac{\partial \psi}{\partial z}, \qquad w = \frac{\partial \varphi}{\partial z} + \frac{\partial \psi}{\partial x}. \qquad (8)$$

Now using (8) and substituting in Equations (5) and (7) we obtain the following wave equation for M_1 satisfied by φ and ψ as

$$(\lambda + 2\alpha + 4\mu_L - 2\mu_T + \beta) \frac{\partial^2 \varphi}{\partial x^2} + (\lambda + \alpha + 2\mu_L) \frac{\partial^2 \varphi}{\partial z^2} + \rho g \frac{\partial \psi}{\partial x} = \rho \frac{\partial^2 \varphi}{\partial t^2}, \qquad (9)$$

$$(\alpha + 3\mu_L - 2\mu_T + \beta) \frac{\partial^2 \psi}{\partial x^2} + \mu_L \frac{\partial^2 \psi}{\partial z^2} - \rho g \frac{\partial \varphi}{\partial x} = \rho \frac{\partial^2 \psi}{\partial t^2}, \qquad (10)$$

$$\mu_L \frac{\partial^2 v}{\partial x^2} + \mu_T \frac{\partial^2 v}{\partial z^2} = \rho \frac{\partial^2 v}{\partial t^2} \qquad (11)$$

and similar relations in M_2 with $\rho, \lambda, \alpha, \mu_L, \mu_T, \beta$ replaced by $\rho', \lambda', \alpha', \mu'_L, \mu'_T, \beta'$

SOLUTION OF THE PROBLEM

To solve the Equations (9) to (11) assume the following

$$\varphi = F(z) \, e^{i\omega(x - ct)},$$
$$\psi = G(z) \, e^{i\omega(x - ct)}, \qquad (12)$$
$$v = H(z) \, e^{i\omega(x - ct)}.$$

Using Equations (12) into (9), (10) and (11) we get a set differential equations for medium M_1 as follows

$$\frac{d^2 F}{dz^2} + h_1^2 \, F + f_1^2 \, G = 0,$$
$$\frac{d^2 G}{dz^2} + l_1^2 \, G + m_1^2 \, F = 0,$$
$$\frac{d^2 H}{dz^2} + k_1^2 H = 0 \qquad (13)$$

Where

$$h_1^2 = \frac{\omega^2 (c^2 - A_1)}{A_2}, \qquad f_1^2 = \frac{i\omega g}{A_2},$$
$$l_1^2 = \frac{\omega^2 (c^2 - A_3)}{A_4}, \qquad m_1^2 = \frac{i\omega g}{A_4},$$
$$k_1^2 = \frac{\omega^2 (c^2 - A_4)}{A_5},$$

$$A_1 = \frac{(\lambda + 2\alpha + 4\mu_L - 2\mu_T + \beta)}{\rho},$$
$$A_2 = \frac{(\lambda + \alpha + 2\mu_L)}{\rho},$$
$$A_3 = \frac{(\alpha + 3\mu_L - 2\mu_T + \beta)}{\rho},$$
$$A_4 = \frac{\mu_L}{\rho},$$
$$A_5 = \frac{\mu_T}{\rho}.$$

And the set differential equations for medium M_2 as follow

$$\frac{d^2 F}{dz^2} + h'^2_1 \, F + f'^2_1 \, G = 0,$$
$$\frac{d^2 G}{dz^2} + l'^2_1 \, G + m'^2_1 \, F = 0,$$
$$\frac{d^2 H}{dz^2} + k'^2_1 H = 0 \qquad (14)$$

where

$$h'^2_1 = \frac{\omega^2(c^2 - A'_1)}{A'_2}, \qquad f'^2_1 = \frac{i\omega g}{A'_2},$$

$$l^2_1 = \frac{\omega^2(c^2 - A'_3)}{A'_4}, \qquad m^2_1 = \frac{i\omega g}{A_4},$$

$$k^2_1 = \frac{\omega^2(c^2 - A'_4)}{A'_5},$$

$$A_1 = \frac{(\lambda l + 2\alpha l + 4\mu'_L - 2\mu'_T + \beta l)}{\rho},$$

$$A_2 = \frac{(\lambda l + \alpha l + 2\mu'_L)}{\rho},$$

$$A_3 = \frac{(\alpha l + 3\mu'_L - 2\mu'_T + \beta l)}{\rho},$$

$$A_4 = \frac{\mu'_L}{\rho},$$

$$A_5 = \frac{\mu'_T}{\rho}.$$

Equations (13) and (14) must have exponential solutions in order that F, G, H will describe surface waves, they must become vanishingly small as $z \to \infty$.

Hence for medium M_1

$$\varphi(x, z, t) = [A e^{-P_1 z} + B e^{-P_2 z}] e^{i\omega(x-ct)},$$
$$\psi(x, z, t) = [C e^{-P_1 z} + D e^{-P_2 z}] e^{i\omega(x-ct)}, \qquad (15)$$
$$v(x, z, t) = [E e^{-ik_1 z}] e^{i\omega(x-ct)}$$

where

$$P_1 = \sqrt{\frac{-a - \sqrt{a^2 - 4b}}{2}},$$

$$P_2 = \sqrt{-\frac{a}{2} + \frac{1}{2}\sqrt{a^2 - 4b}},$$

$$a = (l^2_1 + h^2_1), b = (h^2_1 l^2_1 - m^2_1 f^2_1)$$

and for medium M_2

$$\varphi(x, z, t) = [A' e^{-P'_1 z} + B' e^{-P'_2 z}] e^{i\omega(x-ct)},$$
$$\psi(x, z, t) = [C' e^{-P'_1 z} + D' e^{-P'_2 z}] e^{i\omega(x-ct)}, \qquad (16)$$
$$v(x, z, t) = [E' e^{-ik'_1 z}] e^{i\omega(x-ct)}$$

where

$$P'_1 = \sqrt{\frac{-a - \sqrt{a'^2 - 4b'}}{2}},$$

$$P'_2 = \sqrt{-\frac{a'}{2} + \frac{1}{2}\sqrt{a'^2 - 4b'}},$$

$$a' = (l''^2_1 + h'^2_1), \quad b' = (h'^2_1 l'^2_1 - m'^2_1 f'^2_1).$$

To reduce the constants in the equations (15), (16) to be 5 instead of 10 constants, we follow the following

$$C = \gamma_1 A, \qquad D = \gamma_2 B,$$
$$C' = \gamma'_1 A', \qquad D' = \gamma'_2 B'$$

where, $\gamma_j = \frac{m^2_1}{P_i{}^2 + l^2_1}$, $\gamma'_j = \frac{m'^2_1}{P'_i{}^2 + l'^2_1}$, $i, j = 1, 2$.

After solving equations (15) and (16) we are substituting the values of φ and ψ into equations (8), produced from the values of into u, w, u' and w' which are as follows

$$u = [(i\omega + \gamma_1 P_1)A e^{-P_1 z} + (i\omega + \gamma_2 P_2)B e^{-P_2 z}] e^{i\omega(x-ct)},$$
$$w = [(i\omega\gamma_1 - P_1)A e^{-P_1 z} + (i\omega\gamma_2 - P_2)B e^{-P_2 z}] e^{i\omega(x-ct)}, \quad (17)$$
$$u' = [(i\omega + \gamma'_1 P'_1)A' e^{-P'_1 z} + (i\omega + \gamma'_2 P'_2)B' e^{-P'_2 z}] e^{i\omega(x-ct)},$$

$$w' = [(i\omega\gamma'_1 - P'_1)A' e^{-P'_1 z} + (i\omega\gamma'_2 - P'_2)B' e^{-P'_2 z}] e^{i\omega(x-ct)}.$$

BOUNDARY CONDITIONS

The boundary conditions in the problem are:

(i) The displacement components at the boundary surface between the media M_1 and M_2 must be continuous at all times and positions. This means that:

$$[u, v, w] M_1 = [u, v, w] M_2 \text{ at } z = 0$$

then we obtain

$$\frac{\partial\varphi}{\partial x} - \frac{\partial\psi}{\partial z} = \frac{\partial\varphi'}{\partial x} - \frac{\partial\psi}{\partial z},$$
$$E e^{-ik_1 z} = E' e^{-ik'_1 z}, \qquad (18)$$
$$\frac{\partial\varphi}{\partial z} + \frac{\partial\psi}{\partial x} = \frac{\partial\varphi'}{\partial z} + \frac{\partial\psi}{\partial x},$$

$$(1 - i\gamma_1\beta_1)A + (1 - i\gamma_2\beta_2)B - (1 - i\gamma'_1\beta'_1)A' - (1 - i\gamma'_2\beta'_2)B' = 0,$$

$$E - E' = 0, \qquad (19)$$

$$(\gamma_1 + i\beta_1)A + (\gamma_2 + i\beta_2)B - (\gamma'_1 + i\beta'_1)A' - (\gamma'_2 + i\beta'_2)B' = 0. \qquad (20)$$

(ii) The stress components τ_{13}, τ_{23} and τ_{33} must be continuous at the boundary $z = 0$

$$[\tau_{13}, \tau_{23}, \tau_{33}] M_1 = [\tau_{13}, \tau_{23}, \tau_{33}] M_2, \text{ at } z = 0$$

$$\tau_{13} = \tau'_{13},$$

$$\mu_L\left(\frac{\partial w}{\partial x} + \frac{\partial u}{\partial z}\right) = \mu'_L\left(\frac{\partial w'}{\partial x} + \frac{\partial u'}{\partial z}\right),$$

$$\mu_L[(-\omega^2\gamma_1 - 2i\omega P_1 - \gamma_1 P_1^2)A + (-\omega^2\gamma_2 - 2i\omega P_2 - \gamma_2 P_2^2)B] \qquad (21)$$

$$+ \mu'_L[(\omega^2\gamma'_1 + 2i\omega P'_1 + \gamma'_1 P'^2_1)A' + (\omega^2\gamma'_2 + 2i\omega P'_2 + \gamma'_2 P'^2_2)B'] = 0,$$

$$\tau_{23} = \tau'_{23},$$

$$\mu_T\frac{\partial v}{\partial z} = \mu'_T\frac{\partial v'}{\partial z},$$

$$ik_1\mu_T E - ik'_1\mu'_T E' = 0, \qquad (22)$$

$$E - E' = 0,$$
$$\tau_{33} = \tau'_{33},$$

$$(\lambda + \alpha)\frac{\partial u}{\partial x} + (\lambda + 2\mu_T)\frac{\partial w}{\partial z} = (\lambda' + \alpha')\frac{\partial u'}{\partial x} + (\lambda' + 2\mu'_T)\frac{\partial w'}{\partial z},$$

$$[\lambda(\beta_1^2 - 1) + \alpha(i\gamma_1\beta_1 - 1) + 2\mu_T(\beta_1^2 - i\gamma_1\beta_1)]A$$
$$+ [\lambda(\beta_2^2 - 1) + \alpha(i\gamma_2\beta_2 - 1) + 2\mu_T(\beta_2^2 - i\gamma_2\beta_2)]B$$
$$- [\lambda'(\beta'^2_1 - 1) + \alpha'(i\gamma'_1\beta'_1 - 1) + 2\mu'_T(\beta'^2_1 - i\gamma'_1\beta'_1)]A'$$
$$- [\lambda'(\beta'^2_2 - 1) + \alpha'(i\gamma'_2\beta'_2 - 1) + 2\mu'_T(\beta'^2_2 - i\gamma'_2\beta'_2)]B' = 0 \qquad (23)$$

Where

$$\beta_j = \frac{p_i}{\omega}, \quad i = j = 1,2,$$
$$\beta'_j = \frac{p'_i}{\omega}, \quad i = j = 1,2.$$

Eliminating the constants A, B, E, A', B' and E' from Equations (18) to (23) we get

$$\text{Det}(a_{ij}) = 0, \quad i, j = 1,2,3,4,5,6 \qquad (24)$$

where

$$a_{11} = 1 - i\gamma_1\beta_1, \quad a_{12} = 1 - i\gamma_2\beta_2, \quad a_{13} = 0, \quad a_{14} = -(1 - i\gamma'_1\beta'_1),$$
$$a_{15} = -(1 - i\gamma'_2\beta'_2), a_{16} = 0,$$
$$a_{21} = \gamma_1 + i\beta_1, \quad a_{22} = \gamma_2 + i\beta_2, \quad a_{23} = 0,$$
$$a_{24} = -(\gamma'_1 + i\beta'_1), a_{25} = -(\gamma'_2 + i\beta'_2), a_{26} = 0,$$
$$a_{31} = 0, a_{32} = 0, a_{33} = 1, a_{34} = 0, a_{35} = 0, a_{36} = -1,$$
$$a_{41} = \mu_1(-\omega^2\gamma_1 - 2i\omega P_1 - \gamma_1 P_1^2),$$
$$a_{42} = \mu_1(-\omega^2\gamma_2 - 2i\omega P_2 - \gamma_2 P_2^2), a_{43} = 0,$$
$$a_{44} = \mu'_1(\omega^2\gamma'_1 + 2i\omega P'_1 + \gamma'_1 P'^2_1)$$
$$a_{45} = \mu'_1(\omega^2\gamma'_2 + 2i\omega P'_2 + \gamma'_2 P'^2_2), a_{46} = 0,$$
$$a_{51} = 0, \quad a_{52} = 0, \quad a_{53} = ik_1\mu_T, \quad a_{54} = 0, \quad a_{55} = 0,$$
$$a_{56} = -ik'_1\mu'_T,$$
$$a_{61} = \lambda(\beta_1^2 - 1) + \alpha(i\gamma_1\beta_1 - 1) + 2\mu_T(\beta_1^2 - i\gamma_1\beta_1), \qquad (25)$$
$$a_{62} = \lambda(\beta_2^2 - 1) + \alpha(i\gamma_2\beta_2 - 1) + 2\mu_T(\beta_2^2 - i\gamma_2\beta_2),$$
$$a_{63} = 0,$$
$$a_{64} = -[\lambda'(\beta'^2_1 - 1) + \alpha'(i\gamma'_1\beta'_1 - 1) + 2\mu'_T(\beta'^2_1 - i\gamma'_1\beta'_1)],$$
$$a_{65} = -[\lambda'(\beta'^2_2 - 1) + \alpha'(i\gamma'_2\beta'_2 - 1) + 2\mu'_T(\beta'^2_2 - i\gamma'_2\beta'_2)],$$
$$a_{66} = 0.$$

From Equation (24), we get the velocity of surface waves in common boundary between two fibre-reinforced elastic anisotropic semi-infinite solid media under the influence of gravity, since the wave velocity c obtained from (24) depends on the particular value of ω which indicates to the dispersion of the general wave form and on the gravity field, imposing a certain changes in the waves form.

PATICULAR CASES

Stoneley waves

It is the generalized form of Rayleigh waves in which we assume that the waves are propagated along the common boundary of two semi-infinite media M_1 and M_2. Therefore, Equation (24) determines

the wave velocity equation for Stoneley waves in anisotropic fibre-reinforced solid elastic media under the influence of gravity.

Clearly from Equation (24), it is follows that wave velocity of the Stoneley waves depends upon the parameters for fibre-reinforced of the material medium, gravity and the densities of both media. Since the wave velocity Equation (24) for Stoneley waves under the presence circumstances depends on the particular value of ω and creates a dispersion of a general wave form.

Further, Equation (24), of course, is in complete agreement with the corresponding classical result, when the effect of gravity and parameters of the fibre-reinforcement are ignored.

Rayleigh waves

To investigate the possibility of Rayleigh waves in anisotropic fibre-reinforced elastic media, we replace medium M_2 by a vacuum, in the preceeding problem. Since the boundary z = 0 is adjacent to vacuum, it is free from surface traction. So the stress boundary condition in this case may be expressed as:

$$\tau_{31} = \tau_{33} = 0 \quad at\ z = 0,$$
$$\tau_{31} = 0 \quad at\ z = 0,$$

which reduces to

$$\mu_L\left(\frac{\partial w}{\partial x} + \frac{\partial u}{\partial z}\right) = 0, \qquad (26)$$

$$(\omega^2\gamma_1 + 2i\omega P_1 + \gamma_1 P_1^2)A + (\omega^2\gamma_2 + 2i\omega P_2 + \gamma_2 P_2^2)B = 0,$$

$$\tau_{33} = 0 \quad at\ z = 0$$

which tends to

$$(\lambda + \alpha)\frac{\partial u}{\partial x} + (\lambda + 2\mu_T)\frac{\partial w}{\partial z} = 0,$$

$$[\lambda(\beta_1^2 - 2i\gamma_1\beta_1 - 1) - \alpha(i\gamma_1\beta_1 + 1) + 2\mu_T(\beta_1^2 - i\gamma_1\beta_1)]A$$
$$+ [\lambda(\beta_2^2 - 2i\gamma_2\beta_2 - 1) - \alpha(i\gamma_2\beta_2 + 1) + 2\mu_T(\beta_2^2 - i\gamma_2\beta_2)]B = 0. \qquad (27)$$

The frequency equation for Rayleigh waves in isotropic elastic medium given in the following form;

$$\text{Det}(a_{ij}) = 0, \quad i, j = 1,2 \qquad (28)$$

where

$$a_{11} = \omega^2\gamma_1 + 2i\omega P_1 + \gamma_1 P_1^2,$$
$$a_{12} = \omega^2\gamma_2 + 2i\omega P_2 + \gamma_2 P_2^2,$$
$$a_{21} = \lambda(\beta_1^2 - 2i\gamma_1\beta_1 - 1) - \alpha(i\gamma_1\beta_1 + 1) + 2\mu_T(\beta_1^2 - i\gamma_1\beta_1),$$
$$a_{22} = \lambda(\beta_2^2 - 2i\gamma_2\beta_2 - 1) - \alpha(i\gamma_2\beta_2 + 1) + 2\mu_T(\beta_2^2 - i\gamma_2\beta_2).$$

Love waves

To investigate the possibility of Love waves in a fibre-reinforced elastic solid media, we replace medium M_2 is obtained by two horizontal plane surface at a distance H-apart, while M_1 remains infinite.

For medium M_1, the displacement component v remains same as in general case given by Equation (15). For the medium M_2, we preserve the full solution, since the displacement component along y-axis that is, v no longer diminishes with increasing distance from the boundary surface of two media.

In this case the boundary conditions are

$$\tau_{13} = 0 \ , \quad \tau_{33} = 0 \ , \quad v = v' \ at \ z = 0.$$
$$\tau_{13} = \tau'_{13} \ , \quad \tau_{33} = \tau'_{33} \ , \quad v' = 0 \ at \ z = -H,$$
$$\tau_{13} = 0 \quad at \ z = 0.$$
$$\mu_L \left(\frac{\partial w}{\partial x} + \frac{\partial u}{\partial z} \right) = 0,$$

$$\left(\omega^2 \gamma_1 + 2 i \omega P_1 + \gamma_1 P_1^2 \right) A + \left(\omega^2 \gamma_2 + 2 i \omega P_2 + \gamma_2 P_2^2 \right) B = 0, \quad (29)$$

$$\tau_{33} = 0 \quad at \ z = 0,$$
$$(\lambda + \alpha) \frac{\partial u}{\partial x} + (\lambda + 2\mu_T) \frac{\partial w}{\partial z} = 0,$$
$$\left[\lambda(P_1^2 - \omega^2) + \alpha(i\,\omega\gamma_1 P_1 - \omega^2) + 2\mu_T(P_1^2 - i\,\omega\gamma_1 P_1) \right] A$$
$$+ \left[\lambda(P_2^2 - \omega^2) + \alpha(i\,\omega\gamma_2 P_2 - \omega^2) + 2\mu_T(P_2^2 - i\,\omega\gamma_2 P_2) \right] B = 0 \quad (30)$$

$$v = v' \quad at \ z = 0.$$

$$E - E' = 0, \tag{31}$$

$$\tau_{13} = \tau'_{13} \quad at \ z = -H.$$

$$\mu_L \left(\frac{\partial w}{\partial x} + \frac{\partial u}{\partial z} \right) = \mu'_L \left(\frac{\partial w'}{\partial x} + \frac{\partial u'}{\partial z} \right),$$

$$\mu'_L \left[(-\omega^2 \gamma'_1 - 2 i \omega P'_1 - \gamma'_1 P'^2_1) \right] e^{P'_1 H} A' + \mu'_L \left[(-\omega^2 \gamma'_2 - 2 i \omega P'_2 - \gamma'_2 P'^2_2) \right]$$
$$\times e^{P'_2 H} B' = 0. \tag{32}$$

$$\tau_{33} = \tau'_{33} \quad at \ z = -H,$$

$$(\lambda + \alpha) \frac{\partial u}{\partial x} + (\lambda + 2\mu_T) \frac{\partial w}{\partial z} = (\lambda' + \alpha') \frac{\partial u'}{\partial x} + (\lambda + 2\mu'_T) \frac{\partial w'}{\partial z},$$

$$\left[\lambda(P_1^2 - \omega^2) + \alpha(i\,\omega\gamma_1 P_1 - \omega^2) + 2\mu_T(P_1^2 - i\,\omega\gamma_1 P_1) \right] e^{P_1 H} A + \left[\lambda(P_2^2 - \omega^2) + \right.$$
$$\alpha(i\,\omega\gamma_2 P_2 - \omega^2) + 2\mu_T(P_2^2 - i\,\omega\gamma_2 P_2) \right] e^{P_2 H} B - \left[\lambda'(P'^2_1 - \omega^2) + \alpha'(i\,\omega\gamma'_1 P'_1 - \right.$$
$$\left. \omega^2) + 2\mu'_T(P'^2_1 - i\,\omega\gamma'_1 P'_1) \right] e^{P'_1 H} A' - \left[\lambda'(P'^2_2 - \omega^2) + \alpha'(i\,\omega\gamma'_2 P'_2 - \omega^2) + \right.$$
$$\left. 2\mu'_T(P'^2_2 - i\,\omega\gamma'_2 P'_2) \right] e^{P'_2 H} B' = 0 \ , \tag{33}$$

$$v' = 0 \quad at \ z = -H, \tag{34}$$

$$E' = 0. \tag{35}$$

In this case the velocity wave it has given from the following equation

$$\mathbf{Det}(a_{ij}) = 0, \quad i,j = 1,2,3,4,5,6 \tag{36}$$

where:

$$a_{11} = \left(\omega^2 \gamma_1 + 2 i \omega P_1 + \gamma_1 P_1^2 \right),$$
$$a_{12} = \left(\omega^2 \gamma_2 + 2 i \omega P_2 + \gamma_2 P_2^2 \right),$$
$$a_{13} = 0, \quad a_{14} = 0, \quad a_{15} = 0, \quad a_{16} = 0,$$

$$a_{21} = \left[\lambda(P_1^2 - \omega^2) + \alpha(i\,\omega\gamma_1 P_1 - \omega^2) + 2\mu_T(P_1^2 - i\,\omega\gamma_1 P_1) \right],$$
$$a_{22} = \left[\lambda(P_2^2 - \omega^2) + \alpha(i\,\omega\gamma_2 P_2 - \omega^2) + 2\mu_T(P_2^2 - i\,\omega\gamma_2 P_2) \right],$$
$$a_{23} = 0, \quad a_{24} = 0, \quad a_{25} = 0, \quad a_{26} = 0,$$
$$a_{31} = 0, \quad a_{32} = 0, \quad a_{33} = 1, \quad a_{34} = 0, \quad a_{35} = 0, \quad a_{36} = -1,$$
$$a_{41} = \mu_L \left[\left(-\omega^2 \gamma_1 - 2 i \omega P_1 - \gamma_1 P_1^2 \right) \right] e^{P_1 H},$$
$$a_{42} = \mu_L \left[\left(-\omega^2 \gamma_2 - 2 i \omega P_2 - \gamma_2 P_2^2 \right) \right] e^{P_2 H},$$
$$a_{43} = 0,$$

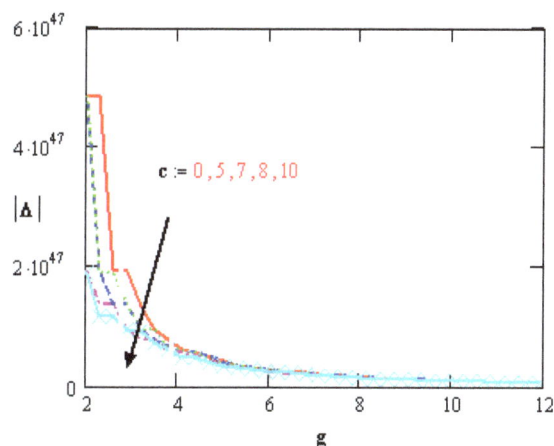

Figure 1. Variation of the secular equation with varies values of c with respect to gravity g.

$$a_{44} = \mu'_L \left[\left(-\omega^2 \gamma'_1 - 2 i \omega P'_1 - \gamma'_1 P'^2_1 \right) \right] e^{P'_1 H},$$
$$a_{45} = \mu'_L \left[\left(-\omega^2 \gamma'_2 - 2 i \omega P'_2 - \gamma'_2 P'^2_2 \right) \right] e^{P'_2 H},$$
$$a_{46} = 0,$$
$$a_{51} = \left[\lambda(P_1^2 - \omega^2) + \alpha(i\,\omega\gamma_1 P_1 - \omega^2) + 2\mu_T(P_1^2 - i\,\omega\gamma_1 P_1) \right] e^{P_1 H},$$
$$a_{52} = \left[\lambda(P_2^2 - \omega^2) + \alpha(i\,\omega\gamma_2 P_2 - \omega^2) + 2\mu_T(P_2^2 - i\,\omega\gamma_2 P_2) \right] e^{P_2 H},$$
$$a_{53} = 0,$$
$$a_{54} = - \left[\lambda'(P'^2_1 - \omega^2) + \alpha'(i\,\omega\gamma'_1 P'_1 - \omega^2) + 2\mu'_T(P'^2_1 - i\,\omega\gamma'_1 P'_1) \right] e^{P'_1 H},$$
$$a_{55} = - \left[\lambda'(P'^2_2 - \omega^2) + \alpha'(i\,\omega\gamma'_2 P'_2 - \omega^2) + 2\mu'_T(P'^2_2 - i\,\omega\gamma'_2 P'_2) \right] e^{P'_2 H},$$
$$a_{56} = 0,$$
$$a_{61} = 0, \quad a_{62} = 0, \quad a_{63} = 0, \quad a_{64} = 0, \quad a_{65} = 0, \quad a_{66} = 1.$$

NUMERICAL RESULTS AND DISCUSSION

The following values of elastic constants are considered (Chattopadhyay et al., 2002; Singh and Singh, 2004), for mediums M_1 and M_2 respectively.

$$\rho = 2660 Kg/m^3 \ , \quad \lambda = 5.65 \times 10^{10} Nm^{-2}, \quad \mu_T = 2.46 \times 10^9 Nm^{-2}, \mu_L = 5.66 \times 10^9 Nm^{-2},$$
$$\alpha = -1.28 \times 10^9 Nm^{-2} \ , \quad \beta = 220.90 \times 10^9 Nm^{-2.}$$

$$\rho = 7800 Kg/m^3 \ , \quad \lambda = 5.65 \times 10^9 Nm^{-2}, \quad \mu_T = 2.46 \times 10^{10} Nm^{-2}, \mu_L = 5.66 \times 10^{10} Nm^{-2},$$
$$\alpha = -1.28 \times 10^{10} Nm^{-2} \ , \quad \beta = 220.90 \times 10^{10} Nm^{-2.}$$

$$c_v = 0.787 \times 10^3 \ J/kg \ K, \quad K = 0.0921 \times 10^3 Jm^{-1} \deg^{-1} s^{-1} \ , T_0 = 293K, c = 1.2 \times 10^4 m^2/s^2 K.$$

The numerical technique outlined above was used to obtain surface wave velocity and with respect to wave number under the effects of gravity and thermal relaxation time parameter in two models. For the sake of brevity some computational results are being presented here. The variations are shown in Figures 1 to 13 respectively.

Figure 1 shows the variation of the secular equation of surface waves, which it decreases with increasing of gravity field until approaching to zero, as well it decreases with increasing of the value of wave speed.

Figure 2 shows the variation of Stoneley wave velocity

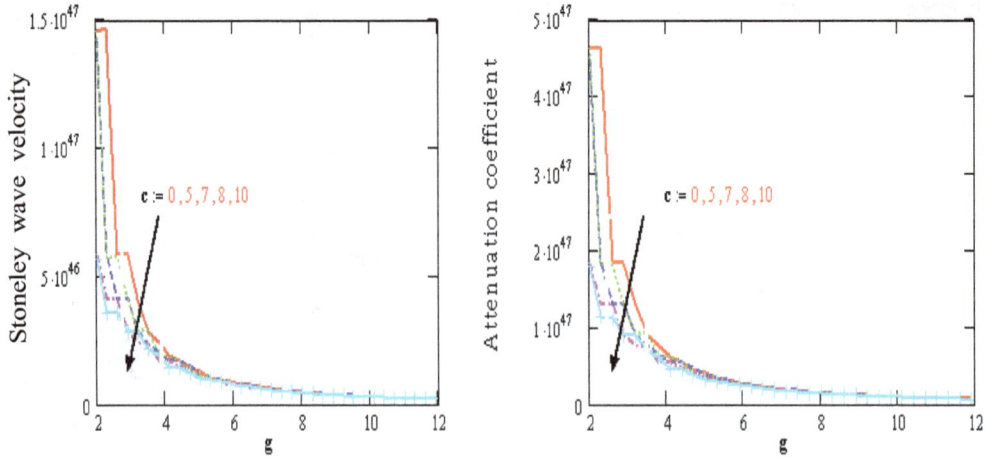

Figure 2. Variation of the Stoneley wave velocity and attenuation coefficient with varies values of c with respect to gravity g.

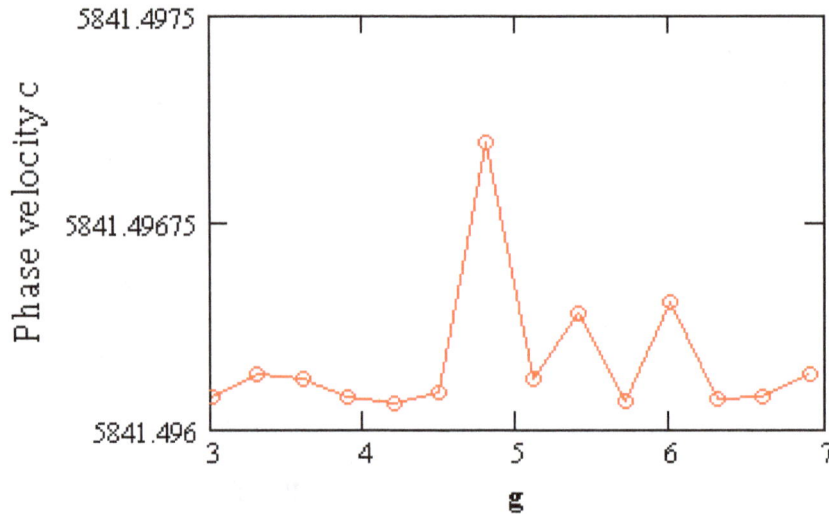

Figure 3. Phase velocity c with respect to gravity g.

and attenuation coefficient of surface waves under the effects of gravity field. The effect of gravity on Stoneley wave velocity which it decreases with increasing of gravity field, as well it decreases with increasing of the value of wave speed. Figure 3 shows the variation of phase velocity of surface waves under the effects of gravity. The value of the phase velocity has an oscillatory behavior with gravity in the whole range of the gravity field. Figure 4 show the variation of Stoneley wave velocity and attenuation coefficient of surface waves under the effect of gravity. The effect of gravity g on Stoneley wave velocity which it decreases with increasing of gravity field, as well it decreases with increasing of the value of wave speed. At a given instant, the velocity of Stoneley waves is finite, which is due to the effect

of gravity. Figure 5 show the the variation secular equation of surface waves under the effect of gravity. The effect of gravity g on secular equation which it decreases with increasing of gravity field, as well it decreases with increasing of the value of wave speed. Figures 6 and 8 show the the variation secular equation for Rayleigh wave under the effect of gravity. The effect of gravity g on secular equation which it decreases with increasing of gravity field, as well it decreases with increasing of the value of wave speed. Figure 7 shows that the variation of phase velocity of Rayleigh wave under the effects of gravity. The value of the phase velocity has an oscillatory behavior with gravity in the whole range of the gravity field. Figures 9 and 10 show the variation of Love wave velocity with respect to depth H, for different values of

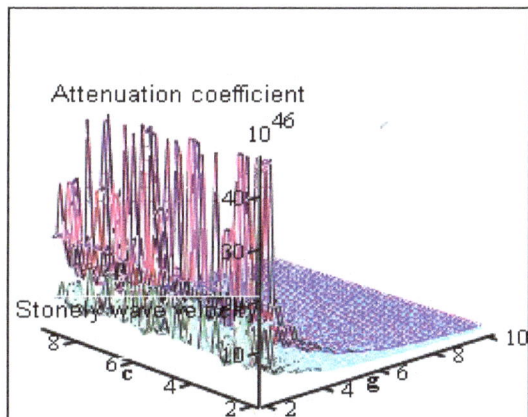

Figure 4. Variation of the Rayleigh wave velocity and attenuation coefficient with varies values of g and c.

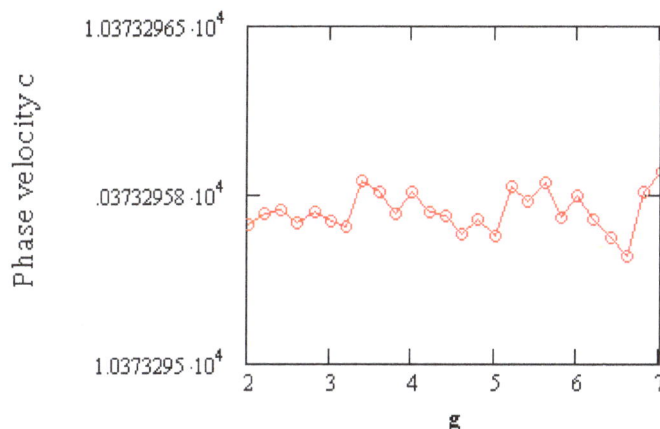

Figure 7. Phase velocity c for Rayleigh waves with respect to gravity g.

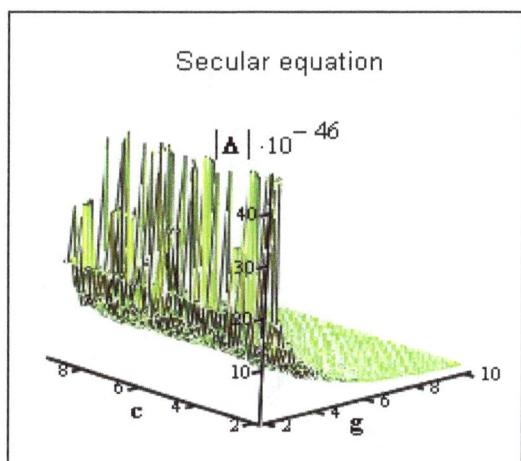

Figure 5. Variation of the secular equation with varies values of g and c.

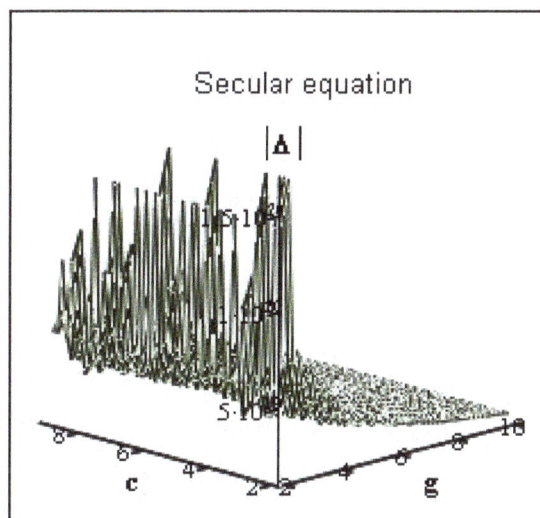

Figure 8. Variation of the secular equation for Rayleigh waves with varies values of g and c.

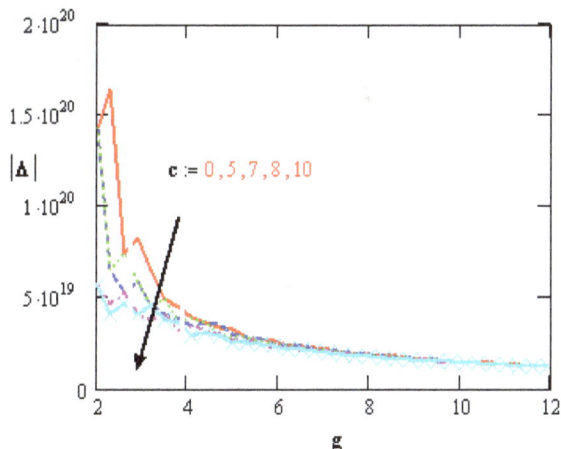

Figure 6. Variation of the secular equation for Rayleigh wave with varies values of c with respect to gravity g.

wave speed and gravity field which it increases with increasing of depth, while it decreases with increasing of wave speed and gravity field, respectively. At a given instant, the velocity of Love wave is finite, which is due to the effect of gravity. Figure 11 shows that the variation of phase velocity of Love wave under the effects of gravity. The value of the phase velocity has an oscillatory behavior with gravity in the whole range of the depth, while it decreases with increasing of gravity field as well it increases with increasing of the value of depth.

Finally, Figures 12 and 13 show the variation of Love wave velocity, attenuation coefficient and secular equation with respect to depth H and phase velocity C which they increase with the increasing of wave speed and depth.

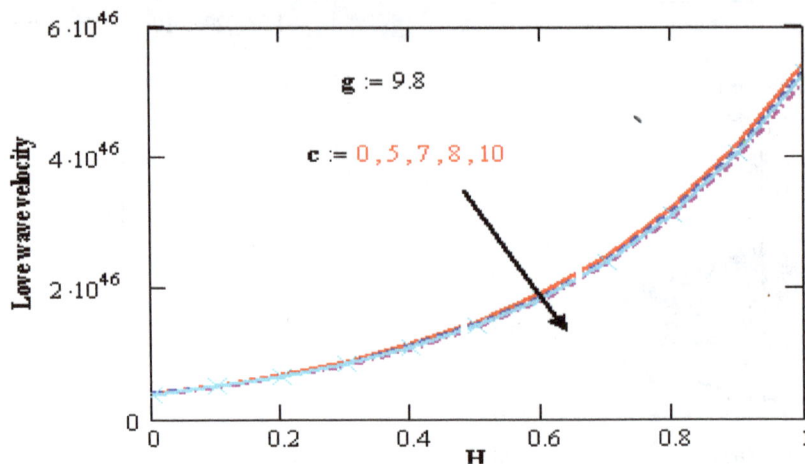

Figure 9. Variation of Love wave with varies values of c with respect to the depth H.

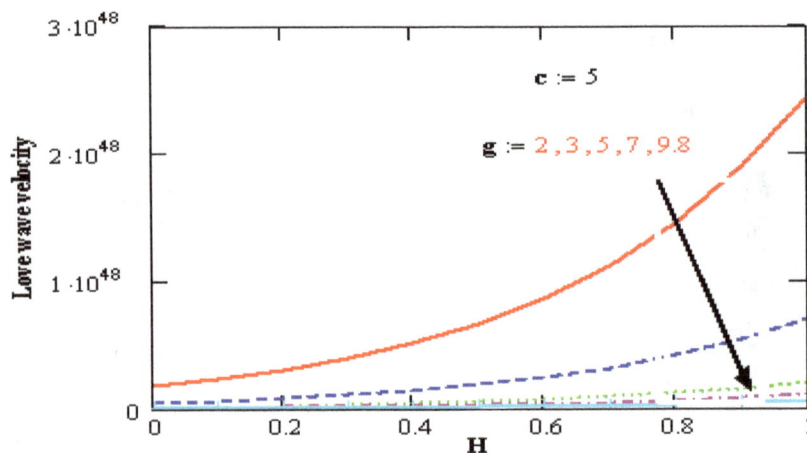

Figure 10. Variation of Love wave with varies values of g with respect to the depth H.

Conclusion

Due to the complicated nature of the governing equations of the elasticity fiber-reinforced theory, the work done in this field is unfortunately limited in number. The method used in this study provides a quite successful in dealing with such problems. This method gives exact solutions in the elastic medium without any assumed restrictions on the actual physical quantities that appear in the governing equations of the problem considered. Important phenomena are observed in all these computations:

1. It was found that for large values of time they give close results. The solutions obtained in the context of elasticity theory, however, exhibit the behavior of speeds of wave propagation.
2. By comparing Figures 1 to 12, it was found that the wave velocity has the same behavior in both media. But with the passage of time and gravity, numerical values of wave velocity in the elastic medium are large in comparison due to the influences of gravity.
3. Special cases are considered as Rayleigh waves, Love wave and surface waves in anisotropic elastic medium, as well in the isotropic case.
4. The results presented in this paper should prove useful for researchers in material science, designers of new materials.
5. Study of the phenomenon of relaxation time and gravity is also used to improve the conditions of oil extractions.

Finally, if the gravity field is neglected, the relevant results obtained are deduced to the results obtained by Sengupta and Nath (2001).

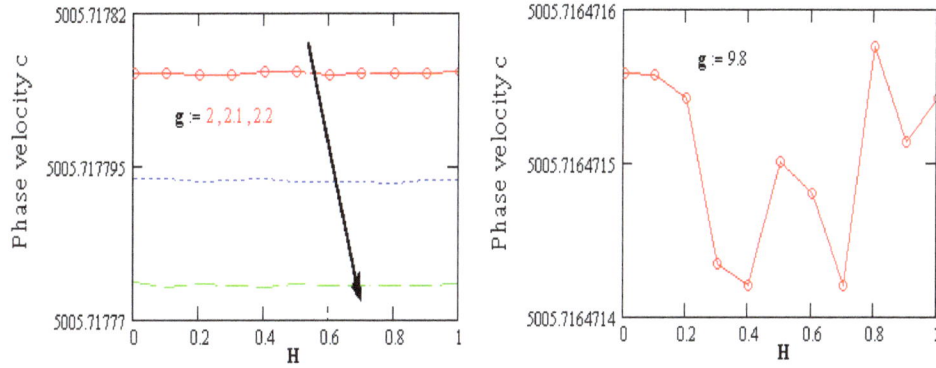

Figure 11. Variation of phase velocity with varies values of g with respect to the depth H.

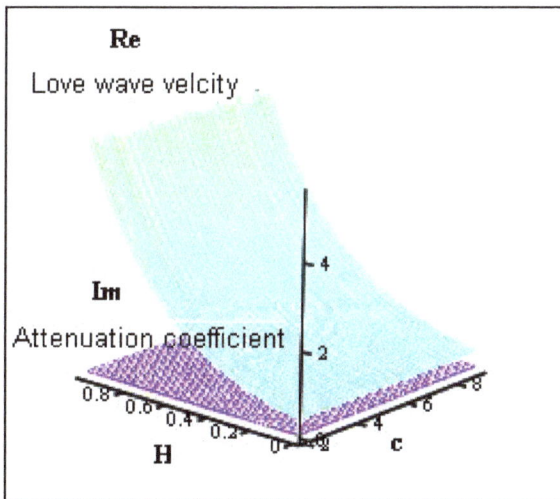

Figure 12. Variation of Love wave velocity and attenuation coefficient with varies values of H and c.

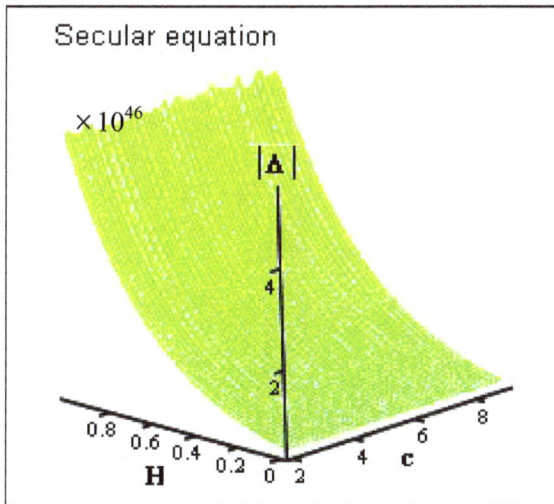

Figure 13. Variation of the secular equation for Love wave with varies values of H and c.

REFERENCES

Abd-Alla AM, Ahmed SM (1999). Propagation of Love waves in a non-homogeneous orthotropic elastic layer under initial stress overlying semi-infinite medium. Appl. Math. Comput. 106:265-275.

Abd-Alla AM (1999). Propagation of Rayleigh waves in an elastic half-space of orthotropic material, Appl. Math. Comput. 99:61-69.

Abd-Alla AM, Abo-Dahab SM, Al-Thamali TA (2012). Propagation of Rayleigh waves in a rotating orthotropic material elastic half-space under initial stress and gravity. J. Mech. Sci. Technol. 26:2815-2823.

Abd-Alla AM, Abd-Alla AN, Zeidan NA (2000). Thermal stresses in a non-homogeneous orthotropic elastic multilayered cylinder. J. Therm. Stresses 23:313-428.

Abd-Alla AM, Abo-Dahab SM (2012). Effect of rotation and initial stress on an infinite generalized magneto-thermoelastic diffusion body with a spherical cavity. J. Therm. Stresses 35:892-912.

Abd-Alla AM, Abo-Dahab SM, Hammad HAH (2011). Propagation of Rayleigh waves in generalized magneto-thermoelastic orthotropic material under initial stress and gravity field. Appl. Math. Model 35:2981-3000.

Abd-Alla AM, Ahmed SM (2003). Stoneley and Rayleigh waves in a non-homogeneous orthotropic elastic medium under the influence of gravity. Appl. Math. Comput. 135:187-200.

Abd-Alla AM, Hammad HAH, Abo-Dahab SM (2004). Rayleigh waves in a magnetoelastic half-space of orthotropic material under influence of initial stress and gravity field. Appl. Math. Comput. 154:583-597.

Abd-Alla AM, SM Ahmed (1996). Rayleigh waves in an orthotropic thermoelastic medium under gravity field and initial stress. Earth Moon Planets 75:185-197.

Abouelregal AE, Abo-Dahab SM (2012). Dual phase lag model on magneto-thermoelasticity infinite non-homogeneous solid having a spherical cavity. J. Therm. Stresses 35:820-841.

Baljeet S (2005). Wave propagation in thermally conducting linear fibre-reinforced composite materials. Arch. Appl. Mech. 75:513-520.

Baljeet S (2007). Wave propagation in an incompressible transversely isotropic fibre-reinforced elastic media. Arc. Appl. Mech. 77:253-258.

Chattopadhyay A, Venkateswarlu RLK, Saha S (2002). Reflection of quasi-P and quasi-SV waves at the free and rigid boundaries of a fibre-reinforced medium. Sādhanā 27:613-630.

Dai HL, Wang X (2006). nStress wave propagation in piezoelectric fiber reinforced laminated composites subjected to thermal shock. Compos. Struct. 74:51-62.

Elnaggar AM, Abd-Alla AM (1989). Rayleigh waves in magneto-thermo-microelastic half-space under initial stress. Earth Moon Planets. 45:175-185.

Espinosa HD, Dwivedi S, Lu HC (2000). Modeling impact induced delamination of woven fiber reinforced composites with contact/cohesive laws. Comput. Methods Appl. Mech. Eng. 183:259-290.

Fu YB, Zhang YT (2006). Continuum-mechanical modelling of kink-band formation in fibre-reinforced composites. Int. J. Solids

Struct. 43:3306-3323.

Huang W, Rokhlin SI, Wang YJ (1995). Effect of fibre-matrix interphase on wave propagation along, and scattering from, multilayered fibres in composites. Transfer matrix approach. Ultrasonics 33:365-375.

Rogerson GA (1992). Enetration of impact waves in a six-ply fibre composite laminate. J. Sound Vib. 158:105-120.

Sang-Ho R, Jeong-Ki L, Jung-Ju L (2007). The group velocity variation of Lamb wave in fiber-reinforced composite plate. Ultrasonics 47:55-63.

Sapan KS, Ranjan C (2011). Surface wave propagation in fiber-reinforced anisotropic elastic layer between liquid saturated porous half space and uniform liquid layer. Acta Geophys. 59:470-482.

Sengupta PR, Nath S (2001). Surface waves in fibre-reinforced anisotropic elastic media. *Sādhanā* 26:363-370.

Singh B, Singh SJ (2004). Reflection of plane waves at the free surface of a fibre-reinforced elastic half space. *Sādhanā* 29:249-257.

Tadashi O (2000). The propagation of Rayleigh waves along an obliquely cut surface in a directional fiber-reinforced composite. Compos. Sci. Technol. 60:2191-2196.

Weitsman Y (1972). On wave propagation and energy scattering in materials reinforced by inextensible fibers. Int. J. Solids Struct. 8:627-650.

Weitsman Y, Benveniste Y (1974). On wave propagation in materials reinforced with bi-directional fibers. J. Sound Vib. 34:179-198.

Weitsrian Y (1992). On the reflection of harmonic waves in fiber-reinforced materials. J. Sound Vib. 26:73-89.

Wu TT, Chai JF (1994). Propagation of surface waves in anisotropic solids, theoretical calculation and experiment. Ultrasonics 32:21-29.

Wu TT, Liu YH (1999). On the measurement of anisotropic elastic constants of fiber-reinforced composite plate using ultrasonic bulk wave and laser generated Lamb wave. Ultrasonics 37:405-412.

14

Application of vertical electrical sounding and horizontal profiling methods to decipher the existing subsurface stratification at river Segen dam site, Tigray, Northern Ethiopia

Abraham Bairu, Yirgale G/her and Gebrehiwot G/her

Tigray Water Resources Bureau, Mekelle, Tigray, Ethiopia.

The study area Segen river dam site is situated in the southeastern zone of Tigray National Regional State in between Hintalo Wajirat and Enderta Weredas. It is geographically located at 37P between 541400 to 542600 UTME Latitude and 1481600 to 1482600 UTMN Longitude about 35 km southwestern part of Mekelle, the capital of Tigray National Regional State. The study was conducted having an objective of the geophysical assessment to provide important subsurface geophysical information useful in evaluating the subsurface geological formations, geological structures, cavities and others for the dam site under investigation. Ten vertical electical sounding (VES) points along the two profile lines and a total of two horizontal profiling (EP) with a Wenner electrode array were collected. The VES results have shown that weak zones at VES 2, 3 and 4 along Profile 1 and at VES 2, 3 and 5 along Profile 2 where the depth goes not more than 20 m deep in both profiles' pseudo cross section except at VES 4 profile which extends up to 30 m deep. Similarly, the electrical resistivity profiling results also have shown that the weak zones extend not more than 20 m of depth. Hence, the result revealed that the investigation requires further core drilling investigations.

Key words: Vertical electrical sounding, electrical profiling, subsurface goephysics, Segen Dam, Tigray, Ethiopia.

INTRODUCTION

The main economic means of Tigray region, located in the northern part of the country is rain fed agriculture. The rainfall is erratic and unreliable. The topography of the area is undulating. Thus with the traditional agricultural practices, natural resources are severely degraded due to human interference as well as natural devastation; the land productivity is declining at alarming rate. As a result, because of moisture limitation and the above reasons, the region is not in a position to cover the annual food requirement of the people.

To alleviate the challenges of food insecurity in the country, promotion of irrigated agriculture was given priority in the strategy of the Nations (Mekuria, 2003). According to Abraham et al. (2005) cited in Nata and Asmelash (2007), irrigation is one of the methods used to increase food production in arid and semi-arid regions.

To avert the shortage of water and promote food security, the Ethiopian government has been involved in the construction of different surface water harvesting structures such as micro-dams, river diversion weirs and

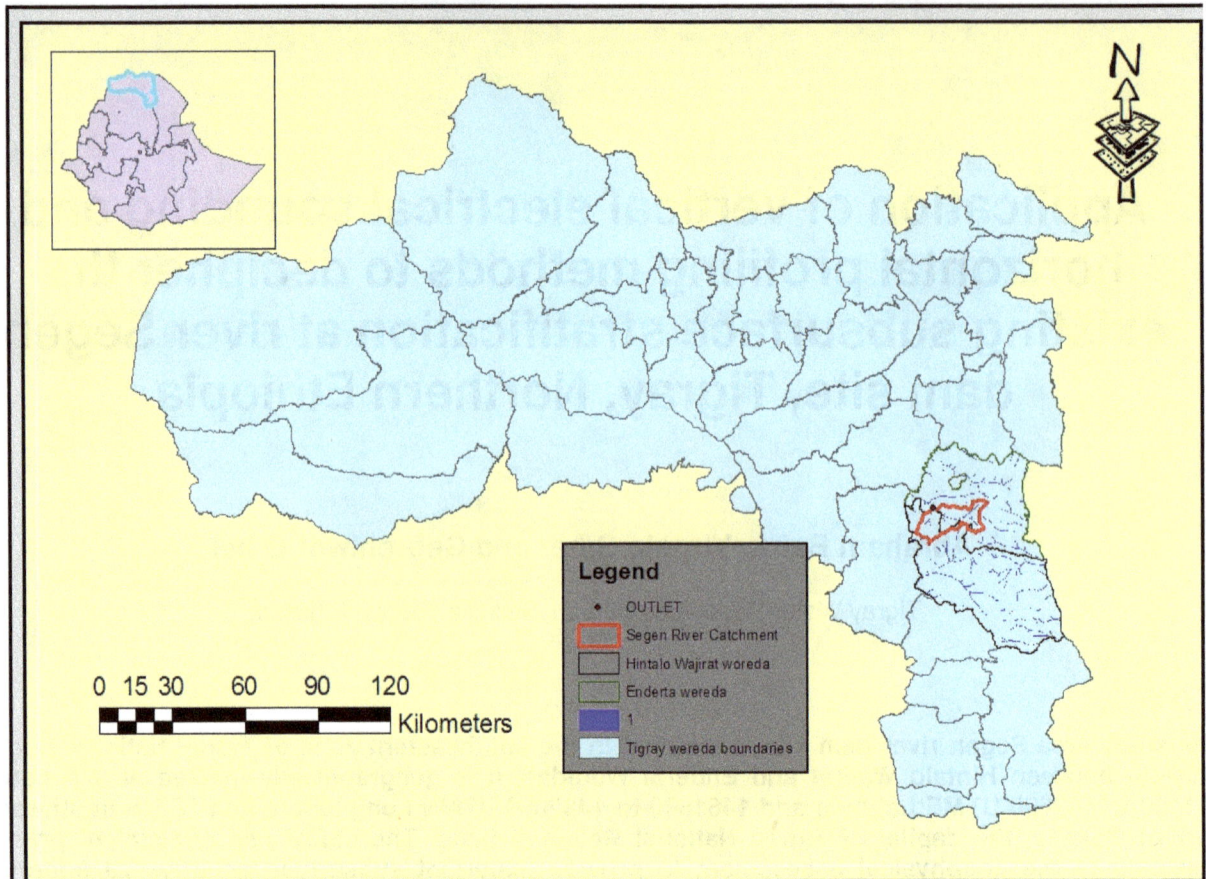

Figure 1. Location map of the Segen river dam site.

ponds in many parts of the country. In the Tigray National Regional State a number of micro-dams and several diversion weirs have been built in the last decades. Preliminary studies (Woldearegay, 2001; Mintesinot et al., 2004) indicate that the constructions of surface water harvesting schemes have economic, hydrological, and environmental benefits.

Electrical methods namely resistivity were developed in early 1900's but are more widely used since the 1970's, due primarily to the availability of computers to process and analyse the data. Electrical resistivity techniques are used extensively in the search for suitable groundwater sources, to monitor types of groundwater pollution, in engineering surveys to locate sub-surface cavities, faults and fissures, permafrost, etc, in archaeology for mapping out the area extent of remnants of buried formations of ancient buildings, amongst many other applications. It is also used extensively in down hole logging (Reynolds, 2000). The electrical resistivity technique of subsurface materials determines the composition of the overburden and depth to bedrock, and thickness of sand, gravel or metal deposits or aquifers, detect fault zones, locate steeply dipping contacts between different earth materials.

The present study intends to determine the geoelectric parameters (layer resistivity, layer thickness, transverse resistance and longitudinal conductance), delineate the weak zones, subsurface geological formations, geological structures, cavities and others.

METHODOLOGY

Location

The Segen river dam site is located about 35 Km southwestern part of Mekelle, the capital of Tigray National Regional State. The dam site is situated between two Weredas of Southeastern Zone of Tigray: the Enderta Wereda and Hintalo Wajirat. It is geographically located at 37P between 541400 to 542600 UTME Latitude and 1481600 to 1482600 UTMN Longitude. It is accessible through the Mekelle – Samre all seasons gravel pack road and seasonal rural roads (Figure 1).

Data collection and analysis

Electrical methods

Both vertical electrical sounding (VES) and electrical profiling (EP) were conducted following two profile lines where one is on the

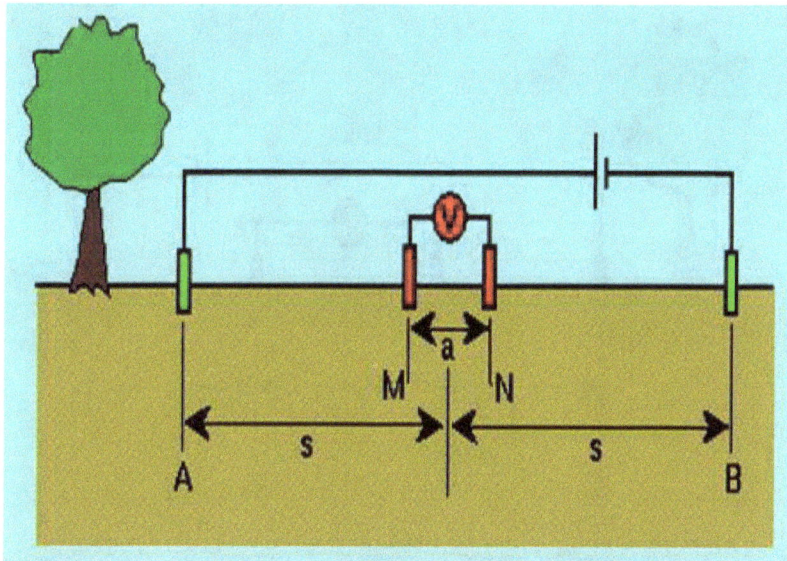

Figure 2. Electrode configurations in Schlumberger array (Mines, 2003).

proposed dam axis and the other is upper stream of the dam axis. Profile lines are aligned in the NE-SW direction. Vertical electrical sounding (VES) with the commonly used Schlumberger configuration was used for the purpose of determining the vertical variation of resistivity. The current and potential electrodes are maintained at the same relative spacing and the whole spread is progressively expanded about a fixed central point (Philip and Michael, 1984). Four electrodes are placed along a straight line on the Earth surface in the same order, A M N B, as in the Wenner array but with AB ≥ 5 MN (Figure 2). For any linear symmetric array A M N B of electrodes, the apparent resistivity (ρa) applying Schlumberger array where AM is the distance on the Earth surface between the positive current electrode A and the potential electrode M. When two current electrodes A and B are used and the potential difference (ΔV) is measured between two measuring electrodes M and N, the apparent resistivity can be written in this form:

$$\rho a = \pi\, \Delta V/I * [\ ((AB/2)2 - (MN/2)2\ )\ /\ MN\ ]\ \text{or}\ \rho a = \pi\, K\, \Delta V/I \qquad (1)$$

The value of the apparent resistivity (ρ) depends on the geometry of the electrode array used, as defined by the geometric factor (K) (Reynolds, 2000).

Electrical profiling (EP) method with Wenner array was used for determining the horizontal or lateral variation of resistivity. In the Wenner array configuration the spacing between successive electrodes remains constant and all electrodes are moved for each reading, this method can be more susceptible to near surface and lateral variations in resistivity and it is sometimes called horizontal electrical profiling (HEP). The four electrodes are collinear and the separations between adjacent electrodes are equal (a) with M,N in between A, B as shown in Figure 3 (Parasnis, 1986). The choice of electrode spacing would primarily depend on the depth of the anomalous resistivity feature (s) to be mapped (Sharma, 1986). The apparent resistivity applying Wenner array configuration can be written in the form:

$$\rho = 2\pi\, a\, \Delta V\, /\, I \qquad (2)$$

Where: ρ is the apparent resistivity, 2πa is the geometric factor (K) and a is the electrode spacing, ΔV is the potential difference and I is the electric current.

Data collection

The type of instrument used for the investigation was ABEM Terrameter SAS 4000/SAS 1000 with appropriate electrodes, cables on reels, and other accessories. Four electrode Schlumberger array was chosen in the VES survey which could provide better interpretation facility and relatively fast data acquisition mechanism. The vertical electrical resistivity sounding was carried out on the profile lines which are on and on the upper stream of the dam axis with the AB/2 and MN/2 spacing ranging from 1.5 to 220 m and 0.5 to 45 m respectively. Ten VES points were collected along the two profile lines, five VES points each. The spacing between two successive VES points VES 1, 2, 3, 4 and 5 in the Profile 1 was in the order of 175, 325, 475, 663 and 783 m, respectively from the 0 station at the Teklehaimanot church. While the spacing between two successive VES points VES 1, 2, 3, 4 and 5 in the Profile 2 was in the order of 175, 325, 475, 650 and 800 m, respectively. A total of two EP with a Wenner electrode array was conducted. It was done along the two profile lines, Profiles 1 and 2 with profile length of 960 and 1025 m respectively with a=5, 10, 20 and 30 m along the proposed dam axis. The horizontal interval or spacing between the two profiles was approximately 50 m (Tables 1 and 2).

Data processing

Based on the fundamental principles and methodologies of the geophysical survey, the collected data are interpreted quantitatively in the case of VES and qualitatively in the case of EP to determine the thickness, nature and lateral variations of the geological formations which are used to obtain a complete geological picture of the area. VES data were entered to the computer and curves were plotted using IPI2win interpretation software. The apparent resistivity and layer thicknesses were converted into useful geological meaning using knowledge of the geological history and direct geological visual observations. EP survey data are entered into the computer and processed using Microsoft Office Excel and Golden Software Surfer V. 8 to determine the lateral variations of the geologic formations.

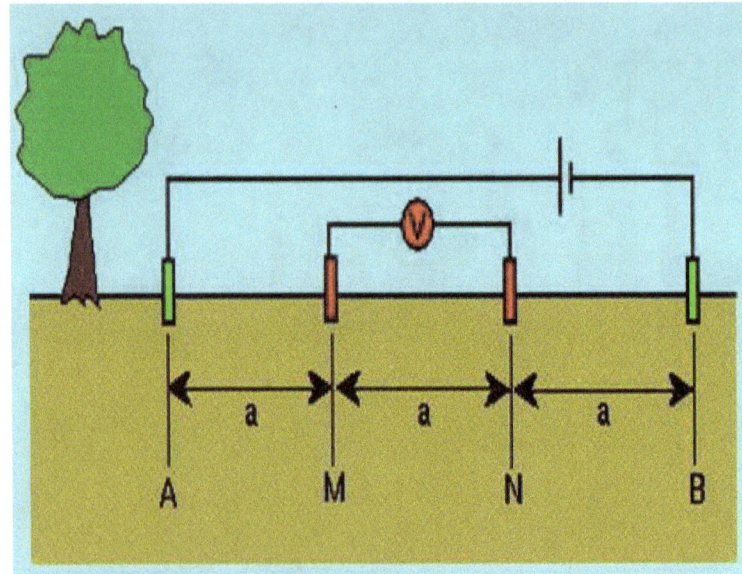

Figure 3. Electrode configurations in Wenner array (Mines, 2003).

Table 1. GPS locations of VES points at both profiles.

Designation	Along Profile 1			Along Profile 2		
	UTME	UTMN	Elevation (m)	UTME	UTMN	Elevation (m)
VES 1	542253	1482359	1852	542311	1482336	1852
VES 2	542152	1482252	1830	542215	1482238	1829
VES 3	542044	1482153	1832	542111	1482126	1829
VES 4	541900	1482038	1824	541997	1481998	1825
VES 5	541814	1481964	1839	541876	1481906	1849

Table 2. GPS location of the start and end coordinates of electrical profiling lines.

Traverse	GPS readings						Remark
	Start			End			
	UTME	UTMN	Elevation (m)	UTME	UTMN	Elevation (m)	
Profile 1	542344	1482485	1888	541657	1481829	1913	Each profile line is approximately
Profile 2	542402	1482475	1892	541720	1481780	1915	50 m away

RESULTS AND DISCUSSION

Vertical electrical sounding (VES)

The VES survey data collected from different locations along the two profile lines are interpreted and presented graphically with its possible geological meanings and the resistivity value, layer thickness and depth tabulated as follows. In terms of resistivity, Igneous rocks such as granite, diorite and gabbro have the highest resistivities while Sedimentary rocks such as shale and sandstone have a lower resistivity compared to Igneous rocks; this is due to the high fluid content in them. Metamorphic rocks on the other hand have intermediate but overlapping resistivities (Felix, 2008). The resistivity values of the 5 VES points along the Profile 1 are stated in such a way that the resistivity values in the first layer ranges from 12.94 to 106 Ohm-m and the depth ranging from 0.75 to 1.41 m. This indicates that the layer consists of clayey sand and shale formations. The second layer has an apparent resistivity values ranging from 17.6 to 214 Ohm-m and a range of depth from 2.16 to 2.91 m. This layer is

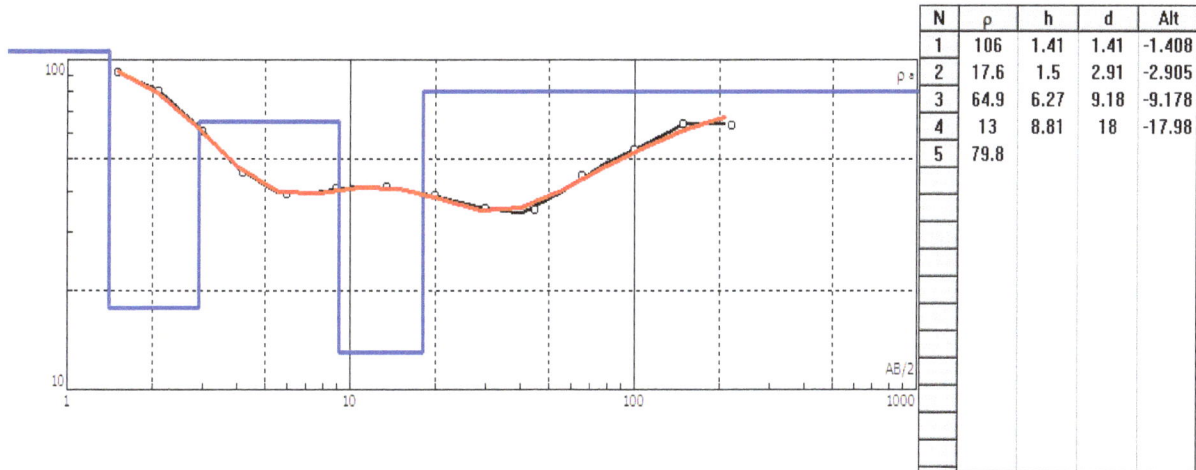

N	ρ	h	d	Alt
1	106	1.41	1.41	-1.408
2	17.6	1.5	2.91	-2.905
3	64.9	6.27	9.18	-9.178
4	13	8.81	18	-17.98
5	79.8			

Figure 4. VES 1 on the Profile 1(RMS = 2.29%).

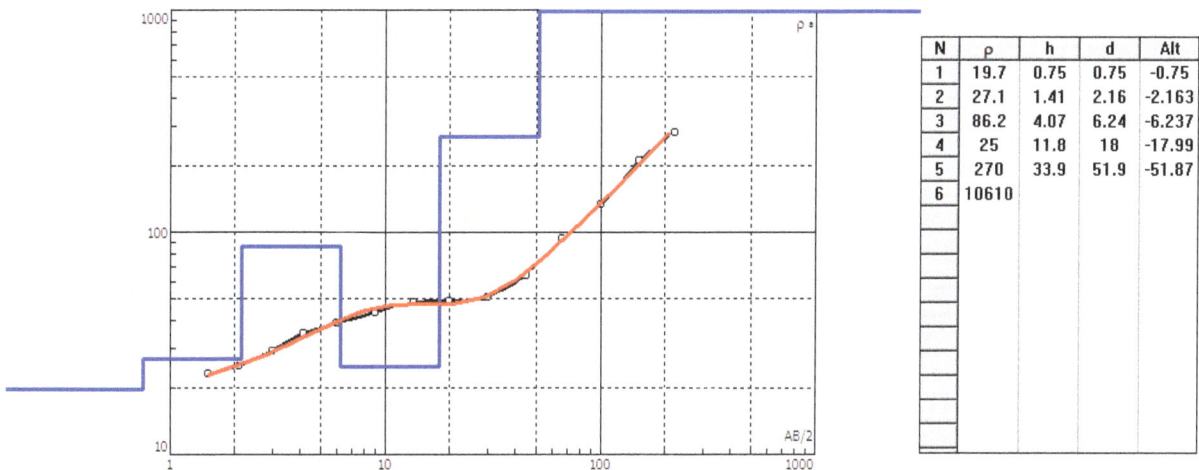

N	ρ	h	d	Alt
1	19.7	0.75	0.75	-0.75
2	27.1	1.41	2.16	-2.163
3	86.2	4.07	6.24	-6.237
4	25	11.8	18	-17.99
5	270	33.9	51.9	-51.87
6	10610			

Figure 5. VES 2 on the Profile 1(RMS = 2.42%).

weathered and fractured limestone. The third layer has an apparent resistivity between 7.191 and 86.2 Ohm-m and the range of depth is from 6.237 to 9.18 m. This layer is interpreted as water-tighted weathered and fructured limestone. The fourth layer has an apparent resistivity between 13 and 406.4 Ohm-m and the range of depth is from 17.99 to 18 m. This layer represents fresh limestone (Loke, 1999; Keller and Frischknecht, 1966) (Figures 4, 5, 6, 7 and 8).

Similarly, the resistivity values of the 5 VES points along the Profile 2 have shown that the resistivity values in the first layer ranges from 1.06 to 184 Ohm-m and the depth ranging from 0.75 to 2.884 m. This indicates that the layer consists of clay dominant sand associated with cobbles, pebbles and boulder formations. The second layer has an apparent resistivity values ranging from 15.9 to 378 Ohm-m and a range of depth from 2.16 to 6.7 m.

This layer is sand dominated clay associated with cobbles, pebbles and boulder formations. The third layer has an apparent resistivity between 16.14 and 1434 Ohm-m and the range of depth is from 6.24 to 24.35 m. This layer is interpreted as fractured limestone intercalated with shale. The fourth layer has an apparent resistivity between 11.7 and 3440 Ohm-m and the range of depth is from 18 to 94.43 m. This layer represents weathered dolerite (Loke, 1999; Keller and Frischknecht, 1966) (Figures 10, 11, 12, 13 and 14).

The pseudo cross section along the Profile 1 shows that the weak zones mainly stretches from VES 2 to 4 from the surface level to about 5 m deep at VES 2 to 3 and to about 100 m deep at VES 3 to 4 (Figure 9). The pseudo cross section along the Profile 2 has also shown that weak zones at VES 2 to 3 and 5. Vertically under VES 3, there exists a hole-like weak zone which extends

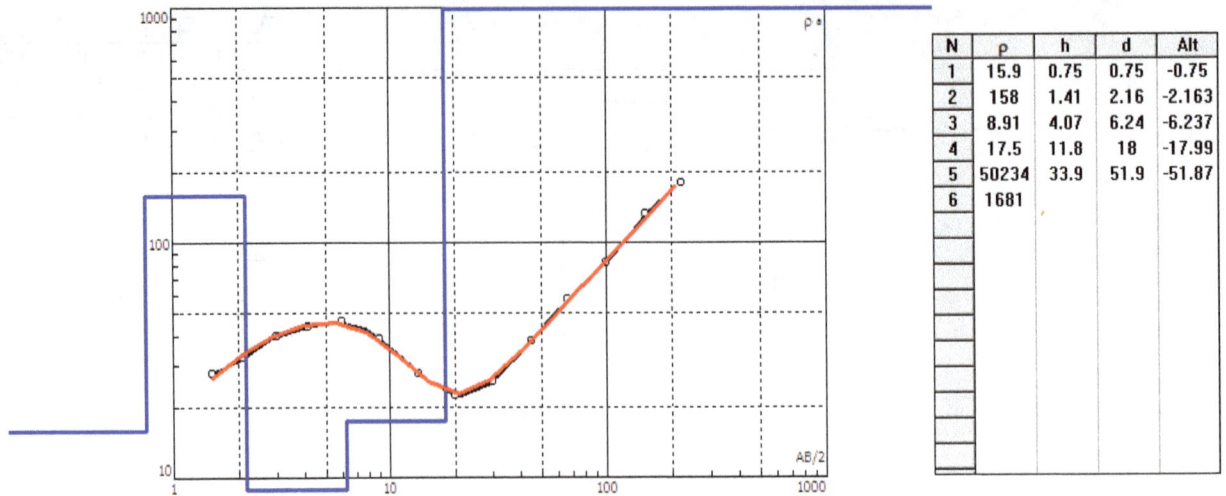

N	ρ	h	d	Alt
1	15.9	0.75	0.75	-0.75
2	158	1.41	2.16	-2.163
3	8.91	4.07	6.24	-6.237
4	17.5	11.8	18	-17.99
5	50234	33.9	51.9	-51.87
6	1681			

Figure 6. VES 3 along the Profile 1(RMS = 2.5%).

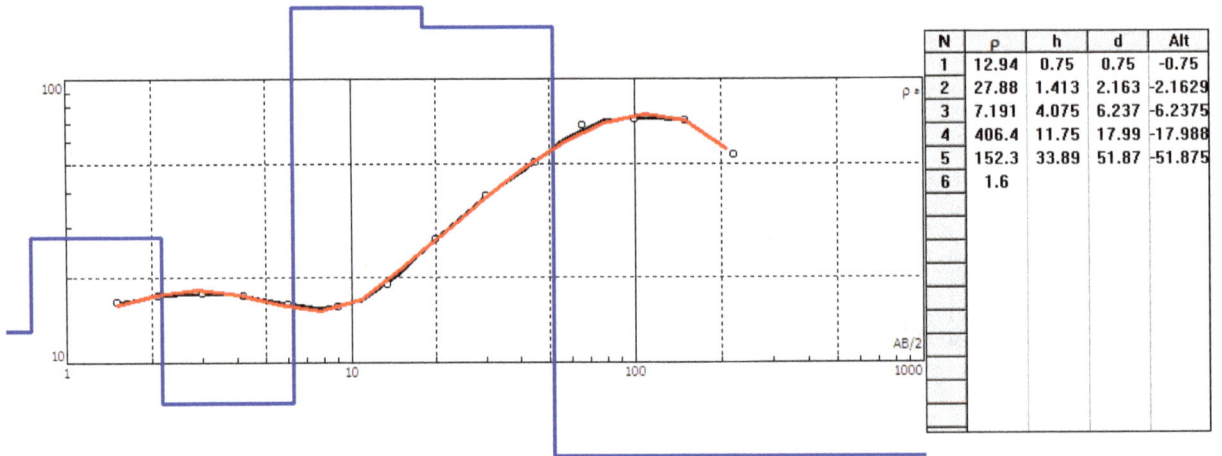

N	ρ	h	d	Alt
1	12.94	0.75	0.75	-0.75
2	27.88	1.413	2.163	-2.1629
3	7.191	4.075	6.237	-6.2375
4	406.4	11.75	17.99	-17.988
5	152.3	33.89	51.87	-51.875
6	1.6			

Figure 7. VES 4 along the Profile 1(RMS = 1.85%).

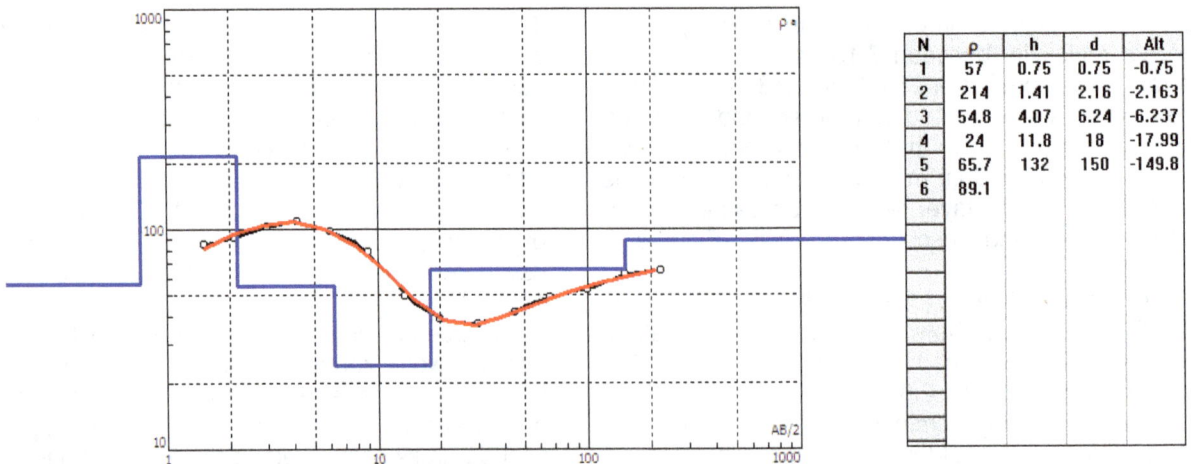

N	ρ	h	d	Alt
1	57	0.75	0.75	-0.75
2	214	1.41	2.16	-2.163
3	54.8	4.07	6.24	-6.237
4	24	11.8	18	-17.99
5	65.7	132	150	-149.8
6	89.1			

Figure 8. VES 5 along the Profile 1(RMS = 2.56%).

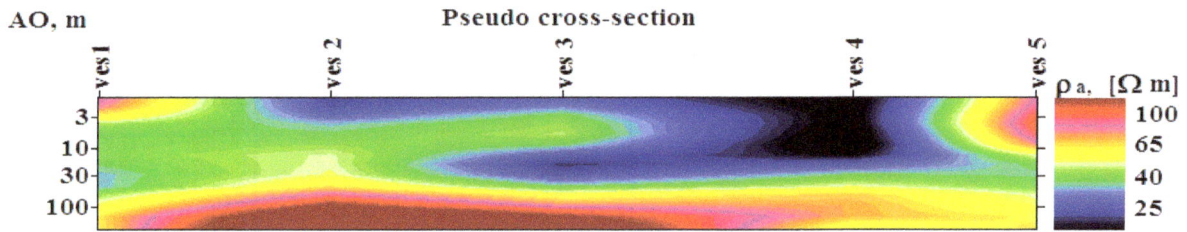

Figure 9. Pseudo cross-section along the Profile 1.

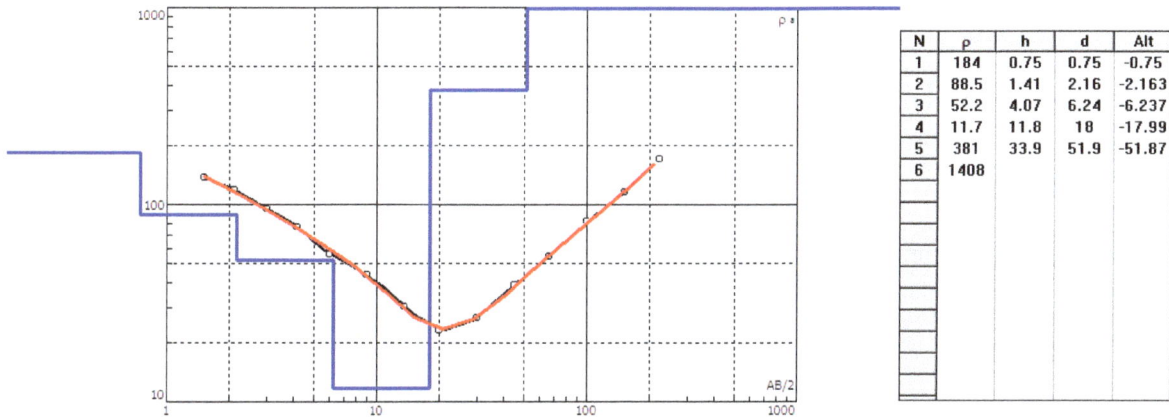

N	ρ	h	d	Alt
1	184	0.75	0.75	-0.75
2	88.5	1.41	2.16	-2.163
3	52.2	4.07	6.24	-6.237
4	11.7	11.8	18	-17.99
5	381	33.9	51.9	-51.87
6	1408			

Figure 10. VES 1 on Profile 2(RMS = 2.27%).

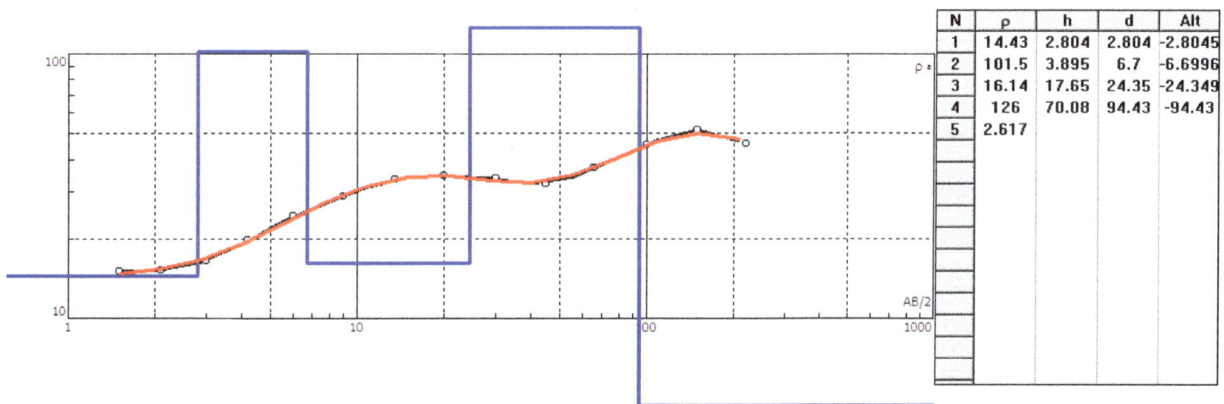

N	ρ	h	d	Alt
1	14.43	2.804	2.804	-2.8045
2	101.5	3.895	6.7	-6.6996
3	16.14	17.65	24.35	-24.349
4	126	70.08	94.43	-94.43
5	2.617			

Figure 11. VES 2 on profile 2(RMS = 1.58%).

approximately from 20 to about 100 m below surface (Figure 15).

Electrical profiling (EP)

Electrical profiling (EP) survey was conducted in the area across the Segen river with Profile 1 on the proposed dam axis with profile length of 960 m and Profile 2 on proposed upper stream of the dam aixs with profile length of 1025 m with sampling intervals of 5, 10, 20 and 30 m along NE-SW direction. The main target of the EP survey is to identify the relative position and orientation of geological structures and lithologic contacts which may have an importace for the dam construction.

Electrical profiling contour map plotted for Profile 1 at 30 m depth (AB/3 = 30 m) is characterized by heterogeneous resistivity. It shows high resistivities

N	ρ	h	d	Alt
1	10.7	0.75	0.75	-0.75
2	15.9	1.41	2.16	-2.163
3	21	4.07	6.24	-6.237
4	40.4	11.8	18	-17.99
5	7.02	33.9	51.9	-51.87
6	546			

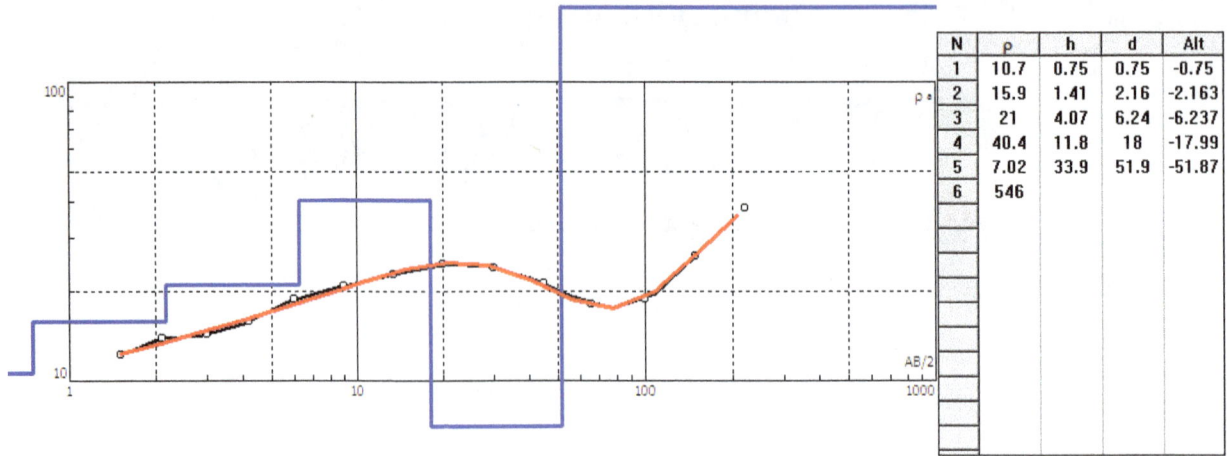

Figure 12. VES 3 on Profile 2(RMS = 2.03%).

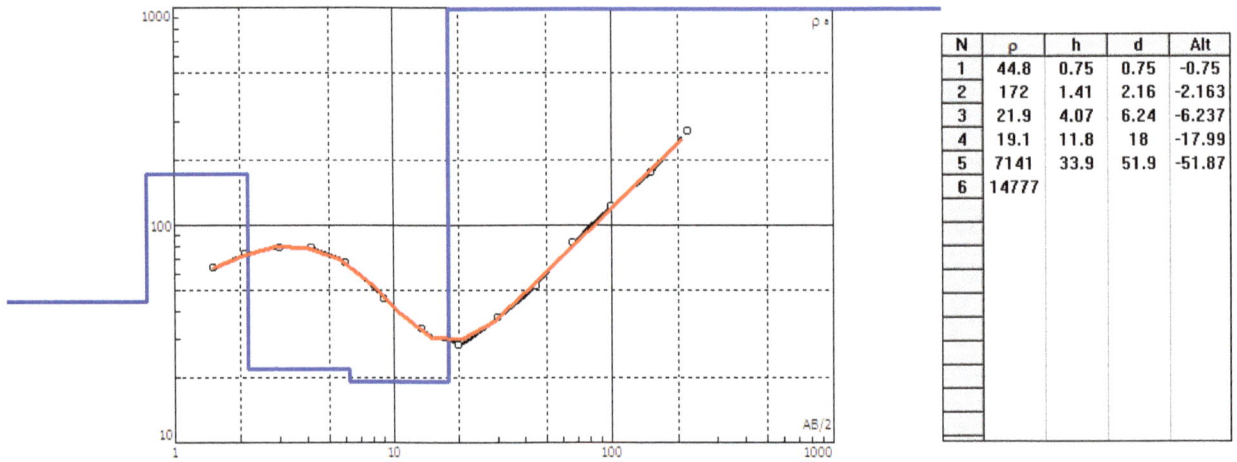

N	ρ	h	d	Alt
1	44.8	0.75	0.75	-0.75
2	172	1.41	2.16	-2.163
3	21.9	4.07	6.24	-6.237
4	19.1	11.8	18	-17.99
5	7141	33.9	51.9	-51.87
6	14777			

Figure 13. VES 4 on Profile 2(RMS =2.3%).

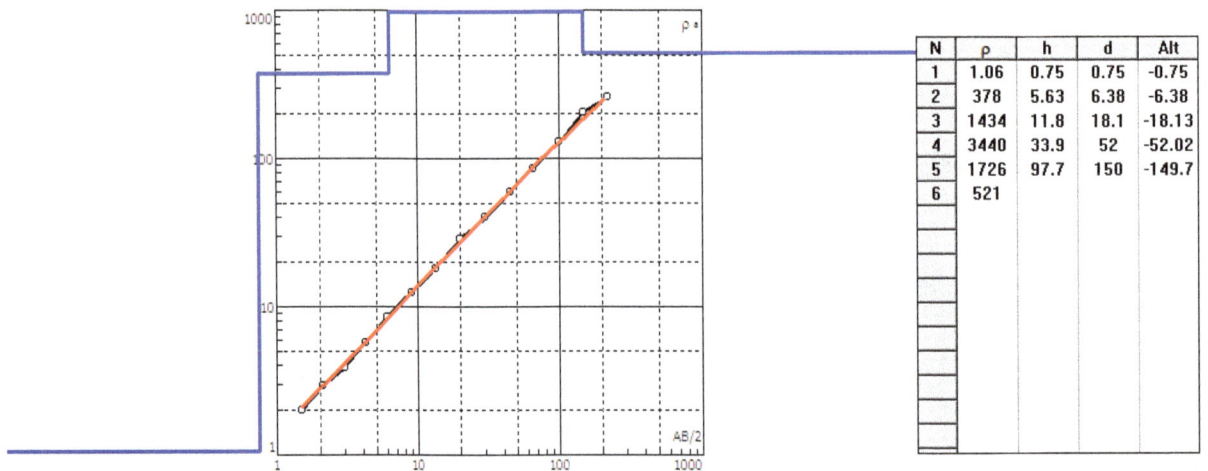

N	ρ	h	d	Alt
1	1.06	0.75	0.75	-0.75
2	378	5.63	6.38	-6.38
3	1434	11.8	18.1	-18.13
4	3440	33.9	52	-52.02
5	1726	97.7	150	-149.7
6	521			

Figure 14. VES 5 on Profile 2(3.65%).

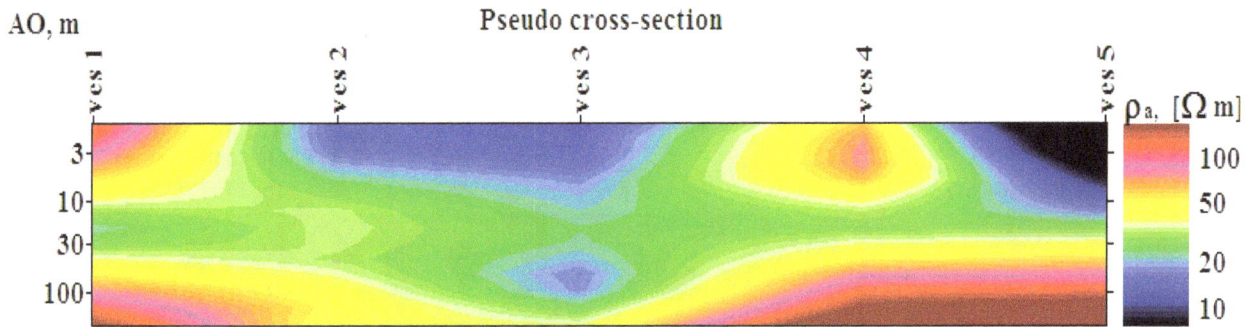

Figure 15. Pseudo cross-section along the Profile 2.

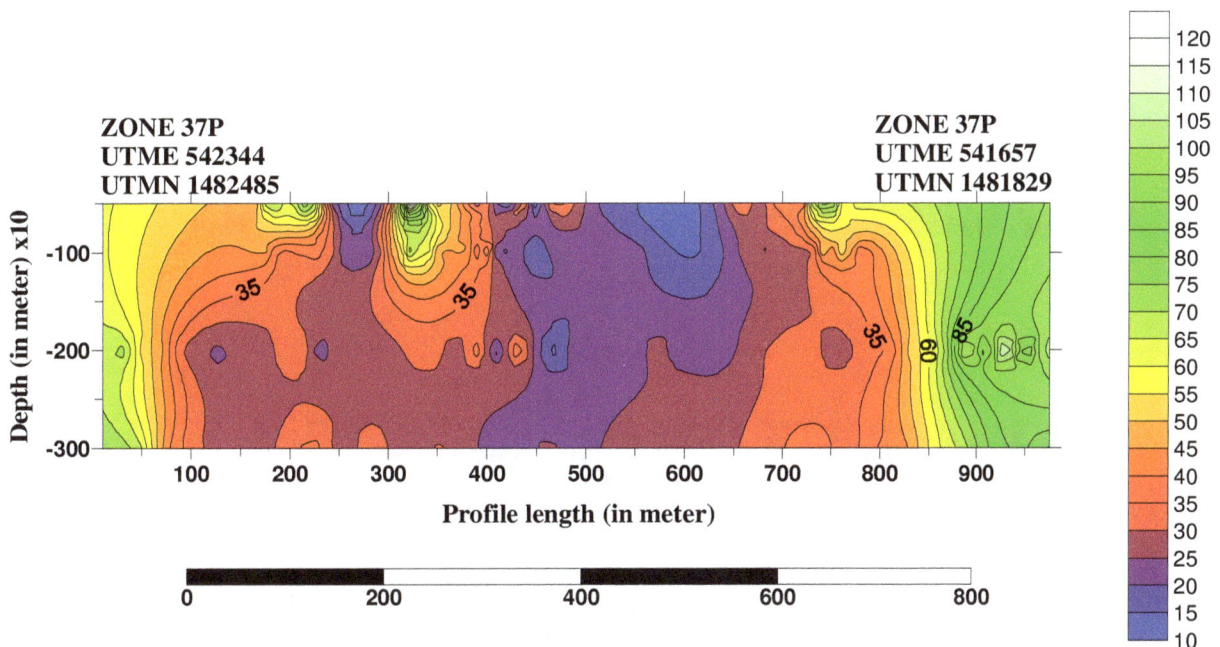

Figure 16. Apparent resistivity contour map of EP Profile 1 showing the lateral variation.

ranging from 35 to 120 Ohm-m at both corners of the profile line. While the resistivity values at the middle of the profile declines from 35 to 10 Ohm-m with the exception of the resisitivity of the profile line at 300 to 400 m from the Teklehaimanot church shows a resistivity value of greater than 35 Ohm-m. The contour map of the Profile 1 shows a "graben-like" borundary between the NE – SW direction (Figure 16).

The electrical profiling contour map plotted for Profile 2 at 30 m depth is also charaterized by heterogeneous resistivity: highest resistivity values at the corners and smallest resistivity values at the middle of the profile line (Figure 17). The horizontal or surfacial contour maps plotted between the Profiles 1 and 2 at a = 5, 10, 20 and 30 m also shows that smallest resistivity values at the middle while the highest values are obtained at the

corners of the profile lines with a "graben-like" structure at the center (Figures 18, 19, 20 and 21).

Conclusion

The vertical electrical resistivity sounding and the electrical resistivity profiling data collected and interpreted above have shown similar trends which is the resistivity values goes inclining from center of the river bed towards both abutments/flanks. The VES results have shown that weak zones at VES 2, 3 and 4 along Profile 1 and at VES 2, 3 and 5 along Profile 2 where the depth goes not more than 20 m deep in both profiles' pseudo cross section except at VES 4 profile which extends up to 30 m deep. Similarly, the electrical

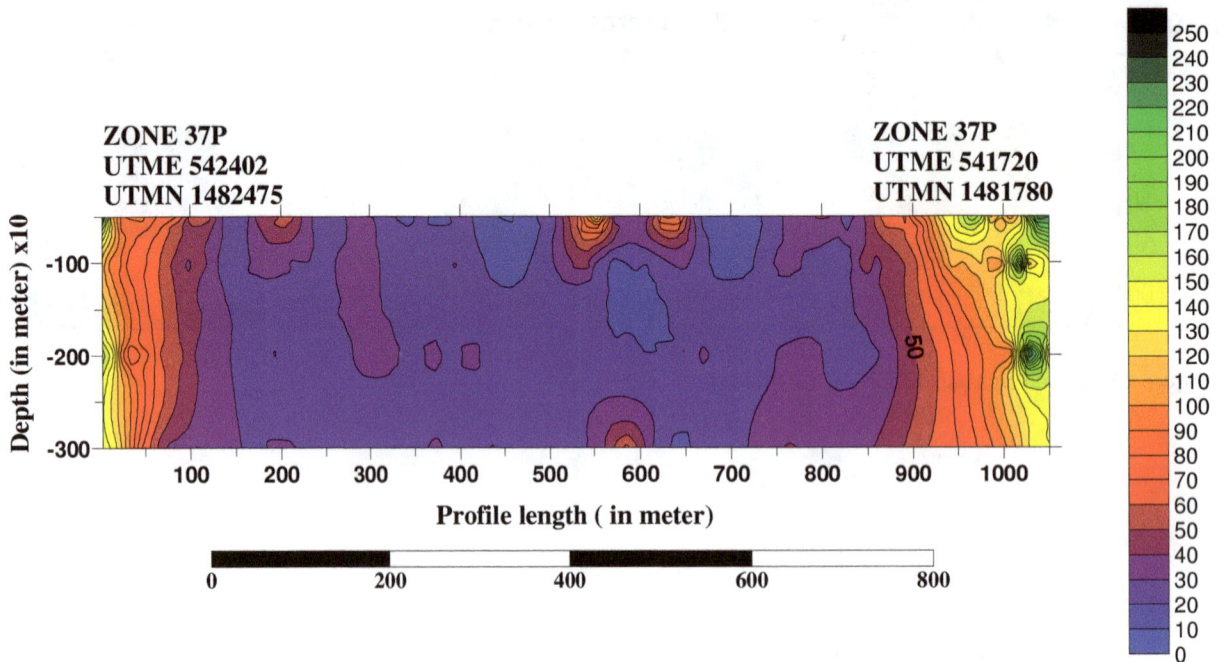

Figure 17. Apparent resistivity contour map of EP Profile 2 showing the lateral variation.

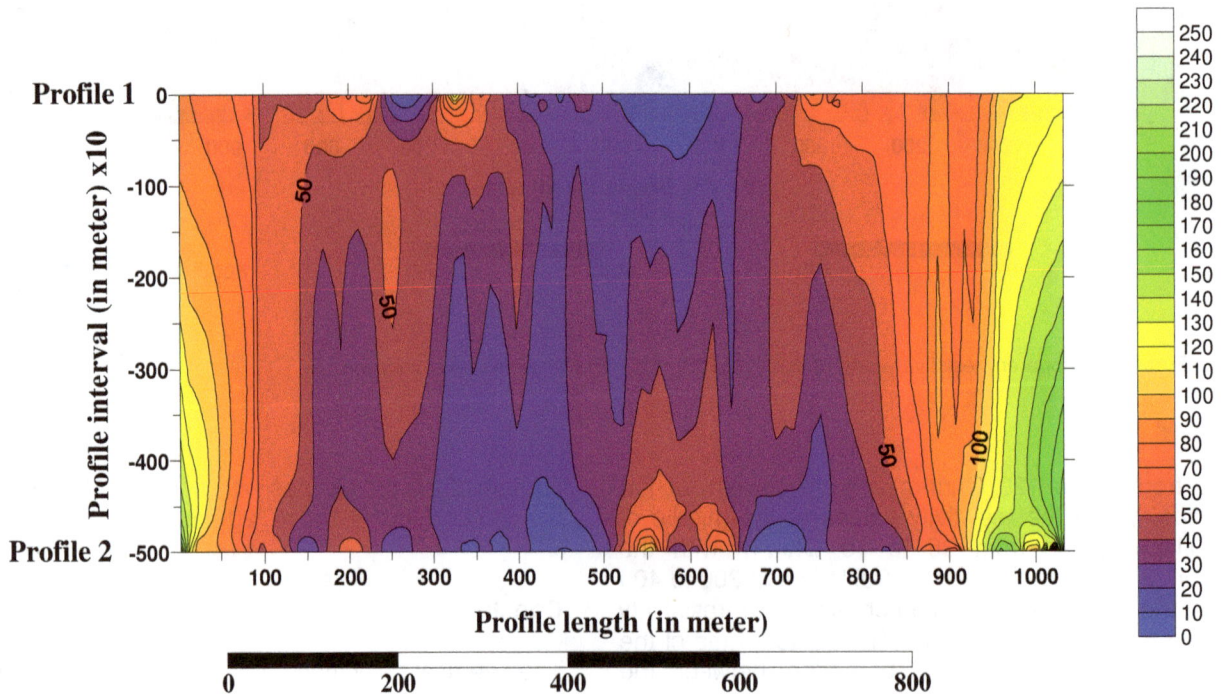

Figure 18. Apparent resistivity contour map between EP Profile 1 and 2 at 5m depth.

resistivity profiling results also have shown that the weak zones extend not more than 20 m of depth. This requires further core drilling investigation before construction is commenced.

RECOMMENDATIONS

Core drilling should be conducted in the proposed dam axis and upper stream of the dam axis where the VES

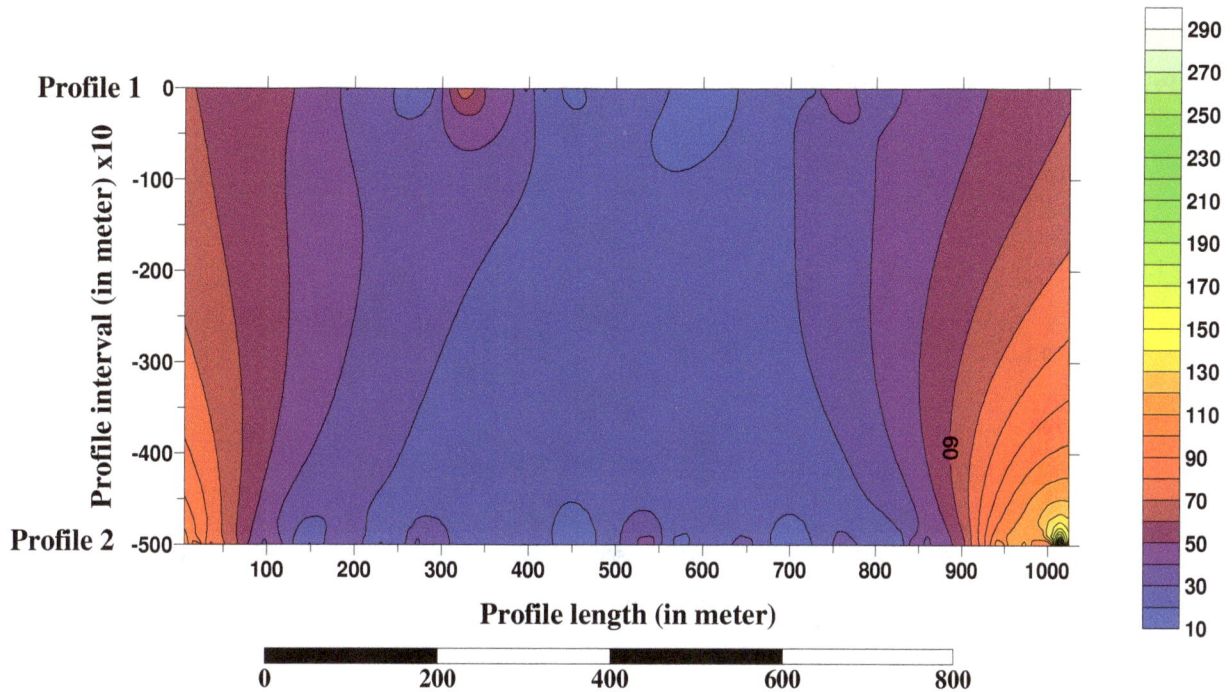

Figure 19. Apparent resistivity contour map between EP Profile 1 and 2 at 10 m depth.

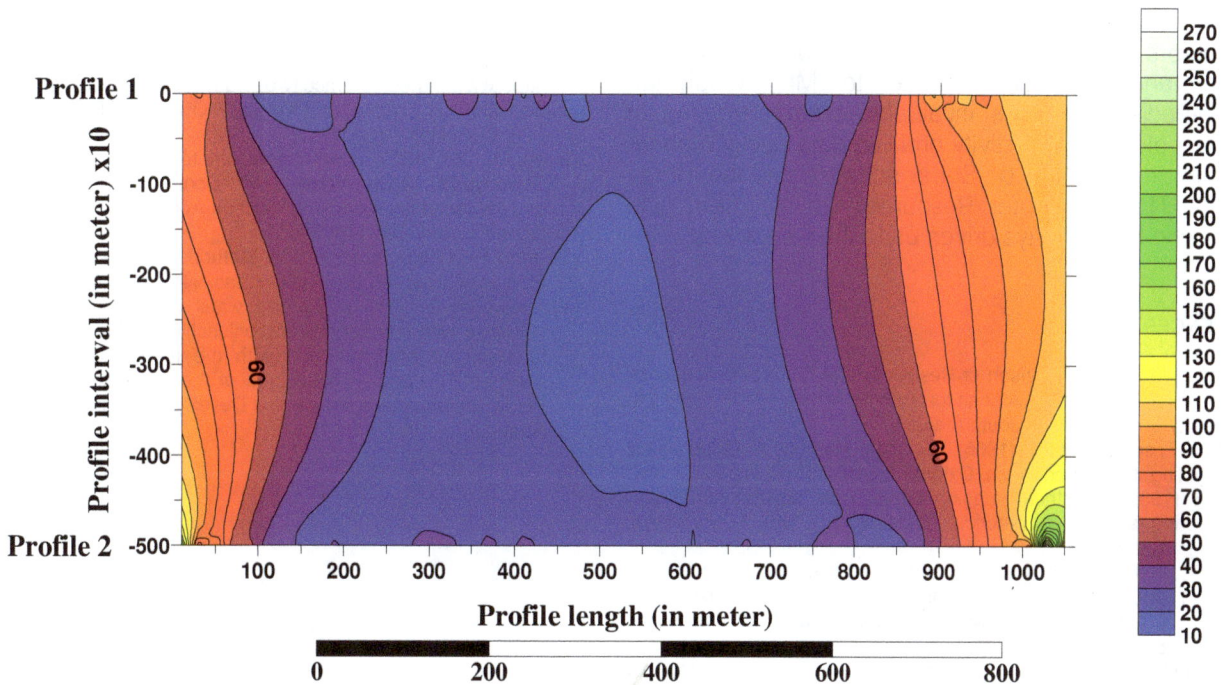

Figure 20. Apparent resistivity contour map between EP Profile 1 and 2 at 20 m depth.

and electrical resistivity profiling were conducted. Specifically, through Profile 1 at VES 2, 3 and 4 which are located at 325, 475 and 663 m and through VES 2, 3 and 5 which are also located at a distance of 325, 475 and 800 m from the starting point at the Teklehaimanot church.

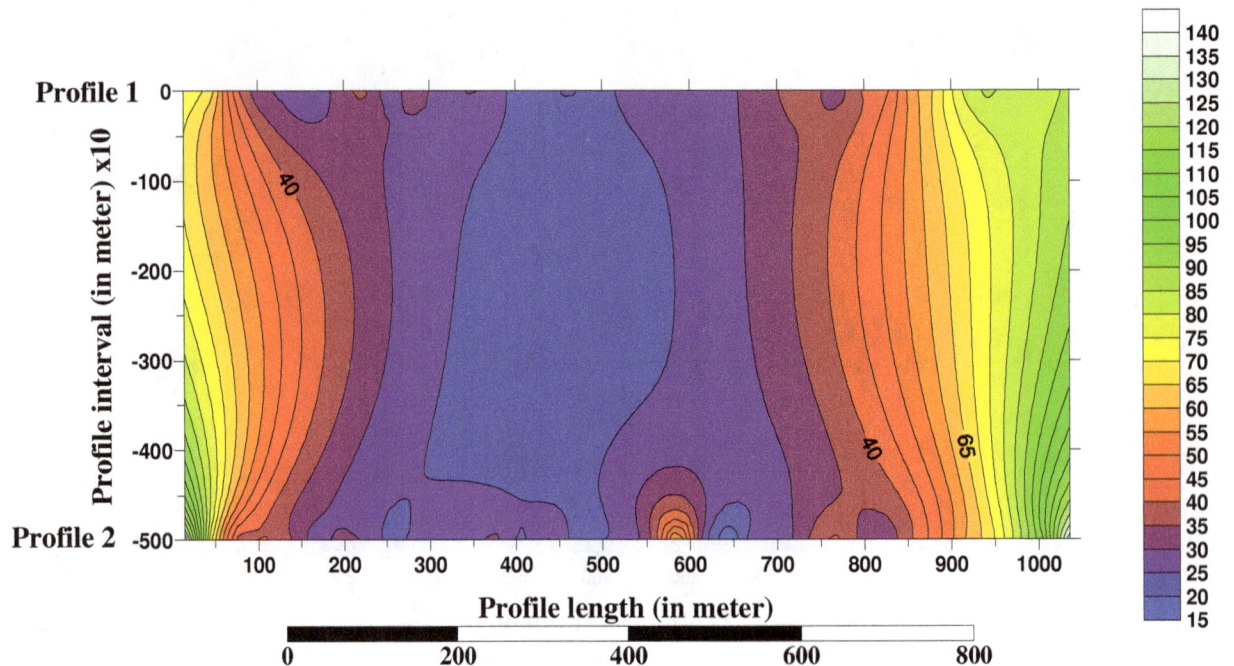

Figure 21. Apparent resistivity contour map between EP Profiles 1 and 2 at 30 m depth.

ACKNOWLEDGEMENTS

The authors of this work, would like to express their most grateful and thankful gratitude to Mr. Abdulwhab Mohammedjemal for his invaluable prompt training in collecting electrical resistivity profiling data using Wenner array and the highlight on Surfer software and the Tigray Regional National Water Resources Development field drivers who have always driven us to the field works.

REFERENCES

Felix OO (2008). A Comparison of Resistivity and Electromagnetics as Geophysical Techniques. African Institute for Mathematical Sciences, University of Cape Town, South Africa.

Keller GV, Frischknecht FC (1966). Electrical Methods in Geophysical Prospecting. Pergamon Press, Oxford. P. 519.

Loke MH (1999). Electrical imaging surveys for environmental and engineering studies. A practical guide to 2-D and 3-D surveys, Penang, Malaysia. P. 57.

Mekuria T (2003). Small Scale Irrigation for Food Security in Sub – Saharan Africa, Report and Recommendation of CTA Studies Used, Jan 20 – 29, Ethiopia, www.cta.int/pubs/wd8031.pdf. Accessed 11/09/2008.

Mines (2003). Colorado School of Mines, Page Site in the Internet (http://www.mines.edu).

Mintesinot B, Kifle W, Lulseged T (2004). Fighting famine and poverty through water harvesting in northern Ethiopia. Comprehensive Assessment Bright Spot Final Report. pp. 63-72.

Nata T, Asmelash B (2007). Recharging Practices for the enhancement of hand dug wells discharge in Debre Kidane watershed, North Ethiopia. 4th International Work Shop on Water Management and Irrigation: Focus on groundwater. Mekelle University, Mekelle, Ethiopia.

Parasnis DS (1986). Principles of Applied Geophysics. Fourth Edition, Chapman and Hall, London. P. 402.

Philip K, Michael B (1984). An Introduction to Geophysical Exploration. Blackwell, London. http://www.lnu.edu.ua/faculty/geology/phis_geo/fourman/library-Earth/AN%20INTRODUCTION%20TO%20GEOPHYSICAL%20EXPLORATION.pdf

Reynolds JM (2000). An Introduction to Applied and Environmental Geophysics. John Wiley and Sons, Inc., New York, USA. http://www.amazon.com/Introduction-Applied-Environmental-Geophysics/dp/0471485365.

Sharma PV (1986). Geophysical Methods in Geology, Second Edition, Elsevier Science Publishing Co., Inc., New York, USA. 2nd Ed.

Woldearegay K (2001). Surface water harvesting and groundwater recharge with implications to conjunctive water management in arid to semi-arid environments (with the model of the Mekelle area, northern Ethiopia). Proceedings of the International Conference on Public Management, Policy and Development, June 3-6, 2001, Addis Ababa.

15

Formation and identification of counter electrojet (CEJ)

Francisca N. Okeke[1], Esther A. Hanson[2], Eucharia C. Okoro[1], Isikwue B. C.[3] and Oby J. Ugonabo[1]

[1]Department of Physics and Astronomy, University of Nigeria, Nsukka, Enugu State, Nigeria.
[2]Centre for Basic Space Science, University of Nigeria, Nsukka, Enugu State, Nigeria.
[3]Department of Physics, University of Agriculture, Makurdi, Benue State, Nigeria.

This study investigates the possible occurrence of counter equatorial electrojet (CEJ) and a quicker method for identification of CEJ. Data from a chain of magnetic observatories of World Data Center for Geomagnetism in Tokyo, Japan, was employed. It is strikingly interesting to observe that most CEJ occurred from morning through new dusk, with almost the same pattern of dHin depression. In Ascension Island (ASC), Huancayo (HUA) and Pondicherry (PND), most ΔH were found to be less than zero, which reveals an indication of full CEJ. Partial CEJ occurrences were observed during some hours at these stations where $\Delta Hin > 0$. It is suggested that IMF turning north indicates CEJ, hence storm effects could also be attributed to CEJ existence. Some of our new findings are at variance with results of some previous workers; hence further work is suggested for further clarification. A quick method of easy identification of CEJ is suggested.

Key words: Electrojet, geomagnetic element, counter electrojet, geomagnetism, interplanetary magnetic field (IMF).

INTRODUCTION

For a long time now, there has been varying opinions about the nature and formation of counter electrojet phenomena. A lot of inconsistencies exist from the results of several workers. There is therefore need to actually attempt to specify cause(s) and formation for counter equatorial electrojet (CEJ) and throw some more light to its origin and formation.

Onwumechili (1997) modified the definition of Mayaud (1977) and defined CEJ as a westward electric current flowing on a very quiet day within a narrow band, centered on the dip equator. It could also be termed reversed equatorial electrojet (EEJ). It is important to note that westward currents and depression of H outside the EEJ zone and those within EEJ zone on magnetically disturbed periods are excluded. When the westward current flows outside the narrow band where the normal EEJ flows eastwards, then it is not definite whether it is CEJ or not. Since it has been observed that negative depressions of Sq(H) in the equatorial zone frequently come from the ring current, magnetosphere – ionosphere coupling, and polar-equatorial coupling, it becomes vital that emphasis must be laid on very quiet periods. This implies that the negative depressions of Sq(H) could be as a result of CEJ as small as -10nT or by disturbance with Ap = 5 (Onwumechili, 1997). He recommended that the study of CEJ should be limited to quiet days with Ap $\leq$ 6.

Gouin (1962) observed a conspicuous depression of the H component of geomagnetic element at local noon, at Addis Ababa. He noted that the H values were well

Table 1. Coordinates of EEJ stations.

Country	Observatory	Code	Geog Lat	Geog. Long.	Geomag. Lat
Ascension Island	Ascension	ASC	7.95S	14.38W	2.36S
Peru	Huancayo	HUA	12.04S	75.32W	1.80S
India	Pondicherry	PND	11.92N	79.92E	2.85N
India	Alibag	ABG	18.68°N	72.87°E	19.50N

below the night level on a very quiet day of January 3rd 1962 with Kp 3+ and Ap of 2. Several workers have examined the cases of various counter electrojet events at different longitudes and observed that the depression of H-field at these longitudes are changing in nature, for example, Rastogi (1974); Mayaud (1977); Marriot et al. (1979); Kane (1976) and Onwumechili (1997). Okeke and Hamano (2000) attributed the pre-noon and after noon maximum in dH to CEJ in some of the abnormal quiet days.

Alex and Mukherjee (2001), found that most frequent and simultaneous occurrence of CEJ at the equatorial stations almost correspond and are found on days of EEJ. Rastogi (1975), ascertained that the solar flare effect on the horizontal component of the geomagnetic field (H) during the period of counter equatorial electrojet current is characterized by a negative crochet (decrease of H-field) at an equatorial electrojet station and a positive crochet (increase of H field) at a low latitude station outside electrojet belt. Gurubaran (2002) suggested that a possible relationship exists between the CEJ field and the noontime D variation observed at low latitudes. Mayaud (1977), Marriot et al. (1979) and Stening (1992) concluded that CEJ are mostly observed in a few hours after dawn and a few hours before dusk and are rarely observed around local noon. In other words, CEJ is never a night phenomenon. Crochet et al. (1979) noted that a very strong day time counter electrojet was observed on Janury 1977 near the magnetic equator in Africa. Also, Rastogi (1999) observed abnormally large westward currents almost the whole of the day time hours on a series of days.

The work of Manoj et al. (2008) concluded that the penetration of electric fields into the equatorial ionosphere is not dependent on the polarity of IMF Bz. This present work examines the formation of CEJ and classifies for the first time nature of type of CEJ that exist.

METHOD OF ANALYSIS

This study employs the steps described by Onwumechili (1997), in identifying CEJ. The first stipulates that a depression of Sq(H) below its night-time level, within a very quiet period, indicates preliminary sign that the current above the observatory has reversed direction. The problem with the above is that it does not indicate which of the two current layers has reversed, since the ionospheric current above the EEJ zone flows in two layers. Hence, there is need for another condition. Again, he introduced a

perturbation ΔH_{in} at a station inside EEJ zone and another ΔH_{out} at a station in the same longitude but just outside the influence of the EEJ.

Hence, $\Delta H_{in} - \Delta H_{out} \, \Re 0$.............. (1)

If the above equation holds, then it implies existence of CEJ. Further, the perturbation by upper current layer which is associated with the worldwide part of Sq gives perturbation of ΔH_u and that of the lower current associated with EEJ gives perturbation ΔH_L. Onwumechili (1997) equally ascertained that WSq is very wide and its current density profile is almost flat, then he summarized as follows:

$\Delta H_{out} \sim \Delta H_U$ and $\Delta H_{in} = \Delta H_L + \Delta H_U$..... (2)

It is clear from Equations (1) and (2) that when $\Delta H_L \, \Re 0$, Equation (1) is satisfied; the implication is that CEJ occurs only when the lower current layer associated with EEJ has reversed westward in part or in whole, that is, $\Delta H_L \, \Re 0$.

In summary, if $\Delta H_{in} \, \Re 0$, it means full CEJ. On the other hand, if $\Delta H_{in} \, \Re 0$, it is partial CEJ. In line with the method described above, the perturbations ΔH_{in}, ΔH_{out}, ΔH_U, and ΔH_L were calculated, and the steps required were taken for the analyses.

Source of data

The data used in the study was obtained from World Data Center for Geomagnetism in Kyoto, Japan. The data consists of hourly values of both H and Z components of geomagnetic intensities recorded for three internationally most quiet days as shown in Table 2 of the year 2000. The work focuses on three equatorial electrojet stations namely Ascension, Huancayo and Pondicherry, and one station located considerably outside the region of the EEJ namely Alibag (Table 1).

RESULTS

ANALYSES OF DATA

Figures 1(a-d) depict diurnal variation of H and Z geomagnetic components at one of the stations, Alibag (ABG) under study, for the quiet days of the year 2000. H-field and Z-field variations in ABG on the quiet days indicated follow the normal enhancement of the H around noon and the Z depression, which is the expected trend.

In Ascension Island (ASC) (Figure 2a-b) a very thin band of current flows between 0:00 h and 5:00 h throughout the month of the year except for July. Subsequently, there is enhancement of ΔH with a peak at about 15:00 h, maximizing at approximately 50nT in all the months of the year except in September, where there

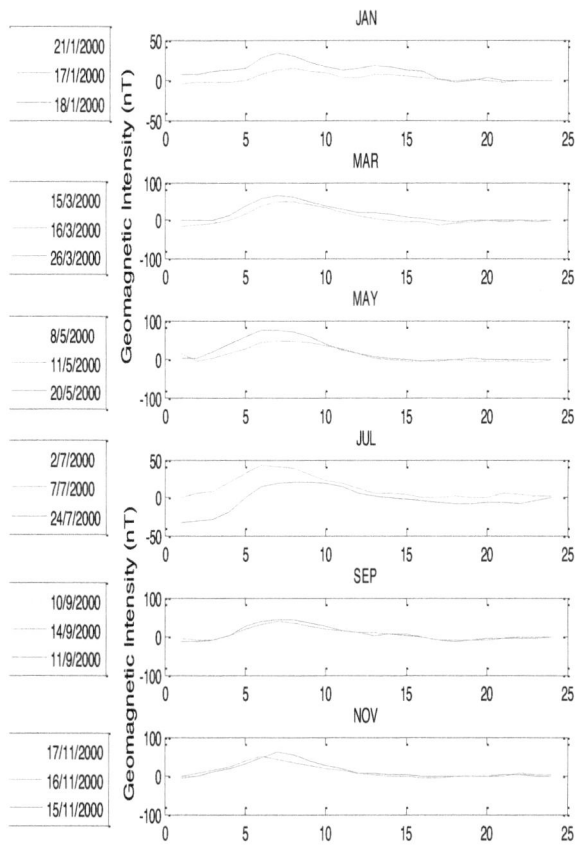

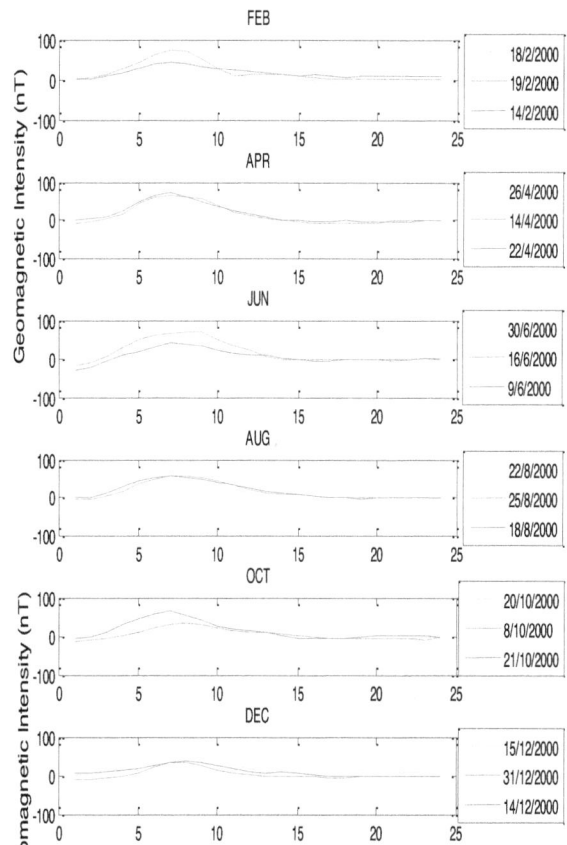

(a)   (b)

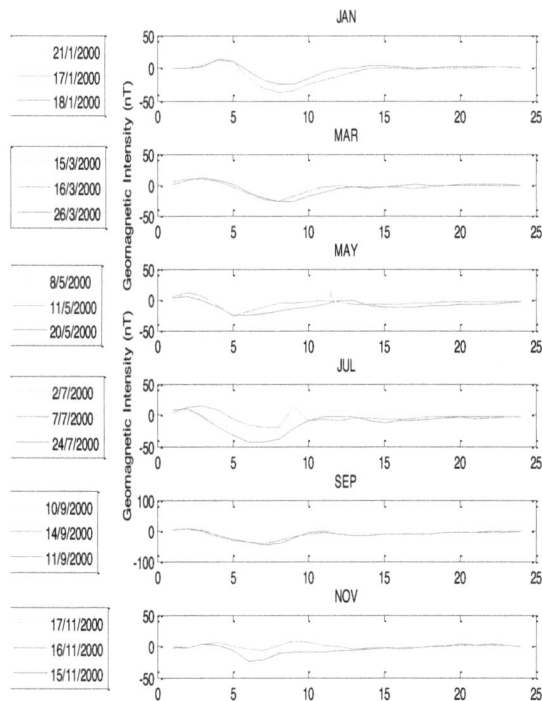

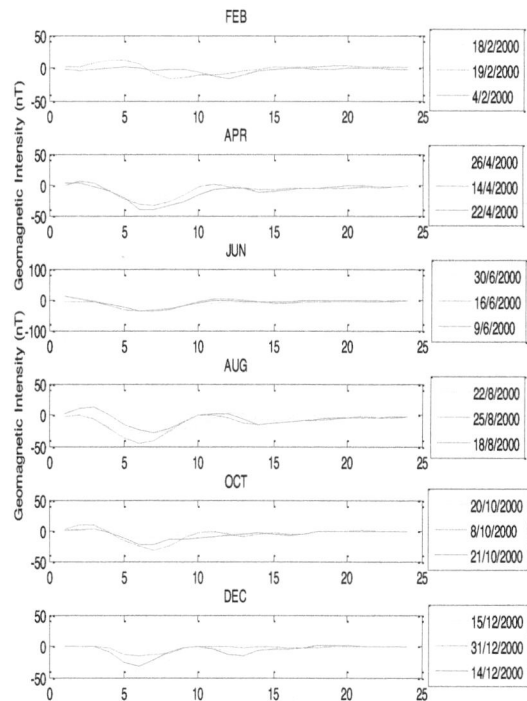

(c)   (d)

Figure 1. (a, b) Diurnal variation of delta ΔH component at ABG in year 2000, (c, d) Diurnal variation of delta Z-component at ABG in year 2000.

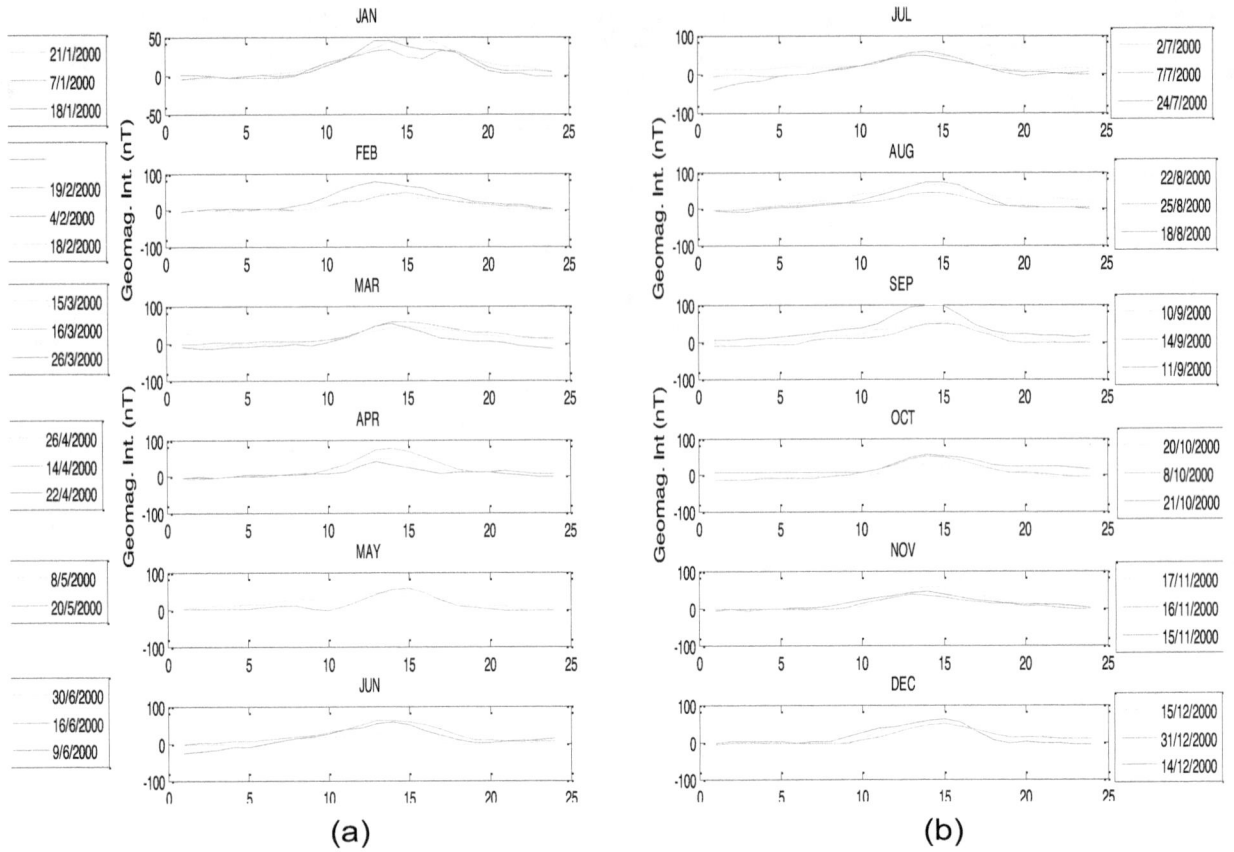

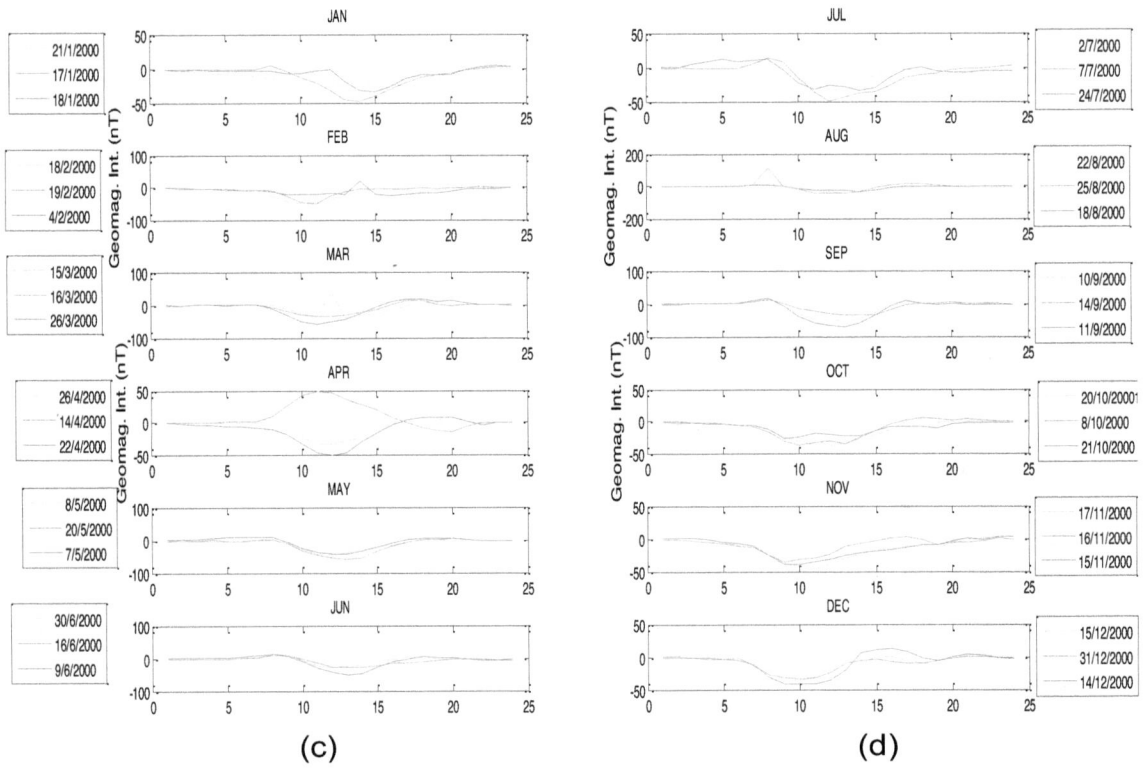

Figure 2. (a, b) Diurnal variation of delta ΔH component at ASC in year 2000, (c, d) Diurnal variation of delta Z-component at ASC in year 2000.

Table 2. International Quiet Days for 2000.

Jan	Feb	Mar	Apr	May	Jun	Jul	Aug	Sept	Oct	Nov	Dec
21	18	15	26	08	30	02	22	10	20	17	15
17	19	16	14	11	16	07	25	14	08	16	31
18	04	26	22	20	09	24	18	11	21	15	14
09	17	27	18	07	25	06	19	09	09	03	20
08	20	04	25	21	17	25	26	22	06	02	30

Table 3. Monthly averages of ΔH-field component of all the stations.

Month	ABG [ΔH_{out}]	ASC [ΔH_{in}]	ASCΔH_{in} - ΔH_{out}	HUA [ΔH_{in}]	HUAΔH_{in} - ΔH_{out}	PND [ΔH_{in}	PNDΔH_{in} - ΔH_{out}
Jan	13.50	13.86	0.36	22.50	9.00	12.12	-1.38
Feb	15.12	19.45	4.33	36.37	21.25	21.3	6.18
Mar	20.83	14.51	-6.32	36.21	15.38	22.2	1.37
Apr	24.07	15.80	-8.27	28.01	3.94	23.4	-0.67
May	17.26	14.10	-3.16	32.06	14.80	22.43	5.17
Jun	21.77	14.64	-7.13	24.78	3.01	20.95	-0.82
Jul	16.90	15.75	-1.15	33.78	16.88	17.56	0.66
Aug	17.07	22.14	5.07	30.32	13.25	20.6	3.53
Sep	17.67	21.50	3.83	36.21	18.54	16.83	-0.84
Oct	17.53	12.32	-5.21	38.83	21.30	17.53	0.00
Nov	14.77	13.07	-1.70	27.04	12.27	15.23	0.53
Dec	9.94	14.06	4.12	27.69	17.75	17.00	7.06

is an enhancement up to 100nT. A depression, which is more conspicuous on January 21 followed by that on January 7 and then that on 18th January with a peak of 5nT, was observed. Nonetheless, a narrow band of current flowed at night-time between 20th and 24th hours, conspicuously in January, February, December and September on 14th.

Interesting is the features on the 15th of March in Figure 2a, where the Sq(H) variation depicts a peculiar pattern. There was no pronounced enhancement, rather, almost a horizontal trend was observed. It could rather be termed a slight depression which cause or route could be attributed to the ring current, magnetosphere-ionosphere coupling or polar-equatorial coupling if not on quiet period as has been chosen in this study. Basically, the Z-component variation in Figure 2c on same day shows enhancement rather than depression so it has given room to further investigate if CEJ has occurred. This is because it has given us information from ΔH that EEJ truly reversed westwards. From Table 3, it is clearly seenthat ΔH_{in} for ASC minus ΔH_{out} is less than zero (ΔH_{in} − ΔH_{out} ℜ0), which indicates the existence of counter electrojet. This occurrence of CEJ at ASC during the noon could be attributed to late morning reversal of E_z, meaning that the reversal could have taken place around noon. Since ΔH_{in} in this station is greater than zero, it is then partial counter electrojet that occurred. More interesting is the fact that, it occurred around the local

noon. This is in variance with findings of Mayaud (1977), Marriot et al. (1979) and Stening (1992). Our findings are with agreement with work of Ezema et al. (1996), who found latitudinal profiles of CEJ at all hours from 07 to 17 h local time.

Incidentally, from Figure 2c and d ΔZ-field shows no reversal on all the days above, except on March 15. Therefore, since it is only the H profile that has reversed in this EEJ station, then it is only the EEJ that has reversed westwards. However, a marked reversal in ΔZ-field was observed on April 14 and slightly on March 15.

Figure 3a-b show a continuous series of enhanced ΔH-component in Huancayo (HUA) occurring between 12:00 h and 20:00 h throughout the year. This trend is preceded by a flat feature in the early hours of all the quiet days under study. A distinct scenario of ΔH depression is noted in the month of December on 31st, 15th and 14th respectively, in order of higher magnitudes of depression. Though there is a westward flow of current from about 1600 to 2400 h, this is associated with a reversal of the ΔZ-component (Figure 3c-d) of same periods in the month of December. While ΔZ-component enhancements are manifested from noon to 18:00 h in the months of January, February and March, the ΔH-components continue to be enhanced at the same time. Figure 4a-d depict geomagnetic intensity at Pondicherry (PND) during the three geomagnetic quietest day under consideration for the year 2000. Results show gradual enhancement of

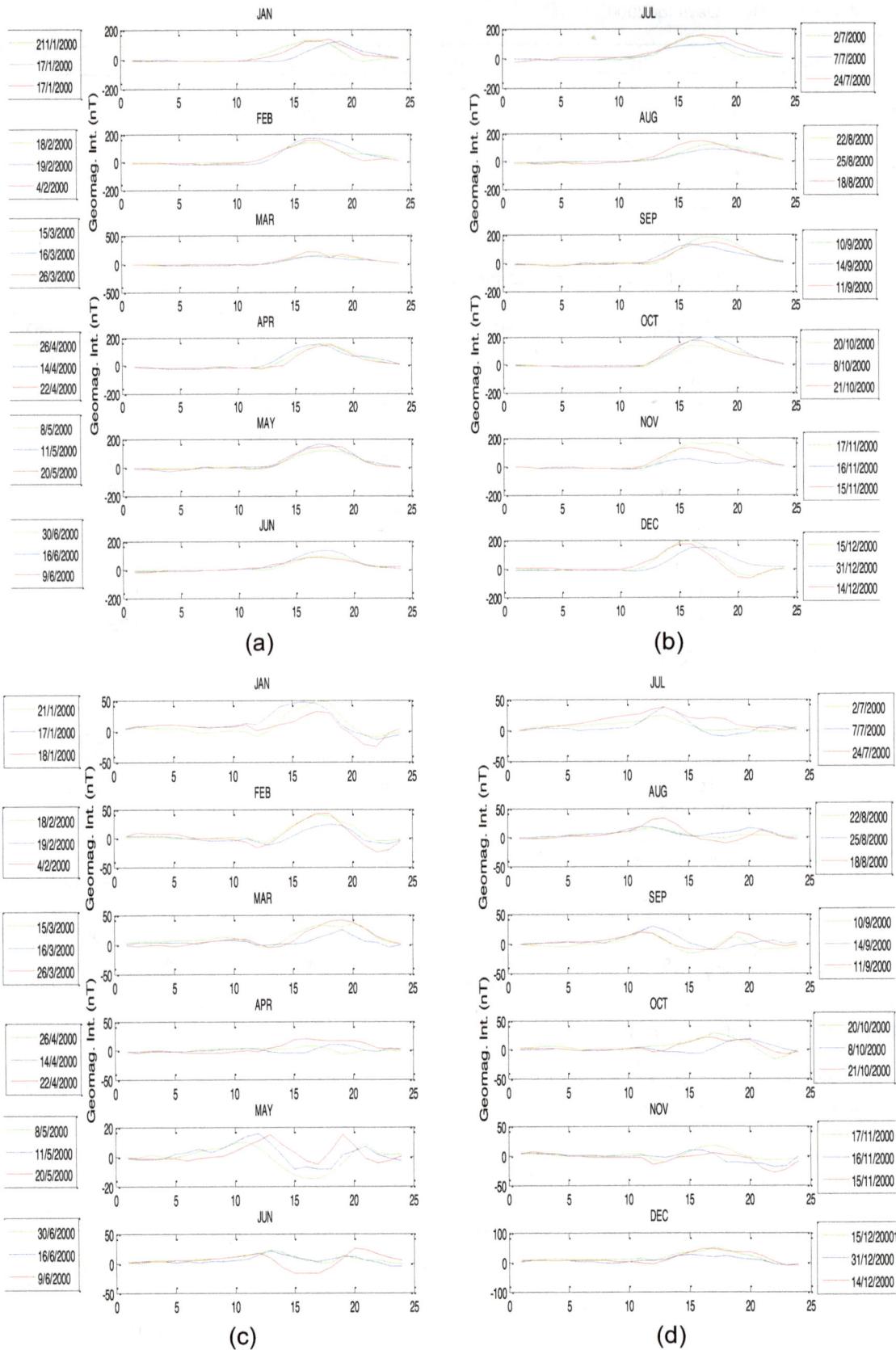

Figure 3. (a, b) Diurnal variation of delta ΔH component at HUA in year 2000, (c, d) Diurnal variation of delta Z-component at HUA in year 2000.

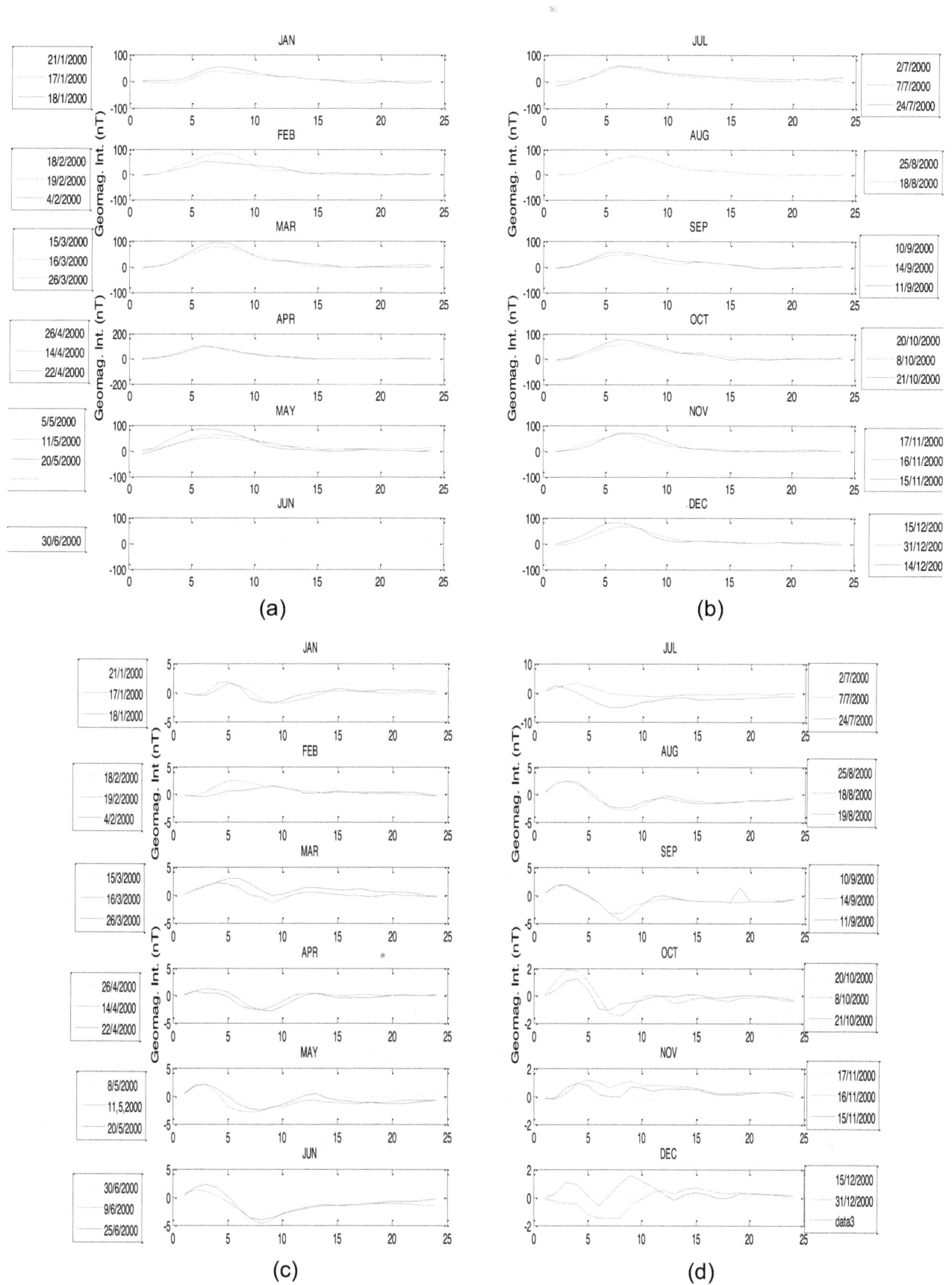

Figure 4. (a, b) Diurnal variation of delta ΔH component at HUA in year 2000, (c, d) Diurnal variation of delta Z-component at HUA in year 2000.

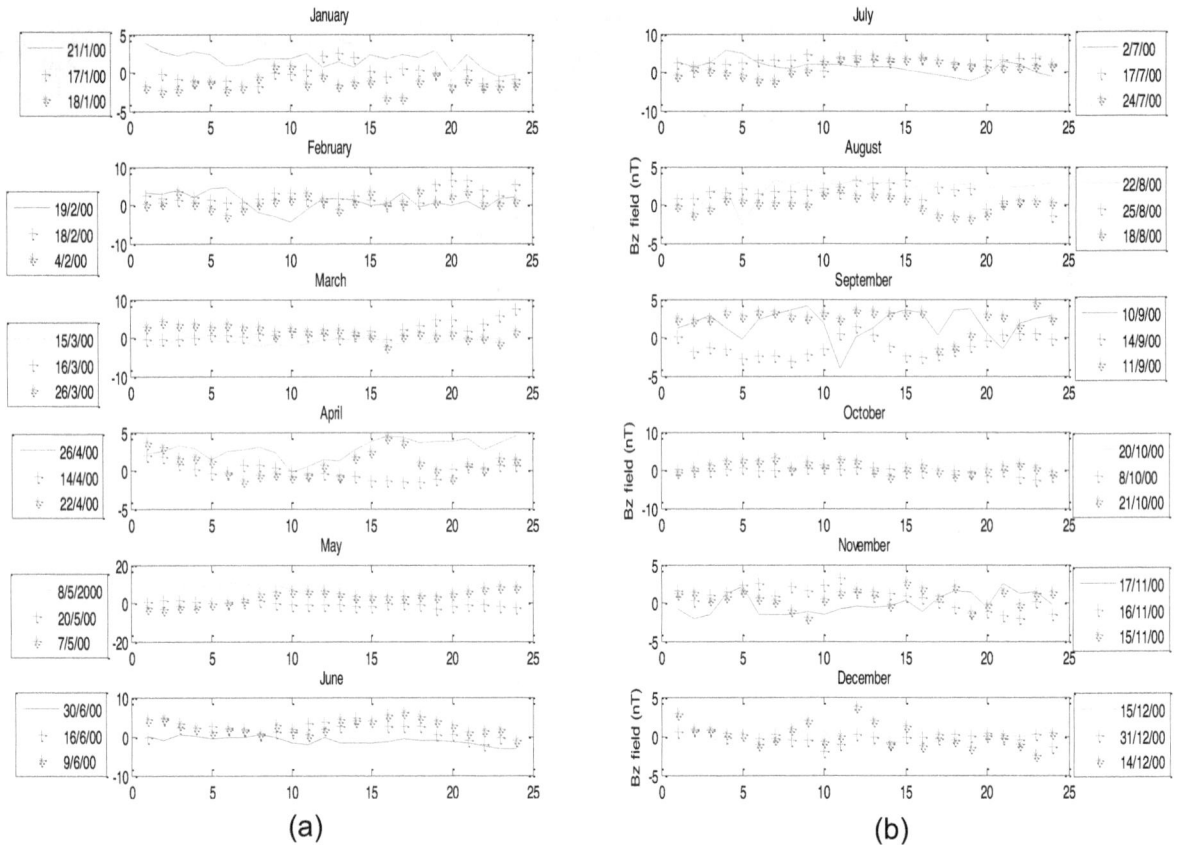

Figure 5. (a) IMF Bz orientation for three quietest days of January - June 2000, (b) IMF Bz orientation for three quietest days of July - December 2000.

ΔH-component from 3:00 h, with a peak at about 8:00 h on all the three most quiet days of the year. This trend is accompanied by depression of ΔZ-component at the same period of time.

From Table 3, it is apparent that the CEJ flow in each of the observatories is not continuous across all the months of the year. For instance, in ASC $\Delta H_{in} - \Delta H_{out}$ ℜ0 holds for the months of March, April, May, June, July, October, November and December. And in PND, Equations (1) and (2) are true for January, April, June and September. The significance being that the lower current layer, which is associated with the EEJ reversed westwards in the months indicated. This also confirms findings from previous workers (Marriot et al., 1979; Hutton and Oyinloye, 1970; Rastogi, 1974), who concluded from their works that afternoon CEJ is most frequent in local summer solstice. Since inconsistency exists both with the concept of definition of CEJ time of occurrence and causes, hence, this study further investigated the relationship between interplanetary magnetic field (IMF) and the CEJ. It was discovered from IMF movement, that as soon as B_z component of IMF turns from South direction to North, then CEJ occurs. Figure 5a and Table 4 support this finding, where northward turning of IMF B_z is observed in the early hours

of the three quietest days of the months of January through May. Also, northward movement of IMF B_z is a post-noon phenomenon as is apparent on February 18 and 19; March 16, April 26 and May 8 of the year 2000. Very remarkable feature of the northward turning was very obvious on April 26, which was the quietest day of that month, where the IMF B_z remained northward throughout the day with greater pronouncements in both pre-noon and post-noon hours.

In Figure 5b, on September 10, the quietest day of the month, the IMF B_z suddenly changed from south to north at 11:00 h and retained its northward orientation for a considerable number of hours. This signature was repeated throughout the day and on September 11, which confirms a CEJ occurrence observed on same dates and time in ASC (Figure 2a-d), in HUA (Figure 3a-d) and in PND (Figure 4a-d). This is a new finding and should be further investigated fully in our next paper.

Conclusion

Common occurrence of ΔH depression with the same obvious pattern could be attributed to the same route cause. The quick method to identify the existence of CEJ

Table 4. IMF Bz magnitudes and orientations for the three quietest days in the year 2000.

Month	1ST QD	Bz Mag. (nT)	Bz Dir	2ND QD	Bz Mag. (nT)	Bz Dir	3RD QD	Bz Mag. (nT)	Bz Dir
Jan	21	1.2	N	17	0.2	N	18	-0.6	S
Feb	18	1.2	N	19	-0.9	S	04	0.5	S
Mar	15	-0.3	S	16	1.3	N	26	1.1	N
Apr	26	1.5	N	14	-0.8	S	22	-0.4	S
May	08	5.4	N	11	0.0	N	20	-1.0	S
Jun	30	-1.2	S	16	1.7	N	09	2.8	N
Jul	02	1.2	N	07	0.6	N	24	-0.1	S
Aug	22	1.1	N	25	-0.3	S	18	-0.2	S
Sep	10	1.1	N	14	-0.2	S	11	3.4	N
Oct	20	0.1	N	08	-1.2	S	21	1.1	N
Nov	17	0.5	N	16	0.3	N	15	0.6	N
Dec	15	0.2	N	31	-0.2	S	14	-0.0	S

is to study IMF movement. As soon as B_z component of IMF turns from South direction to North, the CEJ occurs, which is in variance with prediction of Manoj et al. (2008), who claimed that penetration of electric fields into the equatorial ionosphere is not dependent on the polarity of IMF B_z. It is then easier to examine other contributing factors like the values of ΔH_{in} and the reversal of EEJ and the ΔH values. This study of occurrence of CEJ and causes has constituted an active area of research work as long as the world in which we are is dynamic and not static. The causes of CEJ are yet to be fully explained. Hence, more future work is recommended for more robust results. One of our findings from this work, which revealed that CEJ could equally occur during night-time, is at variance with findings of earlier workers, who on the contrary ascertained that CEJ phenomenon is a morning event. They concluded that CEJ could only occur in the morning hours (Mayaud, 1977; Marriot et al., 1979; Stening, 1992). Hence, further investigation is required in order to confirm this existing controversy.

ACKNOWLEDGEMENTS

We thank and acknowledge the World Data Center, Kyoto, Japan, for providing us with the magnetic data which was employed in this study. The authors are immensely grateful for the assistance rendered by Prof. Nat. Gopalswamy of NASA, who readily made the IMF Bz data available. We also thank the Centre for Basic Space Science, Nsukka, Nigeria, for making available to us the facilities used during the course of this research.

REFERENCES

Alex S, Mukherjee S (2001). Local Time dependence of the equatorial counter electrojet effect via narrow longitudinal belt. Earth Planet Space 53:1151-1161.

Crochet M, Hanwise C, Broche P (1979). HF radar studies of two stream instability during an equatorial counter-electrojet. J. Geophys. Res. 84:5223-5233.

Gouin P (1962). Reversal of the magnetic daily variations at Addis Ababa. Nature 139:1145-1146.

Gurubaran S (2002). The Equatorial counter electrojet: Part of a worldwide current system? Geophys. Res. Lett. 29(a):1337-1340.

Hutton R, Oyinloye JO (1970). The counter electrojet in Nigeria. Ann. Geophys. 26:921-926.

Kane RP (1976). Dilema of the equatorial counter electrojet and the disappearance of Esq. Indian J. Rad. Space Phys. 5:6-12.

Manoj C, Maus S, Lühr H, Alken P (2008). Penetration characteristics of the interplanetary electric field to the daytime equatorial ionosphere. J. Geophy. Res. 113, A12310, doi:10.1029/2008JA013381.

Marriot RT Richmond AD Venkateswaram SV (1979). The quiet time equatorial electrojet and counter electrojet. J. Geomag. Geoelectr. 31:311-340.

Mayaud PN (1977). The equatorial counter electrojet – A Review of its magnetic Aspects. J. Atmos. Terr. Phys. 39:1055-1070.

Okeke FN, Hamano Y (2000). New features of abnormal quiet days in equatorial regions. IJGA 2(2):109-114.

Onwumechili CA (1997). The equatorial electrojet. Gordon and Breach Sci. Publ. UK.

Rastogi RG (1974). Lunar effects in the counter electrojet near the magnetic equator. J. Atmos. Terr. Phys. 36:167-170.

Rastogi RG (1975). On the simultaneous existence of eastward and westward flowing equatorial electrojet current. Proc. Ind. Acad. Sci. 81A:80-92.

Rastogi RG (1999). Morphological aspects of a new type of counter electrojet event. Ann. Geophysicae. 17:210-219.

Stening RJ (1992). The emigma of the counter equatorial electrojet and lunar tidal influences in the equatorial region. Adv. Space Res. 12(6):23-32.

Woodman RF, Rastogi RG, Calderon C (1977). Solar cycle effects on the electric fields in the equatorial ionosphere. J. Geophys. Res. 82(32):5257-5260.

16

Study of nonlinear KdV type equations via homotopy perturbation method and variational iteration method

H. Goodarzian*, T. Armaghani and M. Okazi

Islamic Azad University, Mahdishahr Branch, Mahdishahr, Iran.

The Korteweg-de Vries (KdV) equation is the champion of model equations of nonlinear waves. In fact, it is from numerical experiments of a water wave equation. The objective of this paper is to present a comparative study of He's homotopy perturbation method (HPM) and variational iteration method (VIM) for the semi analytical solution of three different Kortweg-de Vries (KdV) type equations called KdV, K(2,2,) and modified KdV (Burgers) equations. The study has highlighted the efficiency and capability of aforementioned methods in solving these nonlinear problems which has risen from a number of important physical phenomenons.

Key words: Variational iteration method (VIM), homotopy perturbation method (HPM), KdV equation, modified KdV Equation.

INTRODUCTION

It was in 1895 that Korteweg and Vries derived KdV equation to model Russell's phenomenon of solitons (Korteweg and Vries, 1895) like shallow water waves with small but finite amplitudes (Yan, 2001). Solitons are localized waves that propagate without change of its shape and velocity properties and stable against mutual collision (Khattak and Siraj, 2008). It has also been used to describe a number of important physical phenomena such as magneto hydrodynamics waves in warm plasma, acoustic waves in an inharmonic crystal and ion-acoustic waves (Ozis and Ozer, 2006).

Consider three models of KdV equation called KdV, K(2,2) and modified KdV (Korteweg and Vries, 1895) as given respectively by:

$$u_t - 3(u^2)_x + u_{xxx} = 0 \qquad (1)$$

$$u_t + (u^2)_x + (u^2)_{xxx} = 0 \qquad (2)$$

$$u_t + \frac{1}{2}(u^2)_x - u_{xx} = 0 \qquad (3)$$

Equation 1 is the pioneering equation that gives rise to solitary wave solutions. Solitons, which are waves with infinite support, are generated as a result of the balance between the nonlinear convection $(u^n)_x$ and the linear dispersion u_{xxx} in Equations 1-3. Solitons are localized waves that propagate without change of their shape and velocity properties and stable against mutual collisions (Wang, 1996).

$$K(n,n): \quad u_t + (u^n)_x + (u^n)_{xxx} = 0 \qquad (4)$$

Equation (4) is the pioneering equation for compactons. In solitary wave theory, compactons are defined as solitons with finite wave lengths or solitons free of exponential tails (Rosenau and Hyma, 1993). Compactons are generated as a result of the delicate interaction between nonlinear convection $(u^n)_x$ with the genuine nonlinear dispersion $(u^n)_{xxx}$ in Equation 4.

Finally, the modified KdV equation appears in fluid mechanics. This equation incorporates both convection and diffusion in fluid dynamics, and is used to describe the structure of shock waves (Wazwaz, 2006). Hence, KdV type equations have significant roles in engineering and physics. Besides, the analytical solutions of these governing equations may guide authors to know the

*Corresponding author. E-mail: hamed_goodarzian@yahoo.com.

described process deeply and sometimes leads them to know some facts that are not simply understood through common observations, but it is quite difficult to obtain the analytical solution of these problems as these are functioning highly nonlinear.

Many researchers studied the similar kinds of problems in other applications (Abdou and Soliman, 2005; Aboulvafa et al., 2006; He, 1997; Liao, 1992; Malfeit, 1992) and many powerful methods have been proposed to seek the exact solutions of nonlinear differential equations; for instance, Backlund transformation (Ablowitz and Clarkson,1991; Coely, 2001), Darboux transformation (Wadati et al., 1975), the inverse scattering method (Gardner et al., 1967), Hirota's bilinear method (Hirota, 1971), the tanh method (Soliman, 2006), the sine–cosine method (Yan and Zhang, 2000), the homogeneous balance method (Wang, 1996), and the Riccati expansion method with constant coefficients (Yan, 2001).

In this paper, He's variational iteration method (VIM) and homotopy perturbation method (HPM) and Liao's homotopy analysis method (HAM) are used to conduct an analytic study on the KdV, the K(2,2) and the modified KdV equations in order to show all the methods above, are capable in solving a large number of linear or nonlinear differential equations, also all the aforementioned methods give rapidly convergent successive approximations of the exact solution if such solution exists, otherwise approximations can be used for numerical purposes.

HE'S VARIATIONAL ITERATION METHOD

Fundamental

To illustrate the basic concepts of variational iteration method, we consider the following deferential equation (He, 2000):

$$Lu + Nu = g(x) \tag{5}$$

where L is a linear operator, N a nonlinear operator, and $g(x)$ a heterogeneous term. According to VIM, we can construct a correction functional as follows (He, 2000):

$$u_{n+1}(x) = u_n(x) + \int_0^x \lambda \{ Lu_n(\tau) + N\tilde{u}_n(\tau) - g(\tau) \} d\tau \tag{6}$$

where λ is a general Lagrangian multiplier [8, 9], which can be identified optimally via the variational theory, the subscript n indicates the nth order approximation, $\tilde{u}_n$ which is considered as a restricted variation, that is, $\delta \tilde{u}_n = 0$.

The application

Example 1

Considering the KdV equation as:

$$u_t - 3(u)^2_x + u_{xxx} = 0, \quad -\infty < x < +\infty, \quad t > 0 \tag{7}$$

with the following initial condition:

$$u(x,0) = 6x \tag{8}$$

To solve Equations 7 and 8 using VIM, we have the correction functional as:

$$u_{n+1}(x,t) = u_n(x,t) + \int_0^t \lambda \{ u_t - u_{xxt} + uu_x \} d\tau, \tag{9}$$

where uu_x indicates the restricted variations; that is, $\delta(uu_x) = 0$. Making the above correction functional stationary, we obtain the following stationary conditions:

$$1 + \lambda \big|_{\tau=1} = 0 \tag{10a}$$

$$\lambda' = 0 \tag{10b}$$

The Lagrangian multiplier can therefore be identified as:

$$\lambda = -1 \tag{11}$$

substituting Equation 11 into the correction functional equation system (9) results in the following iteration formula:

$$u_{n+1}(x,t) = u_n(x,t) - \int_0^t \{ u_\tau - 3(u)^2_x + u_{xxx} \} d\tau \tag{12}$$

Each result obtained from Equation 12 is $u(x,t)$ with its own error relative to the exact solution, but higher number iterations leads us to obtain results closer to the exact solution. Using the iteration formula (12) and the initial condition as u_0, five iterations were made as follows;

The first iteration results in:

$$u_1(x,t) = 6x(1 + 36t) \tag{13a}$$

The second iteration results in:

$$u_2(x,t) = 6x(1 + 36t + 1296t^2 + 15552t^3) \tag{13b}$$

And finally, the fifth iteration results in:

$$u_5(x,t) = 6x(1 + 36\ t + 1296\ t^2 + 15552\ t^3 + 1679616\ t^4 + 60466176\ t^5) + small\ terms \quad (13c)$$

It is obvious that $u_n(x,t)$ converges to $\dfrac{6x}{1-36t}$ as an exact solution for Equations 7 and 8.

Example 2

We consider the K(2,2) equation:

$$u_t + (u)^2{}_x + (u^2)_{xxx} = 0, \quad x \in R,\ t > 0. \quad (14)$$

$$u(x,0) = x. \quad (15)$$

To solve Equations 14 and 15 using VIM, we have the correction functional as:

$$u_{n+1}(x,t) = u_n(x,t) + \int_0^t \lambda\{u_\tau + (u)^2{}_x + (u^2)_{xxx}\}d\tau \quad (16)$$

Making the above correction functional stationary, we obtain the following stationary conditions:

$$1 + \lambda\big|_{\tau=t} = 0 \quad (17a)$$

$$\lambda' = 0. \quad (17b)$$

The Lagrangian multiplier can therefore be identified as:

$$\lambda = -1 \quad (18)$$

Substituting Equation 18 into the correction functional equation system (16) results in the following iteration formula:

$$u_{n+1}(x,t) = u_n(x,t) - \int_0^t \{u_\tau + (u)^2{}_x + (u^2)_{xxx}\}d\tau \quad (19)$$

Using the iteration formula (19) and the initial condition as u_0, six iterations were made as follows.

The first iteration results in:

$$u_1(x,t) = x(1 - 2t). \quad (20a)$$

The second iteration results in:

$$u_2(x,t) = x\left(1 - 4t + 4t^2 - \frac{8}{3}t^3\right), \quad (20b)$$

And finally the sixth iteration results in:

$$u_6(x,t) = x\left(1 - 2t + 4t^2 - 8t^3 + 16t^4 - 32t^5 + 64t^6\right) + small\ terms \quad (20c)$$

Again, by trying higher iterations we can obtain the exact solution of Equations 14 and 15 in the form of

$$u(x,t) = \frac{x}{1+2t}.$$

Example 3

For the third example, we consider the modified $KdV(mKdV)$ equation as:

$$u_t + \frac{1}{2}\left(u^2\right)_x - u_{xx} = 0, \quad x \in R,\ t > 0, \quad (21)$$

$$u(x,0) = x. \quad (22)$$

The correction functional takes the form of:

$$u_{n+1}(x,t) = u_n(x,t) + \int_0^t \lambda\{u_\tau + \frac{1}{2}\left(u^2\right)_x - u_{xx}\}d\tau, \quad (23)$$

where, again $\lambda = -1$. So, Equation 23 changes to:

$$u_{n+1}(x,t) = u_n(x,t) - \int_0^t \{u_\tau + \frac{1}{2}\left(u^2\right)_x - u_{xx}\}d\tau. \quad (24)$$

Using the iteration formula (24) and the initial condition as u_0, three iterations were made and results are as follows:

$$u_1(x,t) = x(1-t), \quad (25)$$

$$u_2(x,t) = x(1 - t + t^2 - \frac{1}{3}t^3), \quad (26)$$

$$u_3(x,t) = x(1 - t + t^2 - t^3 + \frac{2}{3}t^4), \quad (27)$$

If one try the higher iterations, one can have the exact solution of Equations 21 and 22 in the form of

$$u(x,t) = \frac{x}{1+t}.$$

HE'S HOMOTOPY PERTURBATION METHOD

Fundamental

To illustrate the basic ideas of this method, we consider

the following nonlinear differential Equation (He, 2000):

$$A(u) - f(r) = 0. \quad r \in \Omega \tag{28}$$

Considering the boundary conditions of:

$$B(u, \partial u / \partial n) = 0, \quad r \in \Gamma \tag{29}$$

Where A is a general differential operator, B a boundary operator, $f(r)$ a known analytical function and Γ is the boundary of the domain Ω.

The operator A can be divided into two parts of L and N, where L is the linear part, while N is a nonlinear one. Equation 28 can, therefore, be rewritten as:

$$L(u) + N(u) - f(r) = 0. \tag{30}$$

By the homotopy technique, we construct a homotopy as $v(r, p) : \Omega \times [0,1] \to \Re$ which satisfies (He, 2000):

$$H(v, p) = (1-p)[L(v) - L(u_0)] + p[A(v) - f(r)] = 0, \quad P \in [0,1], \ r \in \Omega \tag{31}$$

where $p \in [0,1]$ is an embedding parameter and u_0 is an initial approximation of Equation 31 which satisfies the boundary conditions. Obviously, considering Equation 31, we will have:

$$H(v,0) = L(v) - l(u_0) = 0, \tag{32}$$

$$H(v,1) = A(v) - f(r) = 0. \tag{33}$$

The changing process of p from zero to unity is just that of $v(r, p)$ from $u_0(r)$ to $u(r)$. In topology, this is called deformation, and $L(v) - L(u_0)$ and $A(v) - f(r)$ are called homotopy. According to HPM, we can first use the embedding parameter p as a "small parameter", and assume that the solution of Equation 31 can be written as a power series in p:

$$v = v_0 + p v_1 + p^2 v_2 + \cdots, \tag{34}$$

setting $p = 1$ results in the approximate solution of Equation 31:

$$u = \lim_{p \to 1} v = v_0 + v_1 + v_2 + \ldots.. \tag{35}$$

The combination of the perturbation method and the homotopy method is called the HPM, which lacks the limitations of the conventional perturbation methods, although this technique can have full advantages of the conventional perturbation techniques.

The series (35) is convergent for most cases. However, the convergence rate depends on the nonlinear operator $A(v)$. The following opinions are suggested by He (2004):

The second derivative of $N(v)$ with respect to v must be small because the parameter p may be relatively large, that is, $p \to 1$.

The norm of $L^{-1} \partial N / \partial v$ must be smaller than one so that the series converges.

Application

Example 1

With the same first example as mentioned previously, the equation is as:

$$u_t - 3(u)^2{}_x + u_{xxx} = 0, \quad -\infty < x < +\infty, \quad t > 0. \tag{36}$$

with the initial condition of:

$$u(x,0) = 6x. \tag{37}$$

Substituting Equation 36 into 31 and then substituting v from Equation 34 and rearranging it as a power series in p, we have an equation system including $n+1$ equations to be simultaneously solved; n is the order of p in Equation 34. Assuming $n = 5$, the system is as follows:

$$\begin{cases} u_{0t} = 0, \\ u_{1t} - 6u_0 u_{0x} + u_{0xxx} = 0, \\ u_{2t} - 6u_1 u_{0x} - 6u_0 u_{1x} + u_{1xxx} = 0, \\ u_{3t} - 6u_0 u_{2x} - 6u_1 u_{1x} - 6u_2 u_{0x} + u_{2xxx} = 0, \\ u_{4t} - 6u_3 u_{0x} - 6u_0 u_{3x} - 6u_2 u_{1x} - 6u_1 u_{2x} + u_{3xxx} = 0, \\ u_{5t} - 6u_3 u_{1x} - 6u_4 u_{0x} - 6u_1 u_{3x} - 6u_2 u_{2x} - 6u_0 u_{4x} + u_{4xxx} = 0 \end{cases} \quad \begin{array}{l} u_0(x,0) = 6x, \\ u_1(x,0) = 0, \\ u_2(x,0) = 0, \\ u_3(x,0) = 0, \\ u_4(x,t) = 0, \\ u_5(x,t) = 0. \end{array} \tag{38}$$

One can now try to obtain a solution for equation system (38), in the form of:

$$u_0(x,t) = 6x, \tag{39a}$$

$$u_1(x,t) = 6x(36t), \tag{39b}$$

$$u_2(x,t) = 6x(1296\,t^2), \tag{39c}$$

$$u_3(x,t) = 6x(46656\,t^3), \tag{39d}$$

$$u_4(x,t) = 6x(1679616\,t^4), \tag{39e}$$

$$u_5(x,t) = 6x(60466176\,t^5). \tag{39f}$$

Having $u_i, i = 0,1,...5$, the solution $u(x,t)$ is as:

$$u(x,t) = \sum_{i=0}^{5} u_i(x,t) = 6x\left[1 + 36t + (36t)^2 + (36t)^3 + (36t)^4 + (36t)^5\right] \tag{40}$$

Trying higher iterations, we can obtain the exact solution of Equations 36 and 37 in the form of $u(x,t) = \dfrac{6x}{1-36t}$.

Example 2

Let us consider the following equation again:

$$u_t + (u)^2{}_x + (u^2)_{xxx} = 0, \quad x \in R,\ t > 0, \tag{41}$$

$$u(x,0) = x. \tag{42}$$

Substituting Equation (41) into (31) and then substituting v from (34), rearranging it as a power series in p and assuming $n = 5$, we have a system of equations including six equations to be simultaneously solved as follows:

$$
\begin{cases}
u_{0t} = 0, \\[4pt]
u_{1t} + 6u_{0x}u_{0xx} + 2u_0 u_{0xxx} + 2u_0 u_{0x} = 0, \\[4pt]
\begin{cases} u_{2t} + 6u_{1x}u_{0xx} + 6u_{0x}u_{1xx} + 2u_1 u_{0x} \\ + 2u_0 u_{1x} + 2u_1 u_{0xxx} + 2u_0 u_{1xxx} \end{cases} = 0, \\[4pt]
\begin{cases} u_{3t} + 6u_{2x}u_{0xx} + 6u_{0x}u_{2xx} + 6u_{1x}u_{1xx} \\ + 2u_2 u_{0xxx} + 2u_0 u_{2xxx} + 2u_0 u_{2x} + \\ 2u_1 u_{1xxx} + 2u_1 u_{1x} + 2u_2 u_{0x} \end{cases} = 0, \\[4pt]
\begin{cases} u_{4t} + 6u_{1x}u_{2xx} + 6u_{2x}u_{1xx} + 6u_{3x}u_{0xx} \\ + 6u_{0x}u_{3xx} + 2u_2 u_{1x} + 2u_0 u_{3xxx} + \\ 2u_3 u_{0x} + 2u_1 u_{2x} + 2u_3 u_{0xxx} + 2u_1 u_{2xxx} \\ + 2u_0 u_{3xxx} + 2u_2 u_{1xxx} \end{cases} = 0, \\[4pt]
\begin{cases} u_{5t} + 6u_{4x}u_{0xx} + 6u_{0x}u_{4xx} + 6u_{2x}u_{2xx} + \\ 6u_{3x}u_{1xx} + 6u_{1x}u_{3xx} + 2u_2 u_{2x} + 2u_1 u_{3x} \\ 2u_0 u_{4x} + 2u_4 u_{0x} + 2u_4 u_{0xxx} + 2u_1 u_{3xxx} \\ + 2u_3 u_{1x} + 2u_3 u_{1xxx} + 2u_0 u_{4xxx} + 2u_2 u_{2xxx} \end{cases} = 0,
\end{cases}
\qquad
\begin{aligned}
& u_0(x,0) = x, \\[10pt]
& u_1(x,0) = 0, \\[10pt]
& u_2(x,0) = 0, \\[14pt]
& u_3(x,0) = 0, \\[20pt]
& u_4(x,0) = 0, \\[24pt]
& u_5(x,0) = 0.
\end{aligned}
\tag{43}
$$

One can now try to obtain a solution for the above equation system in the form of:

$$u_0(x,t) = x, \tag{44a}$$

$$u_1(x,t) = x(-2t), \tag{44b}$$

$$u_2(x,t) = x(4\,t^2), \tag{44c}$$

$$u_3(x,t) = x(-8\,t^3), \tag{44d}$$

$$u_4(x,t) = x(16\,t^4), \tag{44e}$$

$$u_5(x,t) = x(-32\ t^5).\qquad(44f)$$

Having $u_i, i = 0,1,...5$, the solution $u(x,t)$ is therefore as:

$$u(x,t)=\sum_{i=0}^{5}u_i(x,t)=x\left[1-2t+(2t)^2-(2t)^3+(2t)^4-(2t)^5\right]\qquad(45)$$

Higher iterations can make one obtain the exact solution

of Equations 41 and 42 in the form of $u(x,t)=\dfrac{x}{1+2t}$.

Example 3

Let us solve Equations 21 and 22 through HPM; considering $n = 5$; thus, we will have the system equation as:

$$\begin{cases}u_{0t}=0, & u_0(x,0)=x,\\ u_{1t}+u_0u_{0x}=0, & u_1(x,0)=0,\\ u_{2t}+u_1u_{0x}+u_0u_{1x}-u_{1xx}=0, & u_2(x,0)=0,\\ u_{3t}+u_0u_{2x}+u_2u_{0x}+u_1u_{1x}-u_{2xx}=0, & u_3(x,0)=0,\\ u_{4t}+u_3u_{0x}+u_0u_{3x}+u_2u_{1x}+u_1u_{2x}-u_{3xx}=0, & u_4(x,t)=0,\\ u_{5t}+u_3u_{1x}+u_1u_{3x}+u_4u_{0x}+u_0u_{4x}+u_2u_{2x}-u_{4xx}=0 & u_5(x,t)=0.\end{cases}\qquad(46)$$

Trying to solve the system Equation 46 results in:

$$u_0(x,t)=x,\qquad(47a)$$

$$u_1(x,t)=x(-t),\qquad(47b)$$

$$u_2(x,t)=x(t^2),\qquad(47c)$$

$$u_3(x,t)=x(-t^3),\qquad(47d)$$

$$u_4(x,t)=x(t^4),\qquad(47e)$$

$$u_5(x,t)=x(-t^5),\qquad(47f)$$

then,

$$u(x,t)=\sum_{i=0}^{5}u_i(x,t)=x(1-t+t^2-t^3+t^4-t^5).$$
$$(48)$$

Always by adding up the number of iterations one can attain the exact solution of Equations 21 and 22 in the

form $u(x,t)=\dfrac{x}{1+t}$.

Conclusion

The main goals of this study were the assessment of capability of the He's VIM and HPM to solve the KdV type equations. The KdV, K(2,2), and Burgers equations that arise from many important and practical physical phenomenon were examined for rational solutions.

In clear conclusion, two above-mentioned methods were capable to solve this set of problems with successive rapidly convergent approximations without any restrictive assumptions or transformations causing changes in the physical properties of the problems. Also, adding up the number of iterations leads to the explicit solutions for the problems.

Among two methods, VIM is very comprehensible as it reduces the size of calculations and also its iterations are direct and straightforward. HPM do not require small parameters in the equation so that the limitations of the conventional perturbation methods can be eliminated and thereby the calculations are simple and straight forward, though HPM can be more convenient.

REFERENCES

Abdou MA, Soliman AA (2005). Variational iteration method for solving Burger's and coupled Burger's equation. J. Comput. Appl. Math., 181: 245-251.

Ablowitz MJ, Clarkson PA (1991). Nonlinear Evolution Equations and Inverse Scattering. Cambridge University Press. Cambridge, 2:254-262.

Aboulvafa EM, Abdou MA, Mahmoud AA (2006). The solution of nonlinear coagulation problem with mass loss, Chaos Solitons and Fractals, 29: 313-330.

Coely A (2001). Backlund and Darboux Transformations. American Mathematical Society. Providence. Rhode Island, pp. 458-468.

Gardner CS, Green JM, Kruskal MD, Miura RM (1967). Method for solving the Korteweg-deVries equation. Phys. Rev. Lett., 19: 1095-1097.

He JH (1997). A new approach to nonlinear partial differential equations. Comm. Nonlinear Sci. Numer. Simul., 2 (4): 203-205.

He JH (2000). Variational iteration method for autonomous ordinary differential systems. App. Math. Comp., 114: 115-123.

He JH (2004). The homotopy perturbation method for nonlinear oscillators with discontinuities. App. Math. Comp., 151: 287-292.

Hirota R (1971). Exact solution of the Korteweg-de Vries equation for multiple collisions of solitons, Phys. Rev. Lett., 27: 1192-1194.

Khattak AJ, Siraj-ul I (2008). A comparative study of numerical solutions of a class of KdV equation, J. Comput. Appl. Math., 199: 425-434.

Korteweg DJ, Vries G (1895). On the change of form of long waves advancing in a rectangular canal, and on a new type of long stationary wave. Philos. Mag., 39: 422-443.

Liao SJ (1992). The proposed homotopy analysis technique for the solution of nonlinear problems, PhD thesis, Shanghai Jiao Tong University, pp. 58-62.

Malfeit W (1992). Solitary wave solutions of nonlinear wave equations. Am. J. Phys., 60: 650-654.

Ozis T, Ozer S (2006). A simple similarity-transformation-iterative scheme applied to Korteweg-de Vries equation. J. Comput. Appl. Math., 173: 19-32.

Rosenau P, Hyman JM (1993). Compactons: Solitons with finite wavelengths. Phys. Rev. Lett., 70 (5): 564 -567.

Soliman AA (2006). The modified extended tanh-function method for solving Burgers-type equations. Physica A. 361: 394-404.

Yan ZY, Zhang HQ (2000). Auto-Darboux Transformation and exact solutions of the Brusselator reaction diffusion model. Appl. Math. Mech., 22: 541-546.

Yan ZY (2001). New explicit travelling wave solutions for two new integrable coupled nonlinear evolution equations. Phys. Lett., A 292: 100-106.

Wadati M, Sanuki H, Konno K (1975). Relationships among inverse method, backlund transformation and an infinite number of conservation laws. Prog. Theoret. Phys., 53: 419-436.

Wang ML (1996). Exact solutions for a compound KdV-Burgers equation. Phys. Lett. A, 213: 279-287.

Wazwaz AM (2006). The variational iteration method for rational solutions for KdV, K (2, 2), Burgers, and cubic Boussinesq equations. J. Comput. Appl. Math., Article in Press.

17

Predicting geotechnical parameters of fine-grained dredged materials of the gulf of Izmit (NW Turkey) using the slump test method and index property correlations

Kurtulus C.[1] , Bozkurt A.[2] and Endes H.[1]

[1]Department of Geophysics, Kocaeli University, Engineering Faculty, Izmit-Kocaeli, Turkey.
[2]ABM Engineering Co., Izmit-Kocaeli, Turkey.

In this study, an attempt has been done to figure out the geotechnical properties of physically remolded dredged materials of the Gulf of Izmit located in NW Turkey, by using slump test method. For this aim, slum tests were conducted on dredged material specimens using open-ended polyvinyl chloride (PVC) cylinders of 10 cm diameter and height. Water content of the dredged material specimens was obtained in the laboratory using oven method. Bulk wet density, Atterberg limits and phase relations such as bulk unit weight, % solids by weight, void ratio, and engineering behavior properties such as Vane shear strength, and effective stress were determined. Later, the statistical correlations were conducted by regression analysis to obtain the relationships between water content and slump/cylinder height as well as other parameters such as wet bulk density, % solid by weight, void ratio, Vane shear strength and effective stress.

Key words: Vane shear strength, dredged materials, slump test.

INTRODUCTION

Description of dredged material properties is very necessary for any planned dredging operation, and the fundamental geotechnical parameters need to be determined or predicted. While near shore dredging operations, the dredging process generally involves too much material handling, manipulation, and remolding. The natural previously deposited material has been dredged, transported and re-deposited by the time. Therefore, its geotechnical properties are changed. The most fundamental parameters of the dredged materials are grain size distributions, water content, density, specific gravity and percent solid. Engineering properties include shear strength, permeability, viscosity, consolidation and critical erosion.

The laboratory tests of soil classification and description, grain size analysis and distribution, specific gravity of solid particles, water content, bulk density,

shear strength, stress-strain and behavior characteristics provide basic information about the dredged material soil properties. Dredged materials lose their original geotechnical properties while they are being dredged, transported and redeposited. Intergranular bonding and physico-chemical condition affect the materials behavior (Mitchell, 1993; Lee, 2004). As the water content in the dredged material is increased during remoulding and the solid particles become more separated, the material behaves like slurry. If fine-grained particles (silt and clay) composed more than 35% of the total matrix solids, the slurry behaves as viscous material (Spigolon, 1993). High organic content, gas bubbles and fibrous materials affect the shear strength of dredged materials (Klein and Sarsby, 2000; Edil and Wang, 2000). DeMeyer and Mahlerbe (1987) determined the threshold (yield) shear stress σ most clayey slurries less than 10 Pa. Bouziani

Figure 1. Study area. Stars indicate the dredged material collection points (Google earth).

and Benmounah (2013) conducted v-funnel and mini-slump tests with viscosity of Self-Compacting Mortars (SCM), measured at different rotational speeds and linear relationships were obtained between both v-funnel and mini-slump tests and viscosity. Peila et al. (2009) performed the slump cone test on conditioned material to check the mass behavior. Vinai et al. (2008) used a simple slump test to analyze the global characteristics of the conditioned soils.

The dredged materials are used to improve soil structure for agriculture purposes. Some dredged materials may be very good topsoil according to Francingues et al. (2000) and Nelson and Pullen 1990). They are also used for creating embankments (McLellan et al., 1990; Smith and Gailani, 2005), land improvement (Harrison and Luik, 1980; Perrier et al., 1980; Spaine et al., 1978), land creation (Coastal Zone Resources Division, 1978; capping (Palermo et al., 1998), replacement fill and share protection (Comoss et al., 2002).

The advantage of the slump test is that it allows rapid field estimation of a material's water content (with calculated void ratio, density, or other phase relationships) if a previously developed slump-water content curve is available for the given material. The slump test was originally developed for rapid estimation of mine tailing properties in Australia (University of Melbourne, 1996) by observing the height drop (slump) instead of the spread diameter. In this study, slump tests were conducted on 36 dredge material specimens collected along the coastal areas of the Gulf of Izmit (Gürbüz and Gürer, 2008). Index parameters, phase relationships and engineering behavior of the specimens were determined and some correlations were performed.

Study area

The study was performed at the coastal areas of the Gulf of Izmit, located NW Turkey (Figure 1).

Gulf sediment specimens

The dredged material specimens were collected using a small dredging operation from the study area for the purpose of characterizing their geotechnical properties. The grain size distribution graphic of the dredged material specimen collected from Station #1 is given in Figure 2 and the grain size distribution values of specimens collected from 36 stations are given in Table 1.

The slump test method

The slump test is an operation that consists of filling an upright open-ended cylinder with remolded dredged material, getting rid of the excess material at the top; slowly lifting the cylinder and measuring the variation in height (slump) as the material complete its outward flow. The slump test for concrete (ASTM 2000a) has been used for years to measure the consistency of freshly mixed concrete for quality control purposes. For this test a conical upright open ended cylinder is filled with wet concrete and tamped with a rod. After removing the excess material at the top, the cylinder is slowly lifted and the resulted variation in height (slump) of the concrete is measured. The open-ended cylinder size is 150 mm in height, with a 76 mm inside diameter.

Slump test application

Slump tests were conducted on 36 dredged material specimens using an open-ended polyvinyl chloride (PVC) cylinder of 10 cm height and diameter. *In situ* each dredge material specimen was placed into the slump cylinder, leveled off, and allow to flow outward as the cylinder was slowly lifted upward with minimum disturbance to the specimen. After the outward flow has visually stopped, the change in height (slump) as material

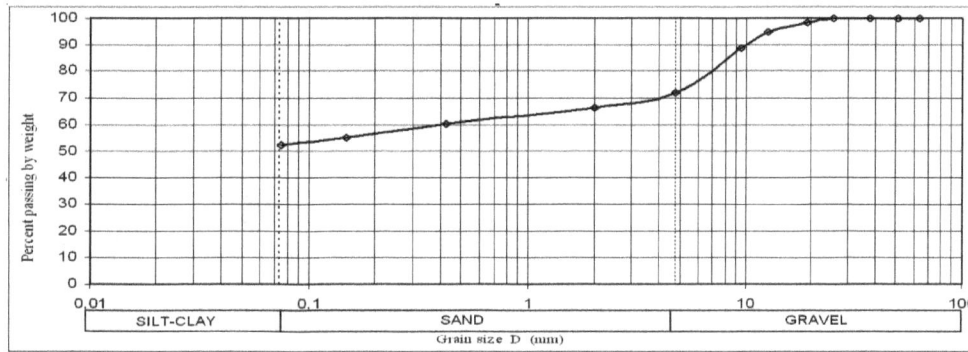

Figure 2. Grain size distribution of dredged specimen collected from Station #1.

Table 1. Grain size distribution of specimens collected from 36 stations.

Station No	Gravel	Sand	Silt+Clay
1	28.31	19.32	52.36
2	30.12	19.61	50.27
3	25.63	21.10	53.27
4	26.58	23.17	20.26
5	29.15	17.94	52.91
6	30.81	17.43	51.76
7	25.94	22.78	51.28
8	28.91	20.45	50.64
9	28.64	15.13	56.23
10	29.48	19.65	50.87
11	29.11	17.39	53.47
12	31.92	15.11	52.98
13	25.48	21.89	52.63
14	26.45	23.15	50.32
15	25.31	20.87	53.82
16	27.96	21.69	50.36
17	21.93	23.76	54.31
18	23.81	23.84	52.36
19	16.34	20.47	63.19
20	21.59	19.52	58.49
21	19.78	25.49	54.73
22	23.62	25.02	51.36
23	12.56	25.11	62.34
24	23.58	25.45	50.96
25	21.36	25.74	52.63
26	22.54	26.79	50.67
27	22.45	22.07	55.48
28	24.51	24.06	51.43
29	24.73	20.59	54.68
30	25.31	24.45	50.23
31	26.94	18.70	54.36
32	38.14	11.74	50.12
33	29.57	19.00	51.43
34	31.96	17.36	50.68
35	19.74	26.47	53.79
36	21.65	28.06	50.29

completes its outward flow is measured. The slump was divided by cylinder height to find normalized slump which is the best predictor (DOER, 2004), (Figure 3). The slump test was performed as quickly as possible to prevent thixotropic effects. The cylinder walls were pre-wetting to prevent material sticking to the walls while lifting the cylinder. The cylinder was removed in less than 7 s after leveling off the top.

Determination of water content (Wn) and bulk wet density (γ)

The water content is the ratio of the weight of water to weight of solids obtained by weighing a specimen in its natural wet state and then again upon drying at 105°C for 24 h. Bulk wet density was obtained using Equation 1.

$$Wn(\%) = 2.10^{11}\gamma^{-4.7128} \tag{1}$$

The relationship between Wn (%) and normalized slump (slump/cyl ht) is given in Figure 4. The water content and normalized slump data show scattering since the coarser-grained soils have less water content variability. The relationship between water content and bulk wet density is illustrated in Figure 5.

Determination of Atterberg limits

The Atterberg limits are a basic measure of a fine-grained soil consisting of the liquid limit (water content at which the soil passes from the liquid to plastic state), the plastic limit (water content at which the soil passes from the plastic to semi-solid state) and the shrinkage limit (water content at which the soil passes from the semi-solid to the solid state). Laboratory testing (ASTM, 2000b) is required to determine the Atterberg limits. A suitable prediction method using water content percent and normalized slump is given by the Equation 2 (DOER-D1, 2004).

$$LL = 52.74 + 0.526Wn - 59.97\ N \tag{2}$$

Figure 3. Sequences of slump test application.

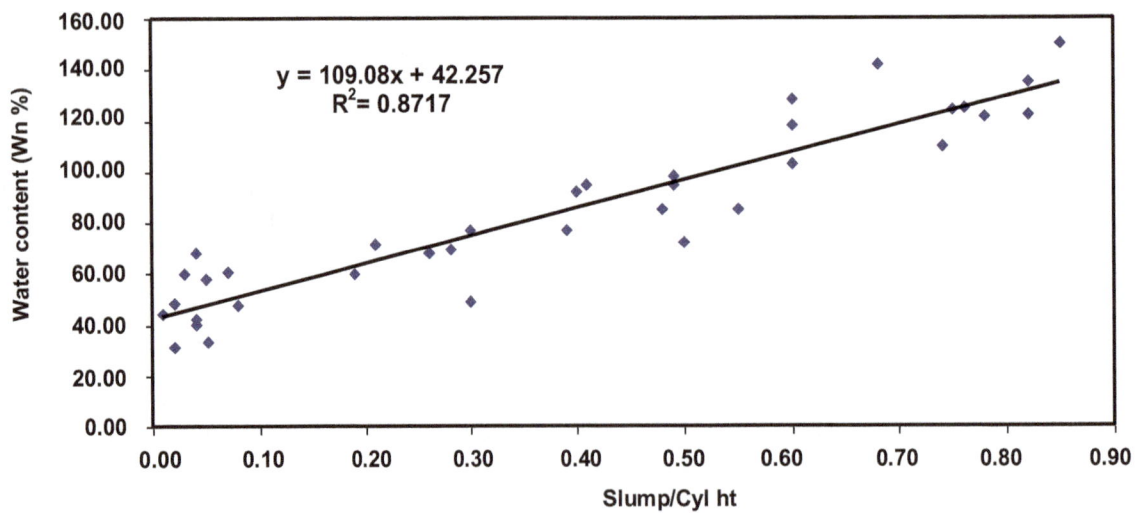

Figure 4. Relationship between water content and normalized slump.

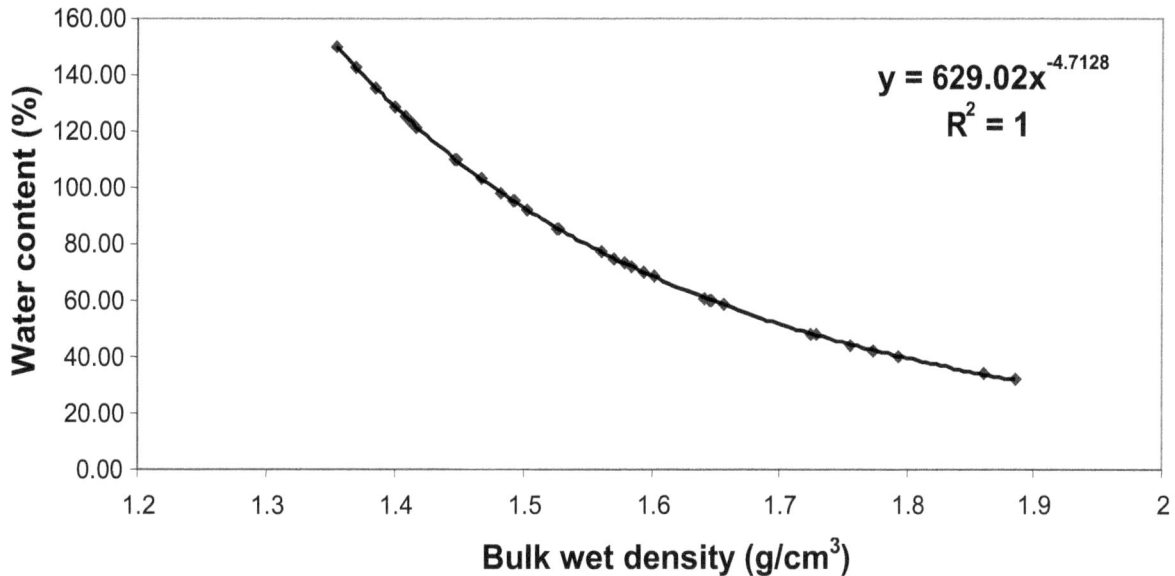

$$y = 629.02x^{-4.7128}$$
$$R^2 = 1$$

Figure 5. The relationship between water content and bulk wet density.

Where, LL: Liquid limit percent, W: Water content percent, N: Normalized slump = (slump / cylinder height). Liquidity index (LI) and Plastic limit percent (PL) are given by Equations 3 and 4.

$$LI = 1.601(Wn/LL) - 0.612 \qquad (3)$$

$$PL = [(LI)(LL) - Wn]/[(LI) - 1] \qquad (4)$$

Plasticity index (PI) is obtained from the difference of LL and PL Equation 5.

$$PI = LL - PL \qquad (5)$$

The Atterberg limits of the dredged material specimens were figured out using Equations (2 to 5). The results are given in Table 2.

Determination of bulk unit weight (saturated wet bulk density), (γ)

The Equation (6) is used to determine the bulk unit weight of the dredged material when only the water content is known (DOER-D1, 2004).

$$\gamma = 233.21\omega^{-0.2051} \qquad (6)$$

Where, γ: Bulk unit weight (N/m^3), Wn: Water content percent .

The result of bulk unit weight of dredged material specimens determined using Equation (6) is given in Table 2.

Determination of void ratio (e)

Void ratio (e) is defined as the ratio of the volume of voids to the volume of solids. If only water content is known, the void ratio can be easily determined conducting a single slump test and using Equation 7.

$$e = 0.028 \, Wn - 0.055N - 0.065 \qquad (7)$$

Where, e: Void ratio, Wn: Water content percent, N: slump / cylinder height.

The void ratio data of dredged material specimens determined using Equation (7) is given in Table 2.

Determination of percent solids

General equation for percent solids by weight is given by Equation (8) (DOER-D1, 2004).

$$\% \, solids \, by \, weight = 10000/(Wn + 100) \qquad (8)$$

Where, Wn: Water content percent.

The percent solids by weight obtained using Equation (8) are given in Table 2. The relation between % solids by weight and water content is shown in Figure 6.

Determination of specific gravity

Specific gravity is the ratio of density of a substance compared to the density of fresh water at 4°C (39°F). Specific gravity compares the mass of a given volume of material to the mass of the same volume of water and

Table 2. Physical properties of dredged material specimens collected from the coastal areas of the Gulf of Izmit.

Specimen No.	Slump/Cyl ht	Solid weight (%)	Water content (%)	Wet bulk density (N/m³)	Liquid limit (%)	Liquidity index	Plastic limit (%)	Plasticity index
1	0.02	75.94744	31.67	1.885489	68.19902	0.131466	26.14076	42.05826
2	0.04	70.28395	42.28	1.773361	72.58048	0.320624	27.98004	44.60044
3	0.07	62.14268	60.92	1.641114	80.58602	0.598296	31.62955	48.95647
4	0.08	67.71398	47.68	1.728704	73.02208	0.433378	28.29723	44.72485
5	0.40	52.04267	92.15	1.503146	77.2229	1.298472	27.21109	50.01181
6	0.41	51.25051	95.12	1.493062	78.18542	1.335769	27.75018	50.43524
7	0.49	50.45154	98.21	1.482968	75.01316	1.484088	27.09455	47.91861
8	0.30	56.43978	77.18	1.560765	75.34568	1.027977	9.780324	65.56536
9	0.60	43.73305	128.66	1.400376	84.43316	1.827618	30.99445	53.43871
10	0.03	62.41808	60.21	1.645201	82.61136	0.554864	32.28662	50.32474
11	0.01	69.30487	44.29	1.75597	75.43684	0.327969	29.08951	46.34733
12	0.02	67.43998	48.28	1.724123	76.93588	0.392684	29.75138	47.1845
13	0.55	53.8648	85.65	1.526658	64.8084	1.503862	23.44472	41.36368
14	0.49	51.18755	95.36	1.492264	73.51406	1.464764	26.5097	47.00436
15	0.50	50.47701	72.48	1.483289	62.205	1.500322	50.95218	11.25282
16	0.60	47.62585	118.45	1.447803	74.60222	1.74801	15.98296	58.61926
17	0.04	57.72339	68.38	1.578215	88.86544	0.707492	18.83166	70.03378
18	0.05	63.15125	58.35	1.656192	80.4336	0.549434	31.42054	49.01306
19	0.04	71.33685	40.18	1.792635	71.47588	0.287998	27.52109	43.95479
20	0.052	74.8111	33.67	1.861147	67.33198	0.188595	25.84592	41.48606
21	0.60	49.20049	103.25	1.467304	71.0675	1.714003	25.99419	45.07331
22	0.68	41.26093	142.36	1.37063	85.04266	2.068047	31.37712	53.66554
23	0.82	42.50436	135.27	1.385568	74.71662	2.286515	27.64887	47.06775
24	0.85	39.96483	150.22	1.355089	80.78122	2.365205	29.91794	50.86328
25	0.26	62.0232	68.35	1.639348	76.6981	0.814741	31.63624	45.06186
26	0.21	59.98081	48.22	1.609749	89.4126	0.677214	35.46784	53.94476
27	0.30	55.16328	82.33	1.543717	57.8769	15.58019	23.31408	34.56282
28	0.39	54.77951	77.12	1.538647	69.91682	1.153943	23.12561	46.79121
29	0.28	60.09254	56.32	1.61134	72.7684	0.928092	34.26943	38.49897
30	0.19	62.3558	48.33	1.644275	72.9057	0.705592	29.06953	43.83617
31	0.82	44.98021	122.32	1.415473	67.90492	2.271949	25.12405	42.78087
32	0.74	46.41233	110.29	1.432914	66.37474	2.048263	24.48139	41.89335
33	0.75	45.41739	124.12	1.420784	73.04962	2.10829	26.96926	46.08036
34	0.78	45.14061	121.53	1.41742	69.88818	2.172012	25.82564	44.06254
35	0.76	44.38132	125.32	1.408214	73.08112	2.133406	26.99093	46.09019
36	0.48	53.95489	85.34	1.527833	68.84324	1.372644	24.57378	44.26946

Specimen No	Void radio	Specific gravity	Water content / Liquid limit	Vane shear (kPa)	Effective stress (kPa)	Undrained Shear strength (Cu) kN/m2	Compression index Cc	Permeability m/sec
1	0.82066	2.591285	0.464376	0.332503	230.2969	92.86217	0.523791	0.003877
2	1.11664	2.64106	0.582526	0.251262	79.27978	38.90342	0.563224	0.012674
3	1.63691	2.686983	0.755962	0.166543	23.26455	10.84738	0.635274	0.053101
4	1.26564	2.654446	0.652953	0.212619	46.34254	23.16091	0.567199	0.020351
5	2.4932	2.705589	1.193299	0.059043	2.716764	0.433202	0.605006	0.247325
6	2.57581	2.707958	1.216595	0.05587	2.48056	0.364911	0.613669	0.277925
7	2.65793	2.706374	1.309237	0.044851	1.756347	0.184464	0.585118	0.310768
8	2.07954	2.694403	1.024345	0.088135	5.571432	1.503198	0.588111	0.131322
9	3.50448	2.72383	1.523809	0.026966	0.860068	0.037992	0.669898	0.828806
10	1.61923	2.689304	0.728834	0.177608	27.62866	13.24592	0.653502	0.051026
11	1.17457	2.651998	0.587114	0.248544	76.4072	37.61104	0.588932	0.015356
12	1.28574	2.66309	0.627536	0.225828	55.86003	27.92821	0.602423	0.0216
13	2.30295	2.688792	1.321588	0.043557	1.680454	0.168427	0.493276	0.18565
14	2.57813	2.703576	1.297167	0.046154	1.834566	0.20161	0.571627	0.27865
15	2.0075	2.676667	1.577204	0.063114	2.587881	0.395421	0.469845	0.11052

Table 2. Contd.

16	2.98116	2.710885	1.474085	0.023172	1.005341	0.054792	0.58142	0.473262
17	1.98352	2.708247	0.824167	0.16129	15.49562	6.564475	0.709789	0.109923
18	1.56605	2.68389	0.725443	0.179042	28.24155	13.58087	0.633902	0.045089
19	1.05784	2.632753	0.562148	0.263701	93.73778	45.20208	0.553283	0.010314
20	0.8749	2.598456	0.50006	0.305527	162.5673	71.40371	0.515988	0.004971
21	2.793	2.705085	1.452844	0.031907	1.076385	0.064069	0.549608	0.370797
22	3.88203	2.726911	1.673983	0.018887	0.552708	0.012573	0.675384	1.187243
23	3.67746	2.718607	1.810441	0.013666	0.382288	0.004603	0.58245	0.981706
24	4.09441	2.725609	1.859591	0.012163	0.337026	0.003205	0.637031	1.430452
25	1.8378	2.688808	0.891156	0.120866	10.72872	4.008339	0.600283	0.081043
26	1.93835	2.692153	0.805256	0.148171	17.28368	7.545382	0.714713	0.098779
27	4.1185	2.745667	0.850426	0.133122	0.070703	1.46E-05	0.430892	2.391568
28	2.07291	2.687902	1.103025	0.073135	3.933439	0.842157	0.539251	0.127336
29	1.8796	2.685143	0.961956	0.102187	7.487823	2.379808	0.564916	0.086991
30	1.60455	2.67425	0.822981	0.142073	15.60096	6.622081	0.566151	0.049257
31	3.31486	2.70999	1.801342	0.013964	0.391457	0.004922	0.521144	0.680927
32	2.98242	2.704162	1.661626	0.019449	0.572314	0.01377	0.507373	0.468257
33	3.36911	2.714397	1.699119	0.017794	0.515283	0.010448	0.567447	0.721164
34	3.29494	2.711215	1.738921	0.016192	0.462103	0.007794	0.538994	0.666602
35	3.40216	2.714778	1.714807	0.017145	0.493479	0.009308	0.56773	0.746438
36	2.29812	2.692899	1.239628	0.0529	2.271074	0.307983	0.529589	0.184358

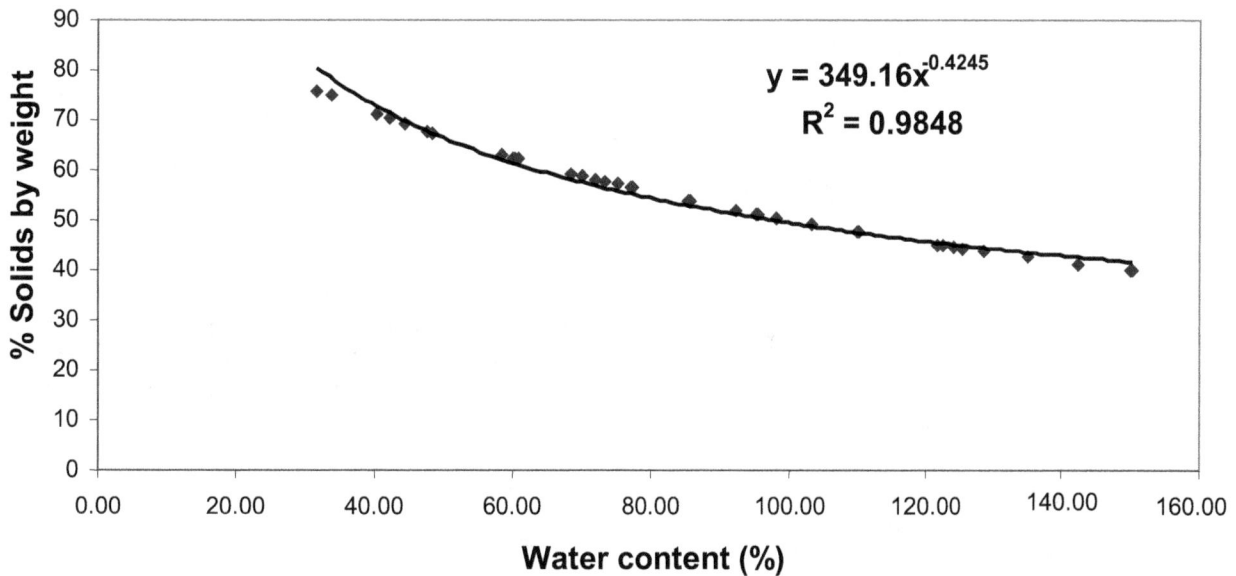

Figure 6. Relationship between % solids by weight and water content.

helps determine types of the minerals. Specific gravity is calculated from Equation 9.

$$Gs = 2.8 - 5.5\,N/W - 6.5/W \qquad (9)$$

Where, Gs: Specific gravity of solids, W: Water content percent, N: Normalized slump=slump/cylinder height. The specific gravity values are given in Figure 2.

Determination of undrained shear strength (c_u)

If only the water content and liquid limit of the dredged material are known, the undrained shear strength is obtained using the approximate Equation 10, (DOER-D1, 2004).

$$LI = \log(170/c_u)/2 \qquad (10)$$

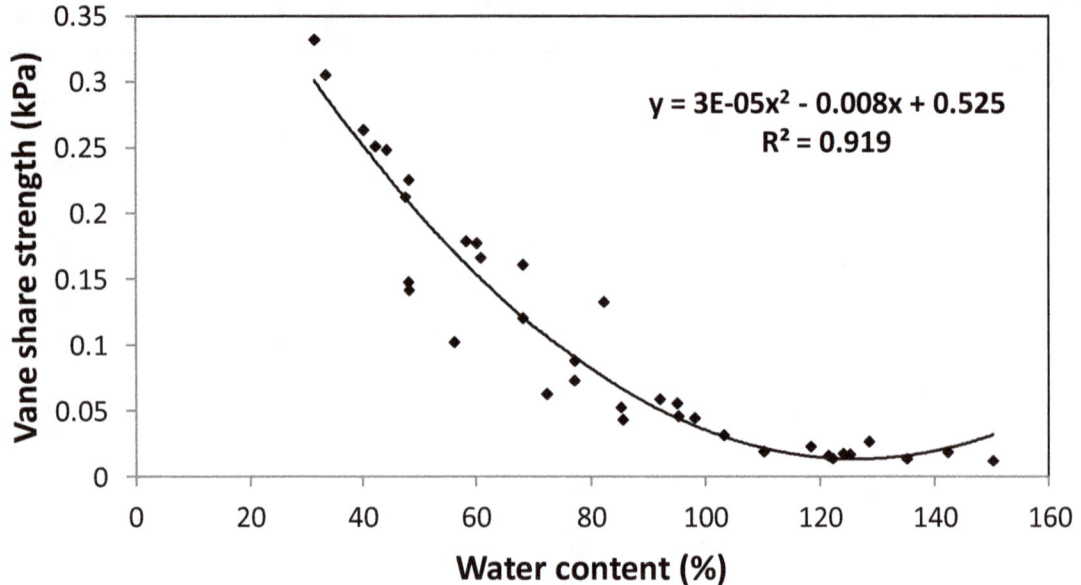

Figure 7. Relation between Vane shear strength and water content (%) / LL.

$$C_u = 170e^{-4.6LI} \qquad (11)$$

Where LI: Liquidity bindex, c_u: Undrained shear strength

$$e = 2.178$$

The soil compression index Cc is given by Equation 12 (Terzaghi and Pack, 1967).

$$Cc = 0.009(LL - 10) \qquad (12)$$

Where Cc= Compression index, LL= Liquid limit percent.

Determination of Vane shear strength

The Vane test is one of the most widely used techniques to estimate the undrained strength of soft clays. It provides an indication of *in-situ* undrained shear strength of fine - grained clays and silts or other fine geomaterials such as mine tailings, organic muck, and substances where undrained strength determination is required. It is a cheaper and quicker method. Vane shear strength is given by Equation 13.

$$Vane\ shear\ strength = 183e^{-2.3714(Wn/LL)} \qquad (13)$$

Where, e: 2.718, Wn/LL: Water content percent / Liquid limit percent.

The Vane shear strength values determined are given in Table 2, and the relation between Vane shear strength and water content (%) / LL is demonstrated in Figure 7.

As depicted in Figure 7, the correlation between Vane shear strength and water content (%) / LL is very good.

Determination of effective stress (σ')

The effective stress is a combination of both the externally applied stresses and the internal pressure of fluid phase(s) and enables the conversion of a multiphase porous medium into mechanically equivalent single-phase continuum. Soil settlement models are developed base on effective stress concept where settlement is always associated with effective stress increase (Terzaghi, 1943; Janbu et al., 1956; Schertmann et al., 1978). If only the water content percent and liquid limit are known, then the effective stress is calculated by Equation 14 (DOER-D1, 2004).

$$\sigma' = 129.77(Wn/LL)^{-4.7044} \qquad (14)$$

Where, σ': Effective stress (kPa), Wn/LL: Water content percent / liquid limit percent.

The calculated effective stress values are given in Table 2, and the relation between effective stress and water content / liquid limit is shown in Figure 8.

Determination of permeability (k)

Permeability is determined by Equation 15.

$$k = 0.0174\ [e - 0.027\ (PL - 0.242PI)\ /\ PI]^{4.29}\ /\ (1 + e) \qquad (15)$$

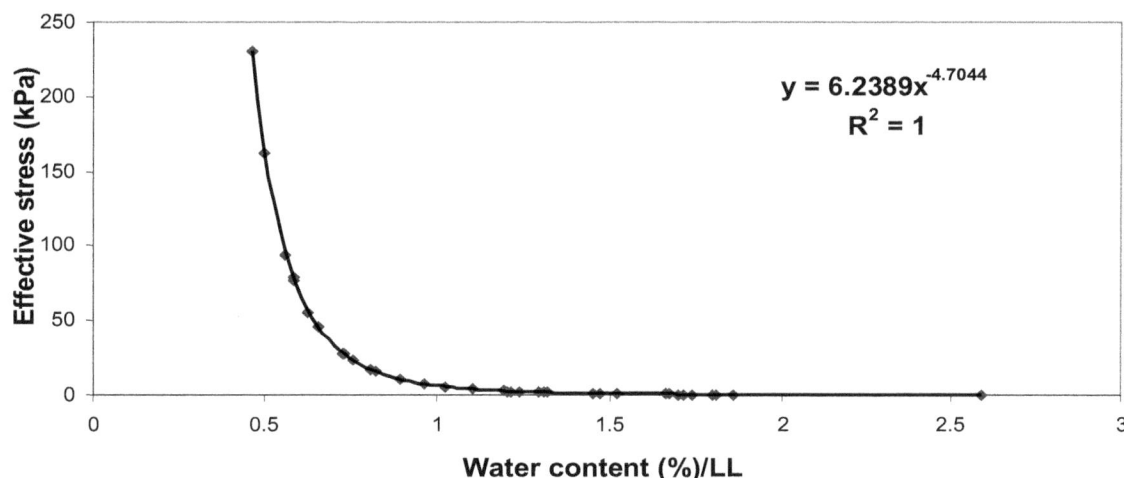

Figure 8. Relation between effective stress and water content / liquid limit.

Where, *k*: permeability, *e*: Void ratio, *PL*: Plastic limit percent, *PI*: Plasticity index.

DISCUSSION

The physical and engineering properties of the dredged materials collected at the coastal areas of the Gulf of Izmit were predicted using correlation equations given by ERDC TN-DOER (2004) in order to figure out their usability in engineering planning and construction aspect. These properties mostly depend upon the site geology. The dredged materials consist of mostly clay and silt which have plasticity index values change in a large range indicating medium to very high plasticity. They have relatively high water content and void ratio but very low shear strength indicating that they are very soft in consistency. These results show similarities with the results obtained by ERDC TN-DOER (2004) and Klein and Sarsby (2000). The water content and bulk wet density obtained by ERDC TN-DOER (2004) are higher than our values. This study is the first study on the marine dredged materials at the coastal areas of the Gulf of Izmit. The results obtained in this study can be utilized by the researchers and the people who are interested in dredged materials.

CONCLUSIONS

In order to determine the physical and geotechnical properties of dredged materials of the Gulf of Izmit, a combination of slump tests were conducted *in-situ* and water content of the dredged material specimens were obtained in the laboratory using the oven method. The other properties such as bulk wet density, Atterberg limits, bulk unit weight, void ratio, percent solids, shear

strength and effective stress were calculated using the formulae given by Doer Technical Note (ERDC TN-DOER-D1, 2004).

Sieve analysis indicated that the dredged material had 12.56 to 30.12% gravel, 11.74 to 28.06% sand and 20.26 to 63.19% clay and silt. Atterberg limit tests demonstrated that the dredged material had liquid limit ranging from 57.86 to 89.41, plastic limit varying from 9.78 to 50.95 and plasticity index ranging from 11.25 to 70.03. The specific gravity of the dredge material varied from 2.59 to 2.72. The natural water content ranged from 31.67 to 150.22% with the void ratio changed from 0.82 to 4.11. The Vane shear strength varied from 0.013 to 0.33 kg/cm^2. The effective stress ranged from 0.07 to 230.29 kg/cm^2. The dredged material was very soft in consistency and exhibit very low shear strength.

Good correlations were determined *(R^2 = 0.87)* between water content and normalized slump, % solids by weight and water content *(R^2 = 0.98)* and Vane shear strength and water content/liquid limit *(R^2 = 0.872)*. Very good correlations were obtained between water content and bulk density and effective stress and water content/liquid limit *(R^2 = 1)*. Water content increases with increasing slum/Cyht, and exponentially decreases with increasing bulk wet density. % solids by weight and Vane shear strength decrease exponentially with increasing water content. Similarly the effective stress also decreases exponentially with increasing water content/LL. The geotechnical parameters using the correlation equations are for the preliminary investigations and do not substitute for standardized laboratory testing requirements.

REFERENCES

ASTM (2000a, b). The standard American Society for Testing and Materials.

Bouziani T, Benmounah A (2013). Correlation Between V-funnel and Mini-slump Test Results with Viscosity, KSCE J. Civil Eng. DOI 10.1007/s12205-013-1569-1, 17(1):173-178.

Coastal Zone Resources Division (1978). Handbook for Terrestrial Wildlife Habitat Development on Dredged Material, Technical Report D-78-37, U.S. Army Engineer Waterways Experiment Station, Vicksburg, MS.

Comoss EJ, Kelly DA, Leslie HZ (2002). Innovative Erosion Control Involving the Beneficial Use of Dredged Material, Indigenous Vegetation and Landscaping Along the Lake Erie Shoreline, Ecological Engineering. 19:203-210.

DeMeyer CP, Mahlerbe B (1987). Optimization of maintenance dredging operations in maritime and estuarine areas, Terra et Aqua, International Journal on Public Works, Ports, and Waterways Developments, Int. Assoc. Dredging Companies. 35:25-39.

Doer Technical note collection, (ERDC TN-DOER-D1: 2004, August).

Edil TB, Wang X (2000). Shear strength and Ko of peats and organic soils, Geotechnics of high water content materials, ASTM STP 1374, T.B. Edil and P.J. Fox, eds., American Society for Testing and Materials, West Conshohocken, PA. 209-225.

Francingues NR, McLellan TN, Hopman RJ, Vann RG, Woodward TD (2000). Innovations in dredging technology: equipment, operations and management, DOER Technical Notes Collection (ERDC TN-DOER-T1), U.S. Army Research and Development Center.

Gürbüz A, Gürer ÖF (2008). Tectonic Geomorphology of the North Anatolian Fault Zone in the Lake Sapanca Basin (Eastern Marmara Region, Turkey). Geosci. J. 12 (3): 215-225.

Harrison W, Luik A (1980). Suitability of Dredged Material for Reclamation of Surface-mined Land, Ottawa, Illinois, Demonstration Project, Technical Report EL-80-7, U.S. Army Engineer Waterways Experiment Station, Vicksburg, MS.

Janbu N, Bjerrum L, Kjaernsli B (1956) "Veiledring ved losning av fundermenteringsoppgaver." Norwegian Geotechnical Institute Publication. (16), Oslo.

Klein A, Sarsby RW (2000). Problems in defining the geotechnical behavior of wastewater sludges, Geotechnics of high water content materials, ASTM STP 1374, T.B. Edil and P.J. Fox, eds., American Society for Testing and Materials, West Conshohocken, PA, 74-87.

Lee LT (2004). Predicting geotechnical parameters for dredged materials using the slump test method and index property correlation, DOER Technical Notes Collection (ERDC TN-DOER-D-X), U.S. Army Engineer Research and Development Center, Vicksburg, Mississippi.

Mclellan TN, Kraus NC, Burke CE (1990). Interim Design for Nearshore Berm Construction, Dredging Research Technical Notes, U. S. (DRP-5-02) Army Engineer Waterways Experiment Station, Vicksburg, MS.

Mitchell JK (1993). Fundamentals of soil behavior, 2nd ed., John Wiley & Sons, Inc., New York.

Nelson DA, Pullen EJ (1990). Environmental Considerations in Using Beach Nourishment for Dredged Material Placement, 113-128. In: R. L. Lazor and R. Medina (eds.) Beneficial Uses of Dredged Material. Technical Report D-90-3, U.S. Army Engineer Research Waterways Experiment Station, Vicksburg, MS.

Palermo MR, Clausner JE, Rollings MP, Williams GL, Myers TE, Fredette TJ, Randall RE (1998). Guidance for Subaqueous Dredged Material Capping, DOER Technical Report (DOER-1), U.S. Army Engineer Research and Development.

Peila D, Oggeri C, Borio L (2009). Using the Slump Test to Assess the Behavior of Conditioned Soil for EPB Tunneling, Environmental and Engineering Science, doi: 10.2113/gseegeosci.15.3.167   15(3):167-174.

Perrier E, Liopis J, Spaine P (1980). Area Strip Mine Reclamation Using Dredged Material A Field Demonstration, Technical Report EL-80-4, U.S. Army Engineer Waterways Experiment Station, Vicksburg, MS.

Schmertmann JH, Hartman JP, Brown PR (1978). "Improved strain influence factor diagrams." Proceedings, American Society of Civil Engineers, 104(GT8): 1131 – 1135.

Smith ER, Gailani JZ (2005). Nearshore Placed Mound Physical Model Experiment, DOER Technical Notes Collection (ERDC TN-DOER-D3), U.S. Army Engineer Research and Development Center, Vicksburg, MS.

Spaine PA, Liopis JL, Perrier ER (1978). Guidance for Land Improvement Using Dredged Material Synthesis Report, Technical Report DS-78-21, U.S. Army Engineer Research Waterways Experiment Station, Vicksburg, MS.

Spigolon SJ (1993). Geotechnical factors in the dredgeability of sediments, Reports 1 and 2, Contract Report DRP-93-3, U.S. Army Waterways Experiment Station, Vicksburg, MS.

Terzaghi K (1943) Theoretical soil mechanics. New York, Wiley Publications

Terzaghi K, Peck RB (1967). Soil Mechanics in Engineering Practice. A Wiley International Edition, P. 729.

Vinai R, Oggeri C, Peila D (2008). Soil conditioning of sand for EPB applications: A laboratory research. Tunnelling and Underground Space Technol. 23(3):308-317.

18

Measurement, prediction and modeling the impact of vibration as the possibility of protection cultural heritage objects

Sebastian Toplak, Andrej Ivanic, Primoz Jelusic and Samo Lubej*

Faculty of Civil Engineering, University of Maribor, Smetanova 17, 2000 Maribor, Slovenia.

This article describes the possibility of using prognostic equations and the theory of fuzzy logic to predict the intensity of vibrations resulting from the use of construction machinery and heavy traffic. Vibrations from construction machinery can be particularly dangerous if they affect heritage buildings. By analyzing the measured results of vibrations at various facilities, we found that using the theory of fuzzy logic and appropriate modeling we can well predict the intensity of vibration caused by heavy traffic and vibration compaction with vibratory rollers.

Key words: Fuzzy logic, ground vibrations, vibrations monitoring, vibrations prognosis, neural networks.

INTRODUCTION

Experts in sustainable planning of roads are facing the problem of how to prepare their proposals for the authorities so that they will be able to support their decisions with numbered facts. A typical example of such a problem is erecting traffic infrastructure in locations where only part of factors can be shown quantitatively and costs estimated accordingly (mainly construction and maintenance costs). Contrary to this, we tend to show numerous consequences, both costs and benefits, qualitatively, which means descriptively, or we completely ignore the obvious consequences of an intervention in space which may cause social harm or benefits, damage to cultural heritage buildings or create the need of displacement. We are not able to adequately quantify costs and benefits not estimate costs. Another problem of sustainable road design is a deficient judgment of benefits and consequences of road's total life cycle from

"ideas" and "construction" to "preservation" and "degradation".

In this paper we focused on and analyzed two very common phenomena occurring during the construction and operation of roads and ancillary facilities: vibrations arising from construction machinery and vibrations arising from freight traffic. The vast majority of construction machinery used in earthworks produces harmful vibrations. Many earthworks, such as piling and vibratory compaction of materials cause vibrations that can be transmitted in soil to nearby facilities. Because of vibrations, generated dynamic forces can cause damage to nearby structures. Old buildings are the most vulnerable of all structures.

An adaptive network fuzzy inference system (ANFIS) is used for predicting the intensity of vibration. ANFIS is considered to be one of the intelligent tools to understand the complex problems (Jang, 1993). Therefore, ANFIS is being successfully used in many industrial areas as well as in research (Faravelli and Yao, 1996; Provenzano et al., 2004; Gokceoglu et al., 2004; Rangel et al., 2005; Kayadelen et al., 2009). Khandelwal (2012) predict the blast induced ground vibration using different conventional vibrations predictors and artificial neural network (ANN) at a surface of coal mine and it was found out that the ANN model based on multiple input parameters have better prediction capability than conventional vibration predictors.

PHYSICAL CHARACTERISTICS OF CONSTRUCTION MACHINERY AND FREIGHT TRAFFIC

Slovenia has no standards of its own in the field of vibration measurements. By joining the European Union, we could take over the well established European standards in this field, such as DIN 4150, Swiss Standard SN 640 312a, British Standard (BS) 7385 and BS 6472. In Europe the standard DIN 4150 is most commonly used for measuring and assessing effects of vibrations on buildings. This standard prescribes the maximum ground vibration velocity which is 3 mm/s for heritage buildings, 5 mm/s for residential buildings and 10 mm/s for industrial facilities. The maximum ground vibration velocity is given with special curves depending on ground vibration frequency.

Many earthworks, such as driving pilots and sheet piling, vibratory compaction of earth materials as well as operation of heavy construction equipment, cause vibrations that can be transmitted to nearby buildings. Dynamic forces generated by these vibrations can cause damage to the buildings. When designing and planning activities at the site we need to assess possible effects of vibrations, and adjust work of vibratory machines so as to minimize their effects on the surrounding buildings (Achmus and Kaiser, 2006). The effects of vibrations caused by construction machinery may vary depending on numerous factors, such as the intensity of sources of vibration, different soil composition and its quality between the sources of vibration and the building, the quality of foundations, building dimensions and the quality of built-in materials. The effects of vibrations are enhanced by the intensity and duration of vibrations and by the frequency and number of vibration events. The effects of vibrations caused by construction machinery may interfere with users of a building and damage the building, since there is shaking and moving that can change structural integrity of the building to such an extent that its stability is at risk. In vibratory compaction of loose soil layers we dynamically compact upper soil layers. In these construction processes vibrations are transmitted through the soil to neighbouring buildings causing damage to them. When planning construction, possible impacts of vibrations should thus be assessed

and resulting risks and building machines selected so that their impact on surrounding buildings is prevented or at least minimized.

Since 2002 the Slovene statistical office has been performing research about environmental costs arising from different environmental purposes in accordance with the European statistical classification of activities relating to the protection of the environment (CEPA). Data are being collected on investment in environmental protection, current expenses for the environment and income from environmental protection activities. CEPA is a general, multi-purpose and functional classification which is used for classifying the activities for environmental protection. According to this classification the protection against noise and vibrations is also recorded. This protection covers the reduction of noise and vibrations caused by road and rail transport, as well as air transport and shipping. According to the above criteria, activities are foreseen, such as monitoring, traffic management, and erection of sound barriers or anti-vibration devices (Statistical Office, 2009). Traffic vibrations are common concerns of society because they often cause problems to people and structures. They constitute an external source and result from heavy traffic such as buses and trucks. Passenger cars and light trucks rarely cause vibrations that are discernible in buildings. Road transport usually causes vibrations in the frequency range from 5 to 25 Hz and ground vibration from 0.05 to 25 mm/s (Hunaidi, 2000). The frequencies and the vibration velocity depend on numerous factors, such as pavement conditions (especially damage and roughness), the speed and weight of vehicles, a vehicle suspension system, the type of soil, the season, the distance between the road and the building and the type of the building.

MEASUREMENTS, DESCRIPTION OF THE EXPERIMENT AND THE RESULTS

Vibration measurements were performed separately for three types of vibrations. We have measured the impact of the vibrating roller on the residential and on heritage object, where motorway was under construction. The second segment of monitoring comprised the measuring of effects at the residential building, during driving and pulling of sheet piling when the round about and the access to a new motorway was under construction. In the third case we measured the impact of heavy traffic on the residential building.

Earthquake intensity, which results from operation of construction machines or traffic, can be measured with seismographs. Seismographs are portable devices that can be placed wherever it is necessary to measure the intensity of vibrations. They measure vibrations which are transmitted to the seismic mass of a geophone and then to three perpendicularly mounted electromagnetic coils.

Table 1. Technical characteristics of dynamic rollers.

Variables	HAMM 3520	AMMAN AC 110
Operating weight (kg)	12480	12100
Rear axle load (kg)	7320	7140
Vibration frequency (Hz)/amplitude (mm)	30/1,19	28-35/1,8-0,8

Vibration components induce voltage and hence electrical impulses in the winding. These impulses are transmitted to the electronic recording with a built-in microprocessor with a certain memory. Measurement values recorded in the memory can be processed analytically with the software. For vibration measurements, we used the measurement equipment of the manufacturer Instanel from Canada, namely the four-channel device Minimate Plus and the eight-channel device Minimate Plus with the associated linear microphone and the geophones. The impact of construction machinery, whose operation causes vibrations, and the impact of road freight transport are also measured by ground vibration velocity. Geophones can be used for these measurements as well as for assessing a building response to vibrations. Ground vibration velocity is usually measured at the source of vibration – on the ground in the immediate vicinity of a construction machine, on the ground in front of the foundation of the building observed, and at the foundation of the building. The results of measurements present the measured components of vibration velocity in all three orthogonal directions.

Effects of vibration from trucks

Heavy vehicle traffic was monitored in two cases during earthworks for the construction of the parking house. We measured vibrations on the gravel road caused by trucks with a total weight of 20 tons in the phase of driving off, and vibrations on the asphalt road caused by trucks with a total weight of 20 tons and running at a speed of 40 km/h.

Effects of vibration from a vibrating roller

The constructor of the motorway section used two types of construction machinery which causes vibrations. In compaction of road section the dynamic roller HAMM 3520 was used. The second dynamic roller was AMMAN AC 110. The following technical characteristics are presented in Table 1. We have measured the effects of vibrations during the operation of one vibrating roller and two synchronously operating rollers HAMM. The second measuring with vibrating roller AMMAN has ben measured the imapct of surface dynamic compaction,

deep dynamic compaction and static compaction.The measuring instruments – Instantel Minimate Plus measuring station were activated manually, with a time interval between individual measurements of 10 s.

Effects of vibration from driving/pulling sheet pilling

The measurement of vibrations during the construction of new bridge in the old kernel of town was performed at 17 measuring points. Vibrations were measured at facilities, which are protected cultural heritage structures. The second part of measurements was performed on the residential building during the construction of a road roundabout. In both cases the sheet piling was used to protect the excavation or construction pit, respectively.

Prognosis of vibration

The prognosis of vibration based on vibration measurements, which were measured and the use of empirical equations by various authors. The use of prognostic equations as an option for reducing the negative effects of vibration.

Prognosis of vibration velocity from trucks

The measurements were analyzed with a model for predicting the intensity of vibration proposed by Watts (1990). The presented model is based on local degradation of the roadway surface, along which vehicles run with a certain speed, and the distance between the moving vehicle and the measuring point. The model is expressed as:

$$PPV = 0.028 \cdot a \cdot (v/48) \cdot t \cdot p \cdot (r/6)^x. \tag{1}$$

Where: PPV = the peak particle velocity (mm/s), a = the maximum degradation of a surface or a defect (mm), v = the measured speed of a vehicle (km/h), t = the coefficient of soil supporting a roadway structure, p = the wheel index, which is over 0.75 for heavy vehicles when one wheel crosses a damaged spot, or 1 in other cases, r = the distance between the measuring point and the moving vehicle.

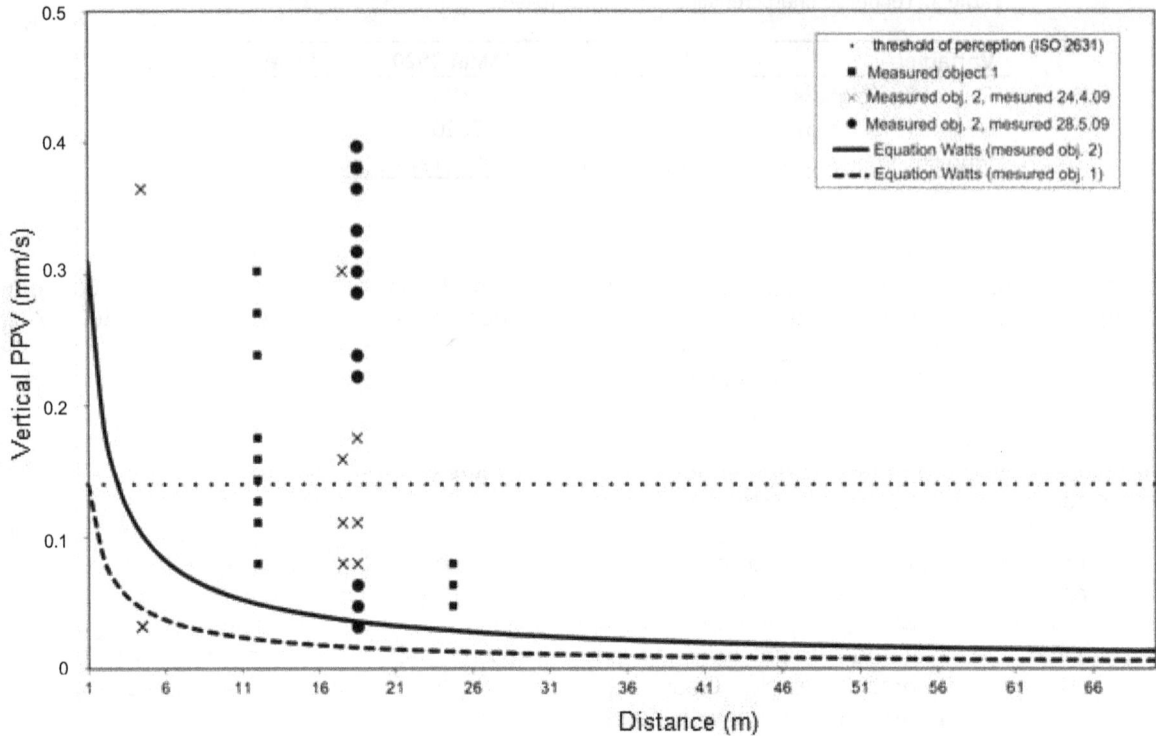

Figure 1. The results of the measured values using the prognostic equation for heavy trucks.

The value of the exponent x determines damping of vibrations and it depends on the site and the distance. The results of the effect of vibrations caused by heavy trucks are given in Figure 1.

Prognosis of vibration velocity from a a vibratory roller

To forecast the potential damage to structures, foundation oscillation velocity, not ground vibration velocity, is used as the base value. Maximum oscillation velocity changes at the transfer to foundations, mainly resulting in the reduction. In the case of resonance, there is a very slight, small increase in oscillation velocity. Due to unknown factors of the transfer of oscillations to the foundations of a building, it is useful to know direct equations to forecast maximal components of foundation oscillation velocity. This is practical because in measuring vibrations measuring devices are placed on foundations and not on the ground, which allows easy calibration of the equations. Such prognostic equations are mainly based on practical experience and are not yet widespread. Since vibratory energy of vibration machines is difficult to assess, a decisive parameter in these prognostic equations is the operating weight of the machine. To predict foundation oscillation velocity arising from vibration rolling we used two equations. The first Equation (2) is suggested by Philipps et al. (2010), and

the second Equation (3) by Achmus et al. (2005).

$$v_{Fi,\max} = 1,1 \frac{\sqrt{G}}{r^{0,7}} \tag{2}$$

$$v_{Fi,\max} = K \frac{\sqrt{G}}{r} \tag{3}$$

Where: $v_{Fi,\max}$ is the maximum foundation oscillation velocity (mm/s), G is the operating weight of the vibrating machine (t) and r is the distance from the vibration source to the foundation (m).

The coefficient K in Equation (2) is 4.31 for a 50% accuracy of results. We used the results of vertical and longitudinal components of foundation oscillation velocity, which are graphically presented in Figure 2, together with the lines of the prognostic Equations (2) and (3) and the relevant literature data (Achmus et al., 2005).

Prognosis of vibration velocity from driving/pulling sheet pilling

Ground and foundation vibration velocity is also taken as the basic physical quantity to predict the potential damage on buildings. In the transfer of vibrations from the ground to the foundations of the building the vibration velocity tends to decrease, yet it can increase in some

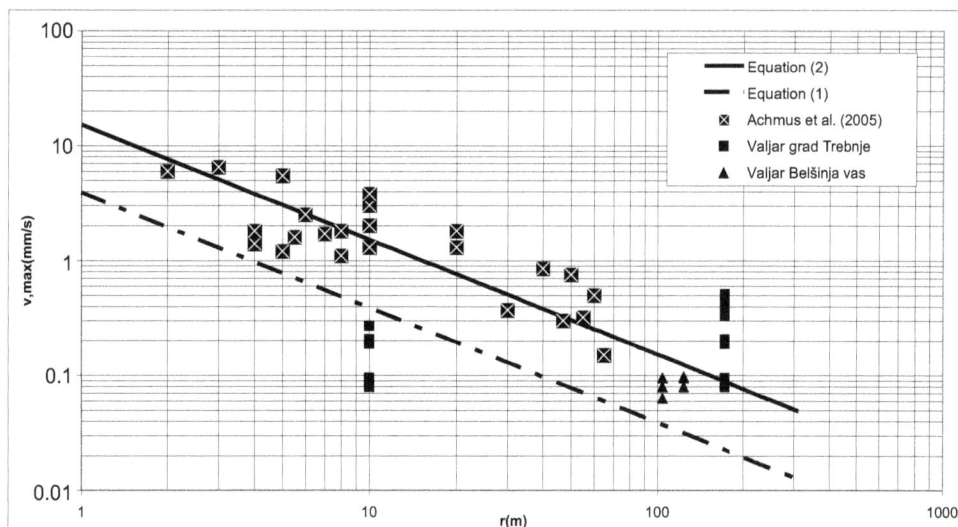

Figure 2. The measured versus predicted values of foundation oscillation velocity.

cases owing to resonance. Especially in pulling sheet piling, sudden changes can occur in vibration velocity in three orthogonal directions. Because of this it is very important to consider the highest measured value in forecasting and evaluating the effect of vibrations. In this case, we used the equation of Achmus et al. (2005), which reads:

$$v_{Fi,max} = K_F \frac{\sqrt{\dfrac{W}{f}}}{r}. \qquad (4)$$

Where: $u_{Fi,max}$ = the maximum foundation oscillation velocity (mm/s), W = the maximum power of the machine in watts, f = the operating frequency of the machine (s-1) and r = the distance from the source of vibration to the measuring point in (m).

The coefficient K_F is 7.9 for a 50% probability of increase, or 18.5 respectively for a 2.25% probability of increase. In analyzing the measured values given in Figure 3, we considered that in driving 7.1 kNm of energy is consumed, while in pulling 2.8 kNm. Vibration frequency is 28 Hz in both cases.

ANFIS MODEL FOR GROUND VIBRATION PREDICTION

The basic structure of the fuzzy inference system (FIS) was introduced by Zadeh (1965). In this type of FIS it is essentially to predetermine the rule structure and the membership functions. Human determined membership functions are subjective and differentiate from person to person. The standard methods, which transform human

knowledge or experience into the fuzzy rules and membership functions, do not exist. Usually there is the collection of input/output data, which we would like to use for constructing the FIS model. The effective method for tuning the membership functions and to minimize the output error measure is the Adaptive-Network-Based Fuzzy Inference System (ANFIS). ANFIS (Jang, 1993) uses a given input/output data to construct a FIS, whose membership function parameters are tuned (adjusted) using either a back propagation algorithm alone or in combination with a least squares type of method. This adjustment allows fuzzy systems to learn from the data they are modelling. ANFIS only supports Sugeno-Takagi-Kang (1985) identification models, which should have only one output parameter. Adaptive network is a superset of all kinds of feed-forward neural networks with supervised learning capability (Rumelhart, 1986). ANFIS is a fuzzy inference system implemented in the framework of adaptive networks and uses the advantages of neural networks and fuzzy logic.

One of the most important stages in the ANFIS technique is data collection. The data was divided into training and checking datasets. Training datasets contains measurements of the vibration caused by trucks and vibratory roller. As an interface for mathematical modelling and data inputs/outputs the MATLAB (2010), a high-level technical computing language, was used.

Vibration caused by trucks

In this case, data is collected using noisy measurements, and the training data cannot be representative of all the features of the data that will be presented to the model. In this model, 15 measurements were used to build a model. Among which, 11 evaluations (70%) were used

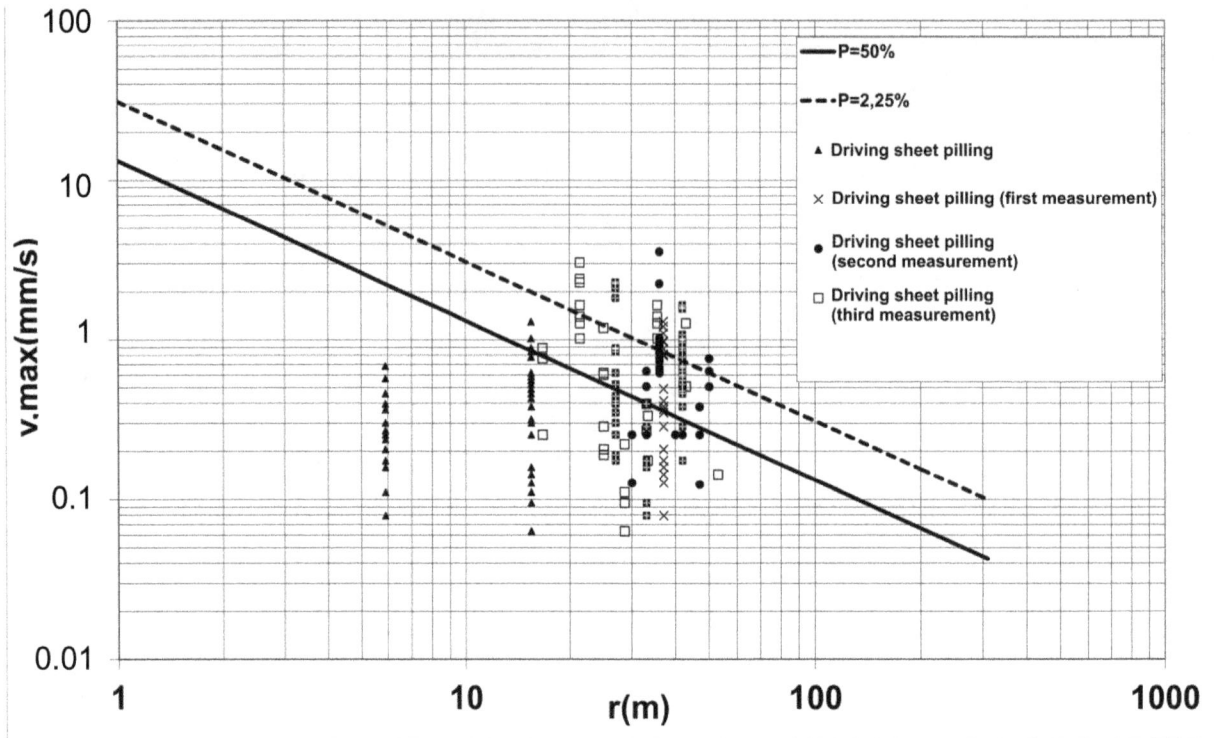

Figure 3. The measured values using the prognostic equation for driving/pulling sheet piling.

Table 2. Training data.

No. measurements	V (km/h)	r (m)	PPV (mm/s)
1	86	7	1.02
2	86	67	0.087
3	71	7	0.766
4	71	67	0.087
5	69	7	0.658
6	69	67	0.092
7	68	67	0.096
8	67	67	0.010
9	65	7	0.809
10	65	67	0.125
11	57	7	0.457

for the training of the ANFIS model, whereas 4 data sets (30%) were chosen for checking the model. Table 2 represents training data and Table 3 represents checking data.

PPV is predicted using ANFIS model based on two parameters (speed of a vehicle and the distance between the measuring point and the moving vehicle). The maximum degradation of a surface is 0.5 mm and the coefficient of soil supporting a roadway structure is 0.1. On the basis of the measured data we build model to predict PPV value for any speed of a vehicle and the

distance between the measuring point and the moving vehicle (Figure 4).

ANFIS use a given input/output data to constructs a FIS whose membership function parameters are tuned (adjusted) using either a back propagation algorithm alone or in combination with a least squares type of method. This adjustment allows a fuzzy systems to learn from the data they are modeling. The hybrid algorithm is described in detail (Jang, 1993). The hybrid algorithm can reduce the error between FIS model and the data that we used to build and verify the model. The aim of ANFIS

Table 3. Checking data.

No. of measurements	V (km/h)	r (m)	PPV (mm/s)
1	70	7	0.502
2	70	67	0.094
3	67	7	0.851
4	67	67	0.103

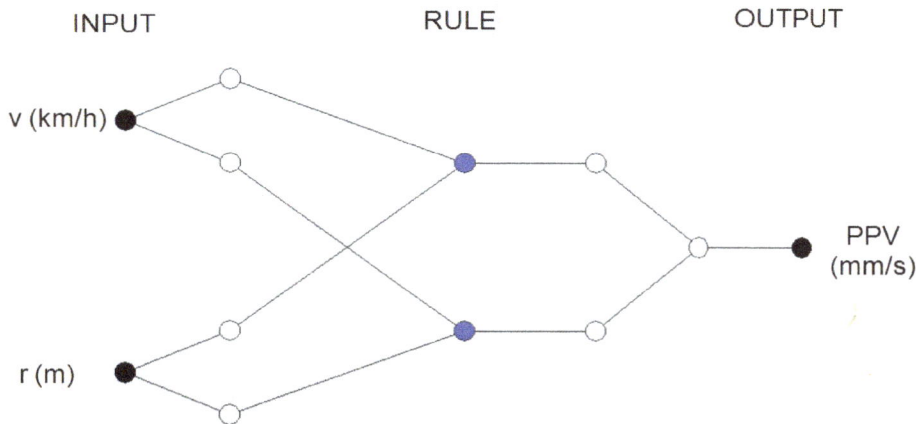

Figure 4. Suggested ANFIS for prediction of ground vibration caused by trucks.

Table 4. The ANFIS structure.

No. of training data sets	11
No. of checking data sets	4
Type	sugeno
No. of input membership functions	2
No. of output membership functions	2
No. of rules	2
And method	prod
Or method	probor
Defuzzification method	wtaver
Implication method	prod
Aggregation method	max

method is to minimize the root mean square error (RMSE) of the model to given attributes. Optimal parameters of the model were achieved when the RMSE is no longer decreasing. ANFIS structure for ground vibration prediction is summarized in Table 4.

Results of ANFIS model - vibration caused by trucks

Results of ANFIS model are shown in Figure 5. Surface shows the influence of speed of a vehicle and the distance between the measuring point and the moving

vehicle on the peak particle velocity. ANFIS method is alternative to existing methods for prediction of ground vibration due to heavy vehicle traffic. However, results need to be generalized as present work is valid only for considered data. Comparison of measured and predicted PPV values for training data is shown in Figure 6.

Vibration caused by vibratory roller

The vibratory energy of machines depends on type of compaction. Tables 5 and 6 contains the measurements

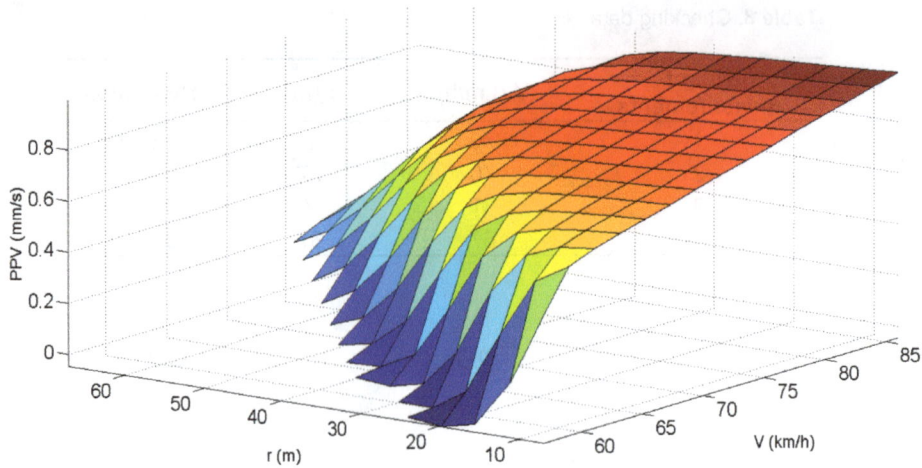

Figure 5. Result of ANFIS model.

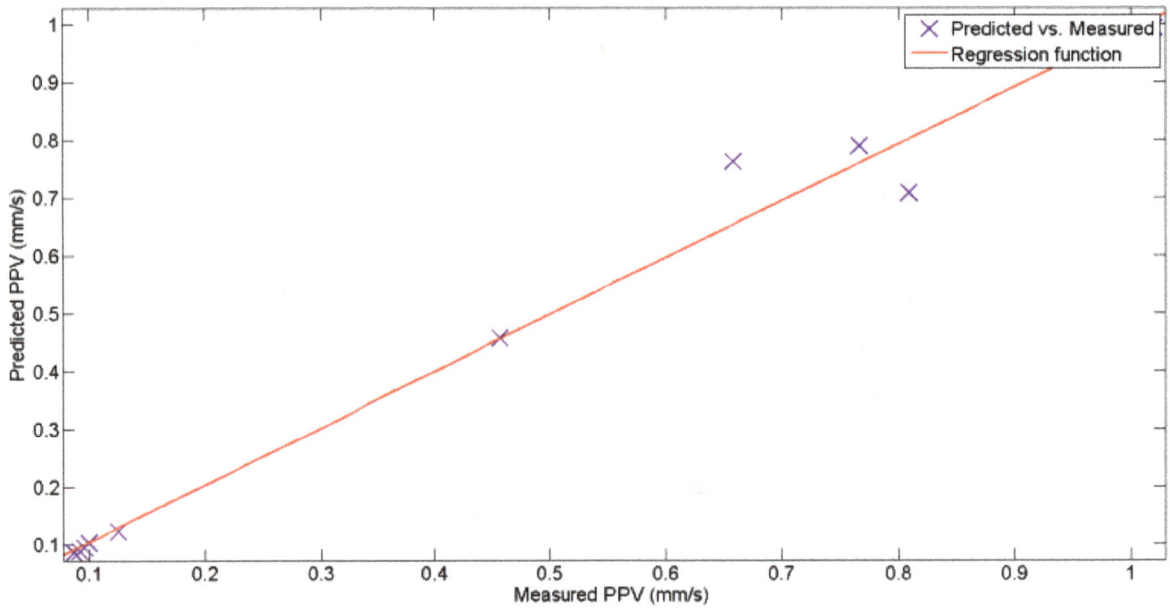

Figure 6. Measured versus predicted PPV values for training data sets.

Table 5. PPV measurement results caused by three types of compaction methods for heritage objects.

Distance (m)	PPV (mm/s) Surface dynamic compaction	PPV (mm/s) Deep dynamic compaction	PPV (mm/s) Static compaction
8.0	3.57	3.51	0.220
8.9	3.28	2.91	0.220
10.7	2.61	1.50	0.311
30.0	2.22	0.93	0.220

of PPV for surface dynamic, deep dynamic and static compaction. Additionally the measurements are obtained on two buildings with different quality of foundations. For each building, type of compaction and various distances

Table 6. PPV measurement results caused by three types of compaction methods for residential objects.

Distance (m)	PPV (mm/s) Surface dynamic compaction	PPV (mm/s) Deep dynamic compaction	PPV (mm/s) Static compaction
5.4	2.01	1.51	0.220
7.1	1.94	1.50	0.220
15.1	0.93	1.26	0.254
16.0	1.02	0.81	0.284

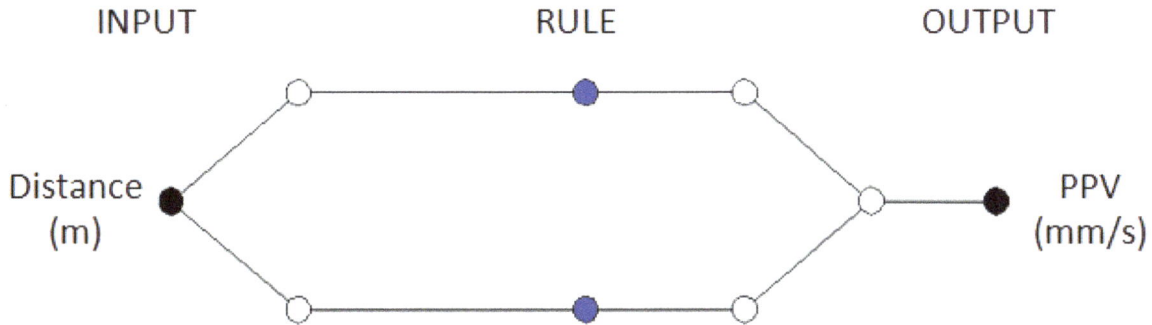

Figure 7. Constructed ANFIS models for prediction of ground vibration caused by compaction.

Table 7. PPV measurements caused by three types of compaction methods.

Number of training data sets	6 x 4
Type	Sugeno
Number of input membership functions	2, Gaussian
Number of output membership functions	2, Constant
Number of rules	2
And method	prod
Or method	probor
Defuzzification method	wtaver
Implication method	prod
Aggregation method	max

from the source of vibrations the PPV were measured. Based on the measurements the simple one parameter ANFIS models were constructed (Figure 7). The PPV value is predicted for two different types of foundations and three types of compaction, therefore six models were constructed with the same structure. Table 7 presents the input data of fuzzy models, which are essential to repeat the calculations.

Results of ANFIS model – vibration caused by vibratory roller

Figure 8 shows the vibration velocity as a function of distance from the source of vibration – based on ANFIS model. Figure 8a shows the graphs for the heritage objects which have generally poorly foundations and have no built-in anti-seismic ties. Graphs on Figure 8b are shown the same physical quantity but for residential objects, which are better quality. A graphical representation of vibration velocity of the distance from the source of vibration indicates that the vibration velocity depends on the type of hardening of soil produced with roller. From the graphs it is evident that surface dynamic compaction and deep dynamic compaction cause almost the same vibration velocity in a small distance from the source of vibration. With increasing distance from the source of vibration surface dynamic compaction leads to

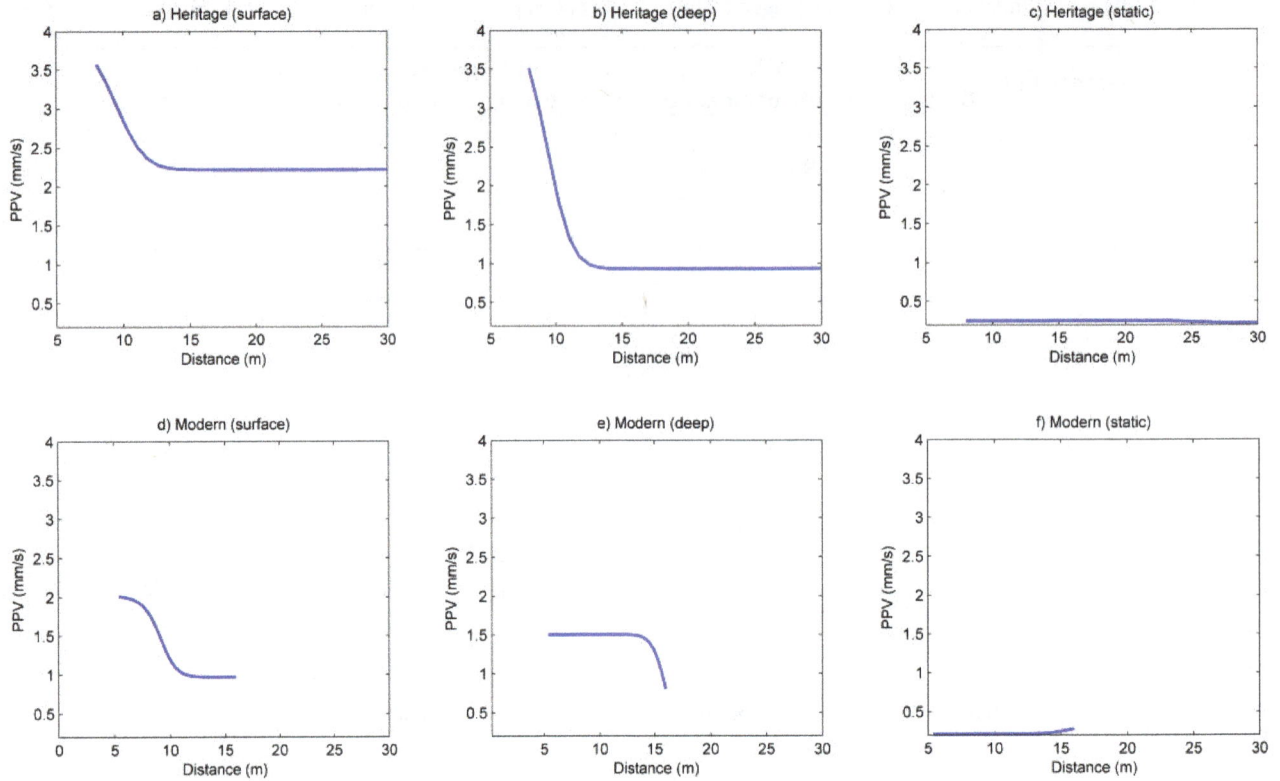

Figure 8. Results of ANFIS model – measuring values which are shown in Tables 5 and 6.

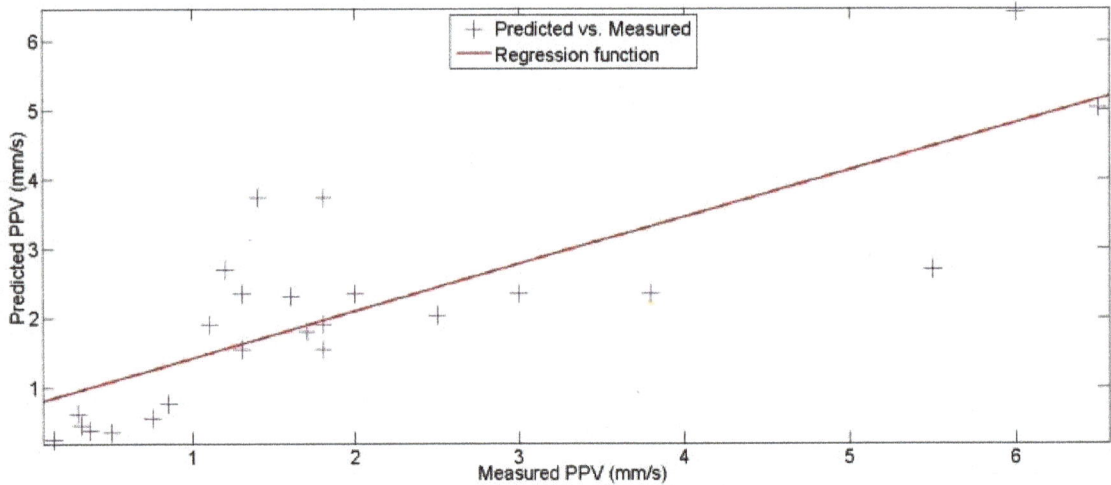

Figure 9. The measured versus predicted ground vibrations due to a road roller activity.

a higher vibration velocity when compared with the deep dynamic compaction. Vibration velocity caused by the static compaction is negligibly small when compared with the vibration velocity induced by surface and deep dynamic compaction. Calculations also confirmed by the fact that the vibration velocity and the distance of the

same intensity and the distance of the source are greater then in residential objects. Additionally, we have constructed an ANFIS, based on measurements provided by Achmus et al. (2005). With the developed model we are able to predict the PPV value for any selected distance. Figure 9 shows the measured versus predicted ground vibrations due to a road roller activity.

CONCLUSIONS AND RECOMMENDATION

The measured foundation oscillation velocities presented in Figure 1 correspond very well with the line of the prognostic Equation (3). We can conclude that the prognostic Equation (3) can be used for the preliminary assessment of foundation oscillation velocity within the scope of risk assessment. Based on the calculation with the prognostic Equation (3), the likely value of foundation oscillation velocity is 0.12 mm/s, while the DIN 4150 standard provides an orientation value of 5 mm/s for horizontal components of oscillation velocity. If this value is not exceeded, we do not expect damage to the building and can exclude the likelihood that horizontal oscillations of structural elements would occur on the top floor of the building with the increase of the oscillation amplitude. For vertical oscillation velocities of the ceiling in the building, DIN 4150 provides an orientation value of 8 mm/s. At a frequency of the vibrating roller of 30 Hz there could be a risk of resonance. However, considering that the transfer factor for wooden ceilings kz ≤ 15 (Funk, 1996), we obtain the maximum oscillation velocity of 1.8 mm/s using the prognostic Equation (4), which means that in no event the guideline value of the standard DIN 4150 is exceeded and no damage is expected to the building. According to the prognostic Equation (4), the measured values of driving, and in some cases of pulling, sheet piling correspond very well with the boundary probability line. The results for pulling sheet piling in a direct vicinity of the measuring point present a border case, a very troublesome one, which is difficult to predict due to the proximity of the source. When planning construction works it is therefore necessary to assess possible effects of vibrations and the resulting risks, and select such building machines whose operation does not affect surrounding buildings or is at least minimized. The measurement of vibrations resulting from heavy trucks has shown that the prognostic Equation (4) sets very strict criteria, which are mostly exceeded in our case.

Data scattering of vibrations is very large, therefore conventional vibration predictors are not able to predict the PPV up to an acceptable limit. Many researchers found out that artificial neural network and neuro-fuzzy technique have superiority in solving problems in which many complex parameters influence the process and results, when process and results are not fully understood. Therefore is very important that experimental data are available. The prediction of ground vibrations due to heavy vehicle traffic is also of this type. Speed of a vehicle and the distance between the measuring point and the moving vehicle are two input parameters for ANFIS model. Built model can be improved with more measurements of ground vibrations. Based on engineering judgment, the proper measurements should be selected for training and checking data sets.

Prediction of vibration caused by the dynamic compaction of soils with vibratory rollers are in good agreement with calculations by prognostic Equations (2) and (3). The elaborated method using techniques FIS and model ANFIS can be applied to assesment of vibration of buildings. The results of experiments show, that application of this methods are possible.

Conflict of Interest

The authors have not declared any conflict of interest.

REFERENCES

Achmus M, Kaiser J (2006). Prognose von Bauwerkserschütterungen infolge Ramm- und Vibrationsverdichtungarbeiten. Danube-European conf. Geotech. Eng. 2:723-728.

Achmus M, Kaiser J, Wörden FT (2005). Bauwerkserschütterungen durch Tiefbauarbeiten: Grundlagen, Messergebnisse, Prognosen. Institut für Grundbau, Bodenmechanik und Energiewasserbau, Universität Hannover, Hannover. 61:1-82.

Faravelli L, Yao T (1996). Use of adaptive networks in fuzzy control of civil structures. Computer-Aided Civ. Infrastructure Eng.11(1):67-76.

Funk K (1996). Expertsystem für Lärm- und Erschütterungsprognosen beim Einbringen von Spundbohlen. Mitteilungen des Curt-Risch-Institut für Dynamik, Schall- und Messtechnik der Universität Hannover. 2(96):1-145.

Gokceoglu C, Yesilnacar E, Sonmez H, Kayabasi AA (2004). Neuro-fuzzy model for modulus of deformation of jointed rock masses. Comput. Geotech. 31:375–383. http://dx.doi.org/10.1016/j.compgeo.2004.05.001

Hunaidi O (2000). Traffic vibrations in buildings. construction technology update No. 39, National Research Council of Canada. 39:1-9.

Jang JSR (1993). ANFIS: Adaptive-Network-Based Fuzzy Inference System. IEEE Trans. Syst. Man Cybernetics. 23(3):665-685. http://dx.doi.org/10.1109/21.256541

Kayadelen C, Gunaydin O, Fener M, Demir A, Ozvan A (2009). Modeling of the angle of shearing resistance of soils using soft computing systems. Expert Syst. Appl. 36:11814-11826. http://dx.doi.org/10.1016/j.eswa.2009.04.008

Khandelwal M (2012). Application of an expert system for assessment of blast vibration. Geotech. Geol. Eng. 30:205-217. http://dx.doi.org/10.1007/s10706-011-9463-4

MATLAB version 7.10.0. (2010). Natick, Massachusetts: The MathWorks Inc.

Philipps G, Stollhoff F, Wieck J (2010). Die vorsorgliche Beweissicherung im Bauwesen: Reihe begründet von Günter Zimmermann. Fraunhofer IRB Verlag. pp. 33-66.

Provenzano P, Ferlisi S, Musso A (2004). Interpretation of a model footing response through an adaptive neural fuzzy inference system. Comput. Geotech. 31:251-266. http://dx.doi.org/10.1016/j.compgeo.2004.03.001

Rangel JL, Iturraran-Viveros U, Ayala AG, Cervantes F (2005). Tunnel stability analysis during construction using a neuro-fuzzy system. Int. J. Numer. Anal. Methods Geomech. 29:1433–1456. http://dx.doi.org/10.1002/nag.463

Rumelhart DG (1986). Learning representations by back-propagating errors. Nature 323:533-536. http://dx.doi.org/10.1038/323533a0

Sugeno M (1985). Industrial applications of fuzzy control. Elsevier Science pub. 1-18.

Watts GR (1990). Traffic induced vibrations in buildings. Research Report 246, Transport and Road Research Laboratory, Department of Transport, UK. 246:1-31.

19

Recent astronomical tests of general relativity

Keith John Treschman

51 Granville Street Wilston 4051 Australia.

This history of experimentation relevant to general relativity covers the time post-1928. Classes of investigation are the weak equivalence principle (equivalence of inertial and gravitational mass and gravitational redshift), orbital precession of a body in gravitational fields (the relativistic perihelion advance of the planets, the relativistic periastron advance of binary pulsars, geodetic precession and Lense-Thirring effect), light propagation in gravitational fields (gravitational optical light deflection, gravitational radio deflection due to the Sun, gravitational lensing, time dilation and atomic clocks) and strong gravity implications (Nordtved effect and potential gravitational waves). The results of experiments are analysed to conclude to what extent they support general relativity. A number of questions are then answered: (a) how much evidence exists to support general relativity, (b) is it a reasonable way of thinking and (c) what is the niche it may occupy?

Key words: general relativity, equivalence principle, orbital precession, gravitational fields.

INTRODUCTION

The special theory of relativity came from the mind of Albert Einstein (1879-1955) in 1905 (Einstein, 1905). In it he proposed that the laws of physics take the same form in all inertial frames and that the velocity of light is constant irrespective of the motion of the emitting body. Previously, Isaac Newton (1642-1727) had supplied the term inertial mass when treating his three laws of motion and gravitational mass in the context of his universal law of gravitation. While Newton had attempted to pursue if these conceptual terms were the same, it was Einstein in 1907 who extended his own notions and declared that acceleration and gravitation were identical, that is, objects of different composition would have identical accelerations in the same gravitational field (Einstein, 1907). This idea is now referred to as the equivalence principle. In a publication in 1916 Einstein broadened his concepts to include an accelerated frame of reference (Einstein, 1916). Within his general theory of relativity he united space and time and presented gravity as a geometrical interpretation of how bodies move in the presence of a mass.

It was claimed that there were three astronomical tests which could act as a litmus examination of general relativity: the anomalous advance of the perihelion of Mercury, the extent to which starlight could be bent as it passes the Sun and the gravitational redshift of light from the Sun. In truth, the gravitational light deflection and the gravitational redshift are derived from the equivalence principle and the Mercury situation from general relativity. This distinction will not be invoked in this paper and the term general relativity will be used to encompass the equivalence principle.

Former work by the current author questioned the early acceptance of the results of these tests of gravitational light deflection in one paper (Treschman, 2014a) and Mercury and gravitational redshift in another (Treschman, 2014b). It was argued in those articles that insufficient evidence existed until the year 1928 for acceptance of general relativity as a reasonable explanation of the data that had been gathered.

AIM OF THIS PAPER

This paper picks up the thread post-1928. It does include the extension a number of other scientists made to general relativity from as early as 1916 and even some experiments that were conducted prior to Einstein's publications which can be interpreted within the worldview of general relativity. The history of several themes is examined to gauge at what level they support general relativity.

In order to ascertain reality, science rests on models, namely, using something known as a proxy for the unknown. Truth is not the issue but how useful is the construct in explaining phenomena and predicting outcomes. The aim in this paper is to place the theory of general relativity in the context of its suitability as a description of the cosmos.

Scientific breakthroughs are often presented as before and after. Yet, acceptance takes a long period of time. Aristarchus (c310-c230 BCE) recorded a heliocentric model which was published much later in 1543 by Nicolaus Copernicus (1473-1543). This was in contrast to the geocentric rendition of Claudius Ptolemy (90-168). Yet, even after the telescopic observations of Galileo Galilei (1564-1642) commencing in 1609, scientists correctly needed more evidence before their world picture was better presented by the earth orbiting the Sun. Interestingly, there are still vestiges of the alternative model today in terms such as "sunrise" and "sunset". The ideas of Isaac Newton (1643-1727) put to print in 1687 had initial difficulty with the notion of action at a distance which had a whiff of magic about it. It is still a practical worldview if one limits the picture to speeds much below that of light and to masses the size of the planets. So, the questions are:

(i) How much evidence exists to support general relativity,
(ii) is it a reasonable way of thinking and
(iii) what is the niche it may occupy?

Answers to these queries are attempted by tracing some selections from the historical record separated into classes based on the type of investigation. The survey of the literature is restricted mainly to journals printed in English.

WEAK EQUIVALENCE PRINCIPLE

Equivalence of Inertial and Gravitational Mass

To elucidate any difference between inertial mass and gravitational mass the Hungarian physicist, Loránd Eötvös (1848-1918), commenced measurements in 1885. He used a torsion balance which consisted of a horizontal rod suspended by a thin fibre and having two masses of different composition but the same gravitational mass at the ends of the rod. He worked firstly with copper and platinum. The rod was oriented parallel with the meridian and had an attached mirror which reflected light into a telescope so that any small twist in the fibre could be observed more easily. The rotation of the Earth created forces on the masses proportional to their inertial masses. The vector sum of the tension in the fibre, the gravitational force and the reaction to the centripetal force would result in a zero torque (beyond the rotation of the rod at the same rate as that of the Earth). For a null movement of the rod, Eötvös could claim a proportionality constant between inertial and gravitational mass.

Continuing with different materials he published his results in 1890 (Eötvös, 1890) in which he claimed an accuracy of 1 in 2×10^7. In 1891 he refined the model to have one of the masses suspended by its own fibre from the rod so that the system could now have measurements in two dimensions. His coworkers from 1906-1909 were Dezsö Pekár (1873-1953) and Jenö Fekete (1880-1943). The later publication by Eötvös (1909) declared an improved accuracy to 1 in 10^8. The final results (Eötvös, 1922) were printed after his death.

Later János Renner (1889-1976) (Renner 1935) who had worked with Eötvös took the results to 2-5 in 10^9 and in another three decades Robert Henry Dicke (1916-1977), Peter G. Roll and R. Krotkov (Roll et al., 1964) had used improved equipment to conclude an accuracy of 1 in 10^{11}. Another avenue for testing the equivalence principle was to probe the motions of the Earth and Moon. Both bodies accelerate in the gravitational field of the Sun. To establish whether the accelerations were different, it was necessary to obtain a more accurate position of the Moon relative to the Earth. It had been proposed to bounce a laser beam off the Moon but the topography would conspire to produce spurious results. Hence, in 1969 on the first human lunar landing, the astronauts of Apollo 11 embedded a retroreflector array on the Moon. This consisted of 100 corner cube prisms in a 10 x 10 array 0.45 m square with each cube made of quartz and dimension 3.8 cm. The design of each prism had a trio of mutually perpendicular surfaces such that an incoming ray is totally internally reflected from three surfaces to generate a deviation of 180°. The array from Apollo 14 in 1971 is similar but the one also in 1971 from Apollo 15 had 300 cubes in a hexagonal array. The Soviet Union landed two rovers on the Moon: Lunokhod 1 from Luna 17 in 1970 and Lunokhod 2 from Luna 21 in

1973. Each of the rovers carried 14 cubes in a triangular formation with 11 cm size apiece in an array 44 x 19 cm (Dickey et al., 1994).

A number of Earth stations have observed a reflected pulse but long term dedication belongs to the Observatoire du CERGA (Centre d'Etudes et de Recherches Géodynamiques et Astronomiques) near Cannes in France with a 1.5 m telescope and the McDonald Laser Ranging System in Texas using a 2.7 m system. The latter was replaced by a dedicated 0.76 cm instrument in 1985. The laser adopted was a neodymium-yttrium-aluminium-garnet one firing a 2×10^{-10} s pulse 10 times per second. In the early 1970s accuracies were at the 25 cm level. This was reduced to 15 cm in the mid 1970s as a result of improvements to the timing system and from 1985 to 2-3 cm. The findings were consistent with general relativity to 1 in 10^4 as well as determining the recession of the Moon from Earth by 3.8 cm yr^{-1} (Gefter, 2005). An improvement to 1 mm accuracy between the Earth and the Moon has been achieved by the 3.5 m arrangement at Apache Point Observatory in New Mexico (Murphy et al., 2008). This requires a 3.3×10^{-12} s exactitude in the one way trip or 6.7×10^{-12} s both ways. The major uncertainty in the distance is due to the libration of the Moon which, on its own, contributes to a spread of 15-36 mm in distance, equivalent to $1.0-2.4 \times 10^{-11}$ s round trip time. Accuracy has improved due to the aperture size of the telescope, altitude of 2880 m, a greater capture of photons and a timing mechanism of atomic standards to 10^{-7} s. Any violation of the equivalence principle would produce a displacement of the lunar orbit along the earth-Sun line with a variation coinciding with the 29.53 days synodic period. This has not occurred to the 0.1% level (Williams et al., 2009).

Gravitational Redshift

Measurements of the gravitational redshift of lines from the Sun followed a tortuous journey. From an apparent tangent of using the lines from Sirius B and then other white dwarfs, scientists unravelled the many factors from which the relativistic redshift emerged. Pursuing another tack, Robert Vivian Pound (1919-2010), Glen Anderson Rebka, Jr (1931-) and Joseph Lyons Snider conceived an imaginative experiment.

Pound and Rebka (1959) reported that a fraction of gamma rays could be emitted from the nuclei of a solid without recoil momentum of the nuclei. They hypothesised that gravitational redshift could be measured from an emitter to a source at a different altitude and register the situation for maximum scattering (Pound and Rebka 1959). The emitter they chose was Co-57 electroplated onto one side of an iron disc. To ensure diffusion of the cobalt into the iron, the disc was heated up to 1000°C for one hour. The absorber was seven units of iron enriched in Fe-57 to 32% electroplated

onto a beryllium disc. The absorption level was one third of the emitted gamma rays. Placed inside a space at the Jefferson Physical Laboratory of Harvard University, the source and absorber were 22.6 m apart. To reduce the absorption of gamma rays by air, helium was run through the tower continuously. The fractional change in frequency was proportional to gh/c^2 where g = 9.8 m s^{-2} is the acceleration due to gravity, h = 22.6 m is the altitude and c = 3.0×10^8 m s^{-1} is the speed of light. The ingenious aspect was to measure the change in energy instead by having gamma rays move against gravity and then with gravity by interchanging the emitter and absorber. Thus, the change in energy down less the change in energy up = $2gh/c^2 = 4.9 \times 10^{-15}$. The authors reported that their experimental result was 1.05 ± 0.10 times the theoretical value (Pound and Rebka, 1960a) for a frequency change of 3.27×10^{-8} s^{-1} for this altitude difference in the gravitational potential of the Earth (Pound and Rebka, 1960b) where the gradient (Hirate, 2012) is 1.1×10^{-16} c^2 m^{-1}. Improvements were effected in 1964 by Pound and Snider and their result was published as 0.999 0 ± 0.007, 6 times the predicted relativistic frequency (Pound and Snider, 1965).

From 1976, spacecraft were involved in this particular test of general relativity. Carrying a hydrogen maser, a 100 kg spin stabilised spacecraft, jointly organised by the National Aeronautics and Space Administration (NASA) and the Smithsonian Astrophysical Observatory, was launched to 10000 km almost vertically. The output frequency of $1.420\ 405\ 751 \times 10^9$ Hz, accurate over 100 s averaging time to 1 in 10^{14}, was compared with another maser on Earth. The agreement with general relativity was calculated to the 7×10^{-5} level (Vessot et al., 1980).

Voyager 1 was launched in 1977, flew by Jupiter in 1979 and reached Saturn in 1980. It carried an ultrastable crystal oscillator. As a result of its close approach to Saturn, a redshift of several hertz was predicted to its 2.3×10^9 Hz downlink sent by its 3.7 m antenna. Comparison was made against the three 64 m stations on Earth which are part of the Deep Space Network: Goldstone in California, near Madrid in Spain and near Canberra in Australia. Each of these stations was referenced to a hydrogen maser frequency standard. The result was in agreement with general relativity to 0.995 6 ± 0.000 4 as a formal uncertainty and ± 0.01 as a realistic uncertainty (Krisher et al., 1990).

Similar communication channels were set for Galileo which was launched in 1989 on a trajectory which included a gravity assist from Venus in 1990 and Earth in 1990 and 1992 before arriving at Jupiter in 1995. During the phase from launch to the first Earth gravity assist, regular frequency measurements of the spacecraft clock were conducted. Personnel from the Jet Propulsion Laboratory reported a 0.5% agreement with general relativity for the total frequency shift and a 1% concord with the solar gravitational redshift (Krisher et al., 1993).

However, it was the Cassini spacecraft on its way to

Saturn which has provided the closest match to general relativity at 0.0023% (Williams et al., 2004). Jointly coordinated by NASA and the Italian Space Agency, Cassini was launched in 1997, and flew by Earth, Venus and Jupiter to orbit Saturn in 2004. In 2002 it was near superior conjunction, with the Earth situated 8.43 astronomical units distant. Interference from the solar corona and the Earth's troposphere could be accounted for by two different uplink frequencies and three different downlink signals with use of Cassini's 4 m antenna. Measurements were conducted on the 18 passages of signals between Earth and Cassini (Bertotti et al., 2003). Each pulsar in a binary system is influenced by the strong gravitational field of the other. From PSR J0737 − 3039 A/B (see later), a redshift parameter of 3.856×10^{-4} s is compared with a relativistic calculation of $3.841\ 8 \times 10^{-4}$ s to give a ratio between them of 1.003 6 (Kramer et al., 2006).

ORBITAL PRECESSION OF A BODY IN GRAVITATIONAL FIELDS

Relativistic Perihelion Advance of the Planets

Between the publication of special relativity in 1905 and general relativity in 1916, Einstein received assistance from Marcel Grossmann (1878-1936) (Einstein and Grossmann, 1913) and Michele Besso (1873-1955) (Janssen, 2002). Grossmann alerted Einstein to how tensor calculus and Riemannian geometry could be applied to general relativity and Besso worked with Einstein on solving some equations which were relevant to the perihelion advance of Mercury. Einstein incorporated into his equations Lorentz transformations named for Hendrik Antoon Lorentz (1853-1828). These involved c the speed of light independent of a reference frame. They showed how measurements of space and time taken by two observers were related. Thus, they gave meaning to how two observers travelling at different relative velocities may make different measures of distance and elapsed time. The Lorentz factor γ (gamma) was defined as

$$\gamma = \frac{1}{\sqrt{1 - \frac{v^2}{c^2}}} \qquad (1)$$

where v is the relative velocity between inertial reference frames. In Einstein's work he used for time dilation for length contraction in the x direction.

$$\Delta t' = \gamma\, \Delta t \qquad (2)$$

and

$$\Delta x' = \frac{\Delta x}{\gamma} \qquad (3)$$

for length contraction in the x direction.

In later experimentation, to ascertain how closely results may be interpreted in the worldview of general relativity, the Lorentz factor was a part of a number of equations and the closer this value is to unity, then general relativity is more supported.

It was in 1916 that Einstein wrote his gravitational field equations applying within a vacuum and chose the Sun as the origin of his coordinate system (Vankov, 1915). He made use of Huygens' principle to formulate the angular deflection of a ray of light at a certain distance from the Sun. Through a series of approximations, he derived a planetary motion equation. As long as the speed of a particle was much less than c the speed of light, Newton's equation could be obtained as a first approximation.

With a switch to planar orbit equations with the polar coordinates r and ϕ as the radius vector and angle respectively, the equations led to the known energy and Kepler's planetary law of areas. One result was:

$$r^2 \frac{d\phi}{ds} = \text{a constant} \qquad (4)$$

where s is displacement. If orbital motion were described, the equation was in agreement with Kepler's third law portraying the relationship between the period of a planet and its distance from the Sun. The curvature of spacetime envisaged by Einstein was an explanation of the Mercury advance as it had further to travel than in flat space due to the distortion created by the mass of the Sun.

To obtain the secular advance of an elliptical orbit Einstein next integrated the equation containing ϕ over the ellipse so that $\Delta\phi$, the change in angle in radians per orbit, is found in terms of a the semi major axis and e the eccentricity. If this is extended to an entire passage, the result in the direction of motion for the period T in s is

$$\Delta\phi = 24\pi^3 \frac{a^2}{T^2 c^2 (1 - e^2)} \qquad (5)$$

With conversion factors of $180/\pi$ to give °, 3 600 for ", a change of period from s to 0.240 844 45 tropical years and 100/orbital period in tropical years producing an answer in " century^{-1}, Einstein calculated a figure of 45" ± 5 century^{-1} for Mercury, the then accepted value for the anomalous advance of the perihelion of Mercury being 42".95 century^{-1}.

By 1943 Gerald Maurice Clemence (1908-1974) had examined meridian observations of Mercury totalling 10 400 in right ascension and 10 406 in declination over the period 1765-1937 and 24 transits of Mercury across the Sun spanning 1799-1940 (Clemence, 1943). From this analysis he adjusted figures for the eccentricity and perihelion of the Earth as well as for the mass of Venus. His new value for the anomalous perihelion advance of Mercury was 43".11 ± 0.45 century^{-1} against the Einstein figure at this time of 43".03 century^{-1}.

With his attention on another planet, Raynor Lockwood Duncombe (1917-2013) scrutinised meridian observations of Venus across 1750-1949 (21009 in right ascension and 19852 in declination) (Duncombe, 1956). After applying corrections to some elements of Venus and the Earth and the mass of Mercury, he deduced, for the first time, results accurate enough for the anomalous advance of the perihelion of Venus. In 1956 this was determined as 8".4 ± 4".8 century^{-1} while the relativity figure was 8".6 century^{-1} (Morton, 1956).

For Earth, HR Morgan dissected studies of the Sun over 1750-1944 from a number of observatories and applied a correction in 1945 to the eccentricity of the planet (Morgan, 1945). He combined with Clemence and Duncombe to determine by 1956 the anomalous advance of the perihelion of Earth as 5".0 ± 1".2 century^{-1} while the Einsteinian amount was 3".8 century^{-1} (Morton *op. cit.*). Kepler's third law of planetary motion for Mercury may be expressed as

$$T^2 = \frac{4\pi^2 a^3}{G(M+m)}$$

(6)

for **G** the universal gravitational constant, **M** is the mass of the Sun and **m** the mass of Mercury. As **m<<M**, it may be omitted. If, then, T^2 is substituted into equation (5), one may express the Einstein derivation into a similar one (Gamalath, 2012) as

$$\Delta\phi = \frac{6\pi GM}{c^2 a(1-e^2)}$$

(7)

For **c** = 2.998 x 10^8 m s^{-1}, **G** = 6.673 x 10^{-11} m^3 kg^{-1} s^{-2}, **M** = 1.989 x 10^{30} kg, and data from a modern almanac (Seidelmann, 2006) the calculations for Mercury, Venus and Earth are juxtaposed against the observed values in Table 1. The calculated values are within the range of the observed figures.

For % difference between the calculated and observed values, the central value gives (43.11 – 42.98)/42.98 x 100 = 0.19%. However, the extreme difference is (43.11 + 0.45 – 42.98)/42.98 x 100 = 1.4%. In a similar way, the values respectively for Venus are 2.3 and 58% and Earth 32 and 62%.

One of the assumptions in Einstein's derivation was that the orbital plane of the planets coincided with the rotational equator of the Sun. This is incorrect but the technology to measure what became known as the

quadrupole moment of the Sun did not exist until the 1980s and particularly into the 1990s. The splitting of spectral lines due to solar oscillations in the 1980s revealed that, with the precision of the measurements, the assumption in the derivation of Mercury's anomalous perihelion advance was acceptable (Campbell and Moffatt, 1983).

A Global Oscillations Network Group GONG was formed in 1995 to produce continuous solar velocity imaging with an aim to ascertain the spherical harmonic functions of the Sun related to its radius and latitude. Six solar observatories in the Canary Islands, Australia, California, Hawaii, India and Chile combined to analyse 33169 splits of spectral lines (Pijpers, 1998). The conclusion was that the results are currently consistent with the figure accepted for Mercury's perihelion advance determined by general relativity. This decision is also supported by the first six months of data obtained from helioseismology measurements taken by the Michelson Doppler Imager aboard SOHO, the Solar Heliospheric Observatory, launched in 1995. An interesting extension to this concept is the use of exoplanets (Zhao and Xie, 2013). Data from the Kepler space observatory launched in 2009 and future missions may give improved accuracy so the periastron advance to these other systems may be added to the information on the solar system planets.

Relativistic Periastron Advance of Binary Pulsars

There are many factors involved in determining the orbits of the planets and the positions of the perihelia. In addition, the *total* change per year in the location of the perihelion of Mercury is as small as 5".7. Fortunately, the same property applicable to the relativistic perihelion advance of the planets may be applied outside the solar system. In addition, within the solar system, the gravitational fields are comparatively weak whereas outside the solar system there are opportunities for some very strong fields. The target is a stellar binary system where at least one of the stars is a pulsar so that the periastron advance may be monitored.

The term binary pulsar is used if one or both objects are pulsars. The first such system was discovered in 1974 by Russell Alan Hulse (1950-) and Joseph Hooton Taylor, Jr (1941-) while conducting a survey at the 305 m Arecibo Observatory in Puerto Rico (Hulse and Taylor, 1975). The technology that existed at this time enabled a computer "to report on any pulsar suspects above a certain sensitivity threshold" (McNamara, 2008). The pulsar had a very short pulsation period of 5.9 x 10^{-2} s in a highly eccentric orbit of e = 0.615 with a period of 0^d.323 0. Its companion is believed to be a neutron star. The pulsar is designated PSR 1913 + 16.

The measurement technique is a comparison between the phases of the radio pulses from the pulsar and those of atomic clocks on the Earth (Will, 1995) to register the

Table 1. Anomalous advance in the perihelia of Mercury, Venus and Earth.

Planet	a x 10^{10} m	e	Orbit in tropical years	$\Delta\phi$ in " per century calculated	$\Delta\phi$ in " per century observed
Mercury	5.791	0.205 6	0.240 844 45	42.98	43.11 ± 0.45
Venus	10.821	0.006 8	0.615 182 57	8.625	8.4 ± 4.8
Earth	14.960	0.016 7	0.999 978 62	3.839	5.0 ± 1.2

small changes over time with the pulse frequency. The Doppler effect alters the arrival time of the pulses. The variation was between 0^d.058 967 and 0^d.069 045 which amounts to 6.7 s over its cycle of 0^d.323 0, that is, 7.75 h (Hulse and Taylor, *op. cit.*). The precision of measurement was such that an initial discrepancy of 2.7 x 10^{-2} s for the period of what was thought to be a single pulsar measured at different times was not considered a false value (McNamara, *op. cit.*). The speed of the orbit is highly relativistic being 10^{-3}c. The relativistic periastron advance of $4°$.226 62 ± 0.000 01 yr^{-1} is 2.7 x 10^3 greater than the 5".7 y^{-1} for the perihelion advance of Mercury. This periastron advance is within 0.8% of the prediction from general relativity (Damour and Taylor, 1991). Also, this system will be revisited later in this paper as monitoring continues for how the companion's gravitational field affects the redshift of the pulses and how the relativistic time dilation is caused by the orbital motion.

A consequence of general relativity, the curvature of spacetime, is implicated in the periastron advance of binary pulsars in the same way as the perihelion advance of the planets. However, in 1918, Einstein proposed that a binary system would lose gravitational wave energy and provided a quadrupole formula for the subsequent damping on the orbital period (Einstein, 1918). However, his results are expressed here from a project which derives Einstein's conclusions (Valença, 2008). Firstly, for **E** energy, **t** time, a condition of **e** = 0, μ reduced mass where $\mu = m_1m_2/(m_1 + m_2)$ for the individual masses, **m** representing the same mass which would be the case if e = 0 and **r** the distance between the two objects, then the change in energy over time is given by

$$\frac{d E (e = 0)}{d t} = \frac{32\mu^2 m^3 G^4}{5c^5 a^5}. \tag{8}$$

Then, a correction is applied for the case when e $\neq$ 0 so that

$$\frac{d E}{d t} = \frac{32\mu^2 m^3 G^4}{5c^5 a^5}(1 + \frac{15}{2}e^2 + \frac{45}{8}e^4)(1 - e^2)^{-7/2}. \tag{9}$$

The change in energy per time may be extended to include a change in the period P denoted as $\dot{P}$ as

$$\frac{d E}{d t} = \frac{-2m_1m_2 G}{3 rP}\dot{P}. \tag{10}$$

From measurements on PSR 1913 + 16, the mass of the pulsar was determined as 1.441 0 ± 0.000 7 M_S (times mass of the Sun) and the companion as 1.387 4 ± 0.000 7 M_S (Will, *op. cit.*). The distance between the pair ranged from 1.1 to 4.8 solar radii. Armed with these data, Taylor, a codiscoverer, and Joel M Weisberg found, in 1989 after 14 years of measurement on the binary pulsar, that the rate of orbital decay was within 1% of that predicted by special and general relativity (Taylor and Weisberg, 1989). By 1995, improvement had reached 0.3% accuracy with a rate of (− 2.402 43 ± 0.000 05) x 10^{-12} ss^{-1}. Once a small effect caused by galactic rotation, the relative acceleration between the binary pulsar and the solar system, is subtracted, the result is (- 2.410 ± 0.009) x 10^{-12} s s^{-1} which is the prediction afforded by general relativity (Will, *op. cit.*). After 30 years of analysis in 1995, Weisberg and Taylor provided consistency between theory and observation at the (0.13 ± 0.21%) level (Weisberg and Taylor, 2005).

A further Arecibo survey operating at 4.30 x 10^8 Hz in 1990 detected another binary pulsar PSR 1534 + 12. The 3.79 x 10^{-2} s pulse of orbital period 3.64 x10^4 s has a rate of decay of 2.43 x 10^{-18} s s^{-1} and periastron advance of $1°$.756 2 yr^{-1}. Due to the strong and narrow pulse, greater precision for this system was expected over time (Wolszczan, 1991). This had been achieved by 1998 with further timing observations with radio telescopes at Arecibo, 43 m Green Bank in West Virginia and 76 m Jodrell Bank and a conclusion that the results were in accord with general relativity to better than 1% (Stairs et al.,1998).

A third binary pulsar PSR 2127 + 11C (Prince et al., 1991) had its relativistic periastron advance measured at $4°$.46 yr^{-1} in 1991 but more work was needed to compare this with general relativity. By 1992, 21 binary pulsars had been studied well enough for their basic parameters to be determined (Taylor, 1992).

A rare situation emerged in 2003. A pulsar discovered with the 64 m radio telescope13 beam receiver (Staveley-Smith et al., 1966) at Parkes Australia was found subsequently to have a companion which is also a pulsar. An improved position was determined with the use of the 20 cm band from interferometric observations with the Australia Telescope Compact Array (Burgay et al., 2003).

Results were published in 2006 after 2.5 years of measurements had been effected on PSR J0737 – 3039A and PSR J0737 – 3039B. Data were gathered at Parkes at 6.80×10^8, 1.374×10^9 and 3.030×10^{10} Hz, 76 m Jodrell Bank Observatory in the UK at 6.10×10^8 and 1.396×10^9 Hz, and 100 m Green Bank at 3.40×10^8, 8.20×10^8 and 1.400×10^9 Hz. A total of 131 416 arrival of pulse times for A with an uncertainty of 1.8×10^{-5} s were received and 507 for B with a maximum uncertainty of 4×10^{-3} s. The system has an orbital period of $0^d.102$ 251 563, respective pulse periods of 2.27×10^{-2} and 2.77×10 s and a periastron advance for A of $16°.90$ yr^{-1} (Lyne et al., 2004).

Four independent tests of general relativity are obtainable with this system. The orbital decay derivative observed was $- 1.252 \times 10^{-12}$ $s\ s^{-1}$, shrinking the distance between the pulsars by 7 mm d^{-1}. The relativistic prediction was $1.247\ 87 \times 10^{-12}$ $s\ s^{-1}$ giving a ratio of observed to expected value of 1.003 (Kramer, op.cit.). Other results relate to gravitational redshift and time dilation.

Geodetic Precession

Yet another property was added to the list for testing general relativity soon after its inception. In 1916 Willem de Sitter (1872-1934) applied relativity theory to the Earth-Moon system. He realised the pair was freely falling in the gravitational field of the Sun. Since the Moon was also orbiting the Earth, he predicted that the Moon ought to undergo a non-Newtonian precession in its orbit (Sitter, 1916). His expected figure was a secular motion of the perigee and the node both of $+ 1".91$century^{-1} (Sitter, 1917). This effect is referred to as geodetic precession.

Shapiro et al. (1988) mined the lunar laser ranging data collected over the period 1970-1986 from the retroreflectors on the Moon. A model of the Moon's motion consisted of two coupled sets of differential equations, one for its orbit and the other for its rotation. Perturbations from the gravitational fields of the Sun, Earth, and other planets as well as torques on the Moon from the Sun and Earth and the drag from tides on the Earth were factored to provide equations as a function of time. An introduced numerical factor h was related to any extra precession of the Moon's orbit about the ecliptic pole that was not included in the predicted relativistic geodetic precession. h would equal zero if it were consistent with general relativity and unity if there were 100% difference from the prediction. From the set of 4 400 echo measurements, their analysis resulted in $h = 0.019 \pm 0.010$ (Shapiro et al., 1988).

According to general relativity the Moon should precess in its orbit by 1.9×10^{-2} s yr^{-1}. A data set of 8 300 lunar laser ranges over the period 1969-1993 yielded a deviation from this amount by $- 0.3 \pm 0.9\%$ (Dickey et al.,

op. cit.). Gravity Probe B Relativity Mission was launched by NASA in 2004 and operated an experiment for 12 months. Its aim was to measure two effects predicted by general relativity: geodetic precession and frame dragging or Lense-Thirring effect. Geodetic precession may be described as a vector perpendicular to the orbital plane whereas frame dragging may be designated as a vector arising from rotation and acting orthogonally to the geodetic precession vector. As the two effects act at right angles to each other, the component vectors could be distinguished.

The satellite was placed in an orbit over both poles of the Earth. The mean altitude was 642 km and the orbital eccentricity was 0.001 4. A telescope was fixed on the bright star IM Pegasi, as were initially four super-conducting niobium coated, 38 mm spherical quartz gyroscopes. Each was surrounded by liquid helium at 2 K where some escaping gas caused the gyroscopes to commence spinning up to an average rate of 72 Hz. The devices were suspended electrically with two spinning clockwise and two counter clockwise. They were tested at maintaining their drift rate accuracy to $5" \times 10^{-4}$ yr^{-1}.The gyroscope is a vector not aligned with the spin axis of the Earth. After one orbit of parallel transport of the Earth, any shift in the axis of a gyroscope would induce a current which enabled the changed to be measured (Will, 2006). The predicted Einstein drift rate was $– 6".606\ 1 \times 10^{-6}$ yr^{-1}. The four results were combined to give a weighted average of $– (6".601 \pm 0.018.3) \times 10^{-6}$ yr^{-1}, giving an accuracy of 0.28% (Everitt et al., 2011). Across the span 1961-2003, 250 000 high precision radar observations from the USA and Russia to the inner planets and spacecraft have been examined. In addition to the perturbations of the planets and the Moon, those of 301 larger asteroids and a ring of small asteroids have been included. The result for γ was $0.999\ 9 \pm 0.000\ 2$ (Pitjeva, 2005). With binary pulsars, if the spin axis is not aligned with the angular momentum axis of the system, geodetic precession should occur. All the candidates that have been discovered so far need a much longer time period of measurement to arrive at definitive answers for this property.

Lense-Thirring Effect

Frame dragging refers to another effect arising from general relativity in which a massive celestial rotating body drags its local spacetime around with it. Whereas geodetic precession operates in the presence of a central mass, frame dragging is postulated to exist as a separate effect if the mass is rotating. This consequence was hypothesised by Josef Lense (1890-1985) and Hans Thirring (1888-1976) in 1918. However, Pfister (2007), in his treatment of the history of this effect, argues from evidence in the Einstein-Besso manuscript 1913, Thirring's notebook of 1917 and a letter from Einstein to

Thirring in 1917 that Einstein pointed to this phenomenon. Frame dragging is a secular precession of an orbiting object which has its orbital plane at an angle to the equator of a central entity which possesses angular momentum. The magnitude of the effect is extremely small compared with geodetic precession.

NASA launched Mars Global Surveyor in 1996 and it was inserted into its orbit in 1997. In the five year period 2000-2005, the orbital plane of the spacecraft was predicted to shift by 1.5 m due to frame dragging and the measured result was 1.6 m, giving a difference from general relativity of the order of 6% (Iorio, 2006).

Twin satellites, Laser Geodynamics Satellite (LAGEOS) launched by NASA in 1976 and LAGEOS II a joint NASA and Italian Space Agency in 1992, are passive reflectors in Earth orbit. Each contains 426 corner cube reflectors, all but four of these made of fused silica glass with the others of germanium for infrared measurements. Their respective orbital parameters are: semi-major axis 12 270 and 12 163 km; eccentricity 0.004 5 and 0.014; inclination to Earth's equator 110° and 52°.65. The expected measure of precession of their line of nodes was 3" x 10^{-4} yr^{-1} which is equivalent to a displacement of 1.9 m in that time. Monitoring was performed by 50 Earth stations as part of the International Laser Ranging Service. From 10^8 laser ranging observations over the period 1993-2003, the measure of the precession of the line of nodes was given as 4".79 x 10^{-2} yr^{-1} against the relativistic prediction of 4".82 x 10^{-2} yr^{-1}. The result of the observation was 99% ± 5 of the predicted value although the authors allow for 10% uncertainty (Ciufolini and Pavlis, 2004).

A later satellite, Laser Relativity Satellite (LARES), was launched by the Italian Space Agency in 2012. It is a spherical, laser ranged passive satellite with 92 retroreflectors made of a tungsten alloy. Its semimajor axis is 7 820 km, eccentricity 0.000 7 and orbital inclination 69°.5. Measurements are ongoing.

One of the difficulties with accurate positioning is the figure of the Earth. To ascertain deviations from spherical symmetry of the Earth's gravity field, Gravity Recovery and Climate Experiment (GRACE) consists of twin satellites of NASA and the German Aerospace Center launched in 2002 in polar orbit, 500 km above the Earth and 220 km between them. They maintain a microwave ranging link which can measure their separation to 1×10^{-5} m. Optical corner reflectors allow their position to be monitored from Earth against the GPS. Gravity Probe B results reported in 2012 gave the frame dragging effect (Everitt et al., op. cit.) as (− 3".72 ± 0.72) x 10^{-4} yr^{-1} compared with the Einstein value of − 3".92 x 10^{-4} yr^{-1}.

LIGHT PROPAGATION IN GRAVITATIONAL FIELDS

Gravitational Optical Light Deflection

The central equation of Einstein which led to his

international fame was that the angle of deviation α of starlight in the vicinity of the Sun with mass M and distance from the centre r be given as

$$\alpha = \frac{4GM}{c^2 r} \qquad (11)$$

where half that value was due to time curvature and the other half from space curvature, an intrinsic part of his general relativity (Einstein, 1916 op. cit.). This amounted to 1".75 at the limb of the Sun. With the technology at the time, confirmation rested on a photographic comparison of the stars near the Sun at a total solar eclipse and the same stellar field six months before or after the eclipse. The deviation for stars a little away from the limb corresponded to 1/60 mm on the plate (Eddington, 1919). Such a small measurement was difficult to ascertain with the precision instruments available in the early part of the twentieth century.

The 1919 British total solar eclipse expedition to Brazil by Andrew Claude de la Cherois Crommelin and to Principe by Arthur Stanley Eddington and Edwin Turner Cottingham demonstrated that starlight was deflected by the Sun. In 1922, with final results published in 1928, an excursion to Wallal in remote Western Australia by the Lick Observatory led by William Wallace Campbell supported the deflection at the limb of the Sun as 1".75 ± 0.09 (Campbell and Trumpler, 1928). A limitation for this technique depends on the ability of a telescope to resolve small angular separations due to refraction as light passes through the system.

$$\text{Angular resolution in arcsecond} = \frac{2.5 \times 10^5 \text{ x wavelength of light in m}}{\text{diameter of mirror in m}}$$

$$(12)$$

For the 33 cm telescope used and visible light, the angular resolution amounted to 0".4. Attempts at repeating the experiment have been performed at a number of total solar eclipses, now nine altogether, and the ones in 1952 and 1973 will be mentioned here.

The National Geographical Society and the Naval Research Laboratory jointly sponsored an expedition to Khartoum in Sudan in 1952 (Biesbroeck, 1953). Disappointingly, wind at the time of the eclipse induced vibrations in the 20 foot (6 m) telescope so that many of the fainter stellar images were not included in the measurement. Nevertheless, one photographic plate exposed for 60 s produced nine measurable stars in the eclipse field and eight in the auxiliary field while a second exposure of 90 s resulted in 11 and eight stars respectively. Two checkplates were secured six months later. The conclusion was 1".70 ± 0.10.

In 1973 the University of Texas mounted a mission to Chinguetti Oasis in Mauritania, Africa (Brune et al., 1976). With a 2.1 m focus, four element astrometric lens, the party prepared for a 6 min 18 s eclipse. Three plates, impregnated with a rectangular scale, were obtained with 60 s eclipse field and 30 s comparison field 10° away in declination. 150 measurable images and 60 comparison field ones were captured. After an elapse of five months, 33 calibration plates were obtained. The result extrapolated to the solar limb of 0".95 ± 0.11 serves to indicate, if general relativity is to be supported, how difficult measurements on photographic plates for the visible region of the spectrum actually is.

Since the launch of the European Space Agency spacecraft Hipparcos (high precision parallax collecting satellite) in 1989, the deflection of light at total solar eclipses has been consigned to a quaint part of history. The 29 cm aperture telescope on board has measured the position of 118 200 stars to a precision of $3" \times 10^{-3}$ for the magnitudes 8 - 9. Any effect on the deflection of starlight by the Sun can now be measured by checking the distance between pairs of stars over time. The advantages inherent in this system were that there was no need for a total solar eclipse, bending by the solar corona could be eliminated, measurements could take place over large angular distances from the Sun and the same instrument was used well calibrated over the entire sky for 37 months. Data were collected on a set of stars chosen within 47 - 133° of the Sun. As an example, the relativistic prediction is that at 90° from the Sun the deflection would be $4".07 \times 10^{-3}$. As a number of theories incorporate some predictions similar to general relativity, nine so called parameterised post-Newtonian parameters have been introduced. Radiation deflected by the gravitational field of the Sun and entering a telescope on Earth is expressed as an amount equal to

$$1".749 \, \frac{(1+\gamma)}{2} \qquad \qquad (13)$$

where γ equals unity in general relativity. The result from Hipparcos was γ = 0.997 ± 0.003 (Froeschlé et al., 1997). An improved astrometric spacecraft from the ESA is Gaia which was launched in December 2013 and took up its residence at the Sun-Earth L_2 Langrangian point in January 2014. The aim of the mission is to record the position of 10^9 objects to a precision of $2".0 \times 10^{-5}$. A future analysis of results based on a similar method as for the Hipparcos data will improve the accuracy of this experiment.

Gravitational Radio Deflection due to the Sun

Since angular resolution is proportional to the reciprocal of the wavelength of light, the longer wavelength radio region provides an improvement over the visible spectrum. It eventually became possible to measure the position of radio sources so precisely with interferometry,

even in the daytime. The blazar 3C279 is a very bright object 12' from the ecliptic and each 08 October it is eclipsed by the Sun. Deflection was measured by two groups in 1969. An Owens Valley Radio Observatory team (Seielstad et al., 1970) in California reported γ = 1.02 ± 0.23 and another Californian band from Goldstone (Muhleman et al., 1970) gave γ = 1.08 ± 0.30. This method was also employed in 1974 with three nearly collinear radio sources, 0116 + 08, 0119 + 11 and 0111 + 02, and a 35 km interferometer baseline (Fomalont and Sramek, 1975). As these radio emitters passed near the Sun, the deflection of their beams was monitored by the National Radio Astronomy Observatory at Green Bank. This comprised three steerable 26 m parabolic antennas with a maximum baseline separation of 2.7 km and a fourth element of 14 m aperture situated 35 km away. The three long baselines are 33.1, 33.8 and 35.3 km. So that the solar coronal refraction may be separated from the contribution from relativity, observations were made simultaneously at two frequencies, 2.695×10^9 and 9.085×10^9 Hz since electron refraction varies as the square of the wavelength.

The deflection at the solar limb was determined as 1".775 ± 0.019 which was 1.015 ± 0.011 times the Einstein value. This corresponds to the parameter γ = 1.030 ± 0.022. The experiment was repeated 12 months later in 1975. The combination of the 1974 and 1975 measurements (Fomalont and Sramek, 1976) produced a limb deflection of 1".761 ± 0.016 corresponding to 1.007 ± 0.009 times the general relativity prediction and γ = 1.014 ± 0.018.

The source 3C279 mentioned earlier in this area is also known as J1256 − 0547. It and three other radio emitters, J1304 − 0346, J1248 − 0632 and J1246 − 0730, were captured by the Very Long Baseline Array in 1990. This comprises 10 parabolic 25 m telescopes across the United States of America. Previous testing had shown that the system could measure relative positions to $1" \times 10^5$ (Fomalont et al., 2009a). The system operated at frequencies of 1.5, 2.3 and 4.3 all $\times 10^{10}$ Hz so that the effect of the solar corona was minimised. Furthermore, the relativistic bending is independent of the wavelength. The result from the four sources combined was γ = 0.999 8 ± 0.000 3 (standard uncertainty) (Fomalont et al., 2009b).

As the length of the baseline in interferometry increases, the accuracy of the determination of γ improves. A major investigation between 1980 and 1990 was conducted by personnel from the National Oceanic and Atmospheric Administration in Rockville, Maryland (Robertson et al., 1991). 74 radio sources collected by 29 very long baseline observatories produced a set of 342 810 observations. Early data used 3 000 km as the baseline, such as from Westford, Massachusetts to Fort Davis, Texas, but later ones operated between 7 000 − 10 000 km, for example, a 7 832 km stretch from Wettzell, Germany to Hartebeesthoek, South Africa. The expected deflection at the Sun's limb is 1".750, at an angle of 90° away from the Sun $4"\times10^{-3}$ and zero deflection at 180°. The scientists concluded a value for γ of 1.000 2 ± 0.002 (standard uncertainty).

Use was made of data collected during 1979-1999 from 87 very long baseline interferometric sites and 541 radio sources (Shapiro et al., 2004). The information was intended to monitor various motions of the Earth but has been analysed to conclude γ = 0.999 8 ± 0.000 4.

Gravitational radar deflection is progressing to the planets. Measurements were taken in 2002 when Jupiter passed within 4' of the quasar J0842 + 1835, in 2008 for Jupiter 1'.4 from J1925 – 2210 and in 2009 for Saturn 1'.3 from J1127 + 0555. More arrays are devoting time to this new avenue and the results are awaiting analysis (Fomalont et al., *op. cit.* 2009b).

Gravitational Lensing

Gravitational lensing refers to the production of an image of a background object presented to an observer by another object between them. The origin of this thought has been traced to eight pages of a notebook Einstein used in 1912 (Renn et al., 1997). In it he indicated the possibility of a double image of the source due to gravitational light bending and suggested that the intensity of these images would be magnified. In 1936 Einstein returned to this idea and wrote about a background star, when bent in the gravitational field of an intermediate star, would be perceived by an observer in line with both of them not as a point-like star but as a luminous circle around the foreground object. From geometry he obtained an expression for the angular radius (later Einstein radius) of the halo (later Einstein ring) in terms of the deviation angle of light passing the lensing star, the distance of the light from the centre of the foreground object and the distance between observer and lensing star. The derivation is explained in detail by Schneider et al. (1992) as

$$\alpha = \left(\frac{4GM}{c^2}\frac{D_1}{D_2D_3}\right) \text{ raised to 0.5 power} \qquad (14)$$

where M is the mass of the lens, D_1, D_2 and D_3 are respectively distances between source and lens, lens to observer and source to observer (Schneider et al., 1992). Einstein also noted again that the apparent brightness of the distant star would be enhanced. It is interesting to note that he saw no hope of a direct observation of this spectacle (Einstein, 1936).

An extension from a star as the lensing object was provided in 1937 by Fritz Zwicky (1937). He theorised that the gravitational fields of a number of foreground nebulae may deflect the light from background nebulae and that this might be used to determine nebular masses

accurately. He also suggested that a search ought to be conducted among globular nebulae for images of globular clusters. In 1964 a proposal was published in which a supernova could be lensed by a galaxy. This would allow very faint, distant objects to produce an image much closer to the observer so measurements could be extended to much greater distances. The wait was until 1979 when the 2.2 m telescope on Mauna Kea belonging to the University of Hawaii recorded two images which, from their identical properties such as the same redshift z = 1.413, were intimated to be the twin QSO 0957 + 561 (Walsh et al., 1979). The galaxy causing the lensing was soon directly recorded along with a third image (Stockton, 1980).

With the Advanced Camera for Surveys (ACS) aboard the Hubble Space Telescope, the Sloan Lens ACS (SLACS) Survey (Bolton et al., 2008) has provided a 2008 list of 131 strong gravitational lens candidates. There are 70 systems with clear evidence for multiple imaging and another 19 probable ones. Selection was made from the spectroscopic database of an absorption dominated galaxy continuum at one redshift and nebular emission lines at a higher redshift. The lines incorporated the Balmer series and O II at 3.727×10^{-12} m and O III at 5.007×10^{-12} m.

An interesting gravitational lens system discovered in 1985 (Huchra et al., 1985) shows how it can add support to the theory of general relativity. It has been resolved by the Hubble Space Telescope to be four quasar images with z = 1.695 surrounding a 15 magnitude spiral galaxy 2237 + 0305 with z = 0.039 4. The four images are concentric but have different levels of brightness. From the application of lens models based on the lensing equation derived by Einstein along with the cosmological interpretation of redshifts, all of the data collected can be explained. The first discovery of an Einstein ring occurred in 1988 (Hewitt et al., 1988) with the radio source MG1131 + 0456 being surrounded by an elliptical ring of emission.

Time Dilation

In 1964 Irwin Ira Shapiro (1929-) proposed that with recent advances in radar astronomy, another test for general relativity would be to measure the time delay between emission and detection of radar pulses bounced off Mercury or Venus when they were near superior conjunction (Shapiro, 1964). The Doppler shift cancels on a round trip. The time delay Δt is given by

$$\Delta t = \frac{4GM_S}{c^3}\frac{1+\gamma}{2}\ln\frac{R_E+R_P+R}{R_E+R_P-R} \qquad (15)$$

where G, M_S, c and γ are as defined previously, R_E, R_P and R are respective distances between the Earth and

Sun, planet and Sun and Earth and planet (Reasenberg et al., 1979). This increase in time amounted to 1.6×10^{-4} s for Mercury when the beam passes by the Sun at two radii from its centre.

Testing began in 1967 and after three years of 1 700 measurements by the Haystack and Arecibo Observatories, Shapiro reported $\gamma = 1.03 \pm 0.04$ (Shapiro et al., 1971). The first measurements made of time dilation with spacecraft were at Mars in 1969. NASA sent a dual mission of Mariner 6 and 7 and the echoes were received with the 64 m telescope at Goldstone where the accuracy of the ranging system was rated as 1×10^{-7} s. The respective data were total time for round trip: 44.72, 42.87 min; distance of beam from centre of Sun: 3.58, 5.90 solar radii; angle Sun-Earth-spacecraft: 0°.95, 1°.56; approximate time delay: 2.0×10^{-4}, 1.8×10^{-4} s; γ 1.003 $\pm$ 0.04, 1.000 $\pm$ 0.012. The combined figure for γ was given as 1.00 $\pm$ 0.03 (Anderson et al., 1975). This 3% uncertainty was lowered to 2% for Mariner 9 in orbit of Mars in 1971 (Reasenberg, op. cit.).

In 1975 NASA launched Viking 1 and Viking 2 which arrived at Mars in 1976. Each spacecraft consisted of an orbiter and lander with radio links to each other. Receiving stations on Earth were the three of the Deep Space Network. By having two set places on the Martian surface, accuracy was reduced to 0.5% (Michael et al., 1977). Two parameters from the two pulsars in a mutual orbit relate to the shape of the time delay and its range. They are given respectively followed by the Einstein comparison and ratio of observed to predicted values: 0.999 74 [0.999 87, 0.999 87] and 6.21×10^{-6} s [6.153 $\times$ 10^{-6} s, 1.009] (Kramer, op. cit.).

Atomic Clocks

In 1967 time was defined by the International Union of Pure and Applied Chemistry in terms of transitions involving the caesium-133 atom. Calibration was initially against ephemeris time where the motion of the Sun or Moon could be the standard. However, tables of motion of these bodies require many factors to be taken into account. Nevertheless, programs now exist that do give an accurate description of time.

Not long after, in 1971, four clocks containing caesium-133 were calibrated against each other and compared with the reference atomic scale at the United States Naval Observatory. As an experiment to test time changes within general relativity, they were flown on a commercial jet firstly eastward around the world. Their time losses amounted to 5.1, 5.5, 5.7 and 7.4 all $\times 10^{-8}$ s to give a mean and standard deviation of $-(5.9 \pm 1.0) \times 10^{-8}$ s against the relativistic prediction with estimated uncertainty of $-(4.0 \pm 2.3) \times 10^{-8}$ s. The westward round the world trip resulted in gains of 2.66, 2.66, 2.77 and 2.84 all $\times 10^{-7}$ s to result in $+(2.73 \pm 0.07) \times 10^{-7}$ s against $+(2.75 \pm 0.21) \times 10^{-7}$ s (Hafele and Keating, 1972).

STRONG GRAVITY IMPLICATIONS

Nordtved Effect

A strong equivalence principle is known as the Nordtved effect after Kenneth Leon Nordtvedt (1939). It treats gravity as a geometric property of spacetime. Measurements described at Appache Point Observatory provide support for relativity to a few parts in 10^5 (Murphy, op. cit.).

Potential Gravitational Waves

As general relativity has dealt with weak fields within the solar system and stronger ones outside, it may be used to see if it will elucidate the situation with exceptionally strong fields. The conversion of rotational energy into gravitational energy would result in orbital decay in a binary pulsar. While decay has been measured, the search for gravitational waves has begun in earnest. A connection between accelerating masses and gravitational waves is hypothesised. However, compared with electromagnetic radiation from accelerating charges, the energy is extremely small. Thus, in their search for gravitational waves, scientists will firstly need to look at massive energy systems.

Towards the end of their existence, double neutron stars spiral inwards, collide and merge with a predicted enormous release of gravitational radiation. This is suggested to be strong enough to identify at the Earth. Detection is currently being attempted by VIRGO in Italy, GEO600 in Germany, TAMA in Japan and LIGO in the USA (Heuvel, 2003). As an example, (Laser Interferometer Gravitational Wave Observatory (LIGO) is on two sites. Each contains two arms four km long with weights suspended at the end of vacuum tubes. Laser beams measure the distances between the loads. The passage of a gravitational wave is expected to change the distance between the weights which would be detected with an interference pattern between the laser beams.

DISCUSSION

A summary of all the previous material is listed in Table 2. The property includes the title in this paper, the experiment performed relevant to that topic, the year of publication (not the year of the experiment) arranged chronologically for that section and percentage difference from relativity as the difference divided by the general relativity value. If there are two figures listed, the first one uses the central figure of the result against the prediction of general relativity. The second value uses the uncertainty, if it exists in the literature, and takes the larger of the difference from general relativity.

Table 2. Percentage difference from relativity for experiments conducted listed under a section, property and year of publication.

Property	Experiment	Year of Publication	% Difference from relativity
Equivalence of Inertial and Gravitational Mass	Torsion balance	1890	5×10^{-6}
	Torsion balance	1909	1×10^{-6}
	Torsion balance	1935	$2\text{-}5 \times 10^{-7}$
	Torsion balance	1964	1×10^{-9}
	Lunar laser ranging	2005	1×10^{-2}
	Lunar laser ranging	2009	1×10^{-1}
Gravitational Redshift	Gamma rays	1960	5, 15
	Gamma rays	1965	0.1, 0.9
	Hydrogen maser on rocket	1980	0.007
	Voyager 1 at Saturn	1990	0.44, 1
	Galileo spacecraft	1993	1
	Cassini spacecraft	2004	0.002 3
	Psr j0737 – 3039a/b	2006	0.36
Relativistic Perihelion Advance of the Planets	Mercury	1943	0.19, 1.4
	Venus	1956	2.3, 58
	Earth	1956	32, 62
Relativistic Periastron Advance of Binary Pulsars	PSR 1913 + 16 orbital decay	1989	1
	PSR 1913 + 16 periastron advance	1991	0.8, 1
	PSR 1913 + 16 orbital decay	1995	0.3
	PSR 1913 + 16 orbital decay + galactic rotation	1995	0
	PSR 1534 + 12 periastron advance	1998	1
	PSR J0737 – 3039A/B orbital decay	2004	0.3
	PSR 1913 + 16 orbital decay	2005	0.13, 0.4
Geodetic Precession	For Moon	1988	1.9, 2
	For Moon	1994	0.3, 2
	Planetary motions	2005	0.01, 0.03
	Gravity Probe B in Earth orbit	2011	0.28
Lense-Thirring Effect	LAGEOS and LAGEOS II in Earth orbit	2004	0.6, 0.7
	Mars Global Surveyor in orbit	2006	6
	Gravity Probe B in Earth orbit	2012	5, 24
Gravitational Optical Light Deflection	Total solar eclipse	1953	2.9, 4
	Total solar eclipse	1976	46
	Hipparchos	1997	0.3
Gravitational Radio Deflection due to the Sun	3C279 owens valley observatory	1970	2, 25
	3C279 goldstone	1970	8, 38
	3 radio sources and interferometry	1975	3, 6
	3 radio sources and interferometry	1976	1.4, 4
	74 radio sources and interferometry	1991	0.02, 0.3
	541 radio sources and interferometry	2004	0.02, 0.06
	4 radio sources and interferometry	2009	2, 5
Gravitational Lensing	Observations in accord with predictions	-	-

Table 2. Contd.

Time Dilation	Radar ranging to Mercury and Venus	1971	3, 7
	Mariner 6 in Mars flyby	1975	0.3, 0.7
	Mariner 7 in Mars flyby	1975	0, 2
	Viking – 2 orbiters and 2 landers at Mars	1977	0.5
	Mariner 9 in Martian orbit	1979	0, 2
	PSR J0737 – 3039A/B – shape of time delay	2006	0.013
	PSR J0737 – 3039A/B – range of time delay	2006	0.9
Atomic Clocks	Flying eastwards around Earth	1972	48, 73
	Flying westwards around Earth	1972	0.7, 3
Nordtved Effect	Lunar laser ranging	2003	$(1) \times 10^{-3}$

As seen from the table, the equivalence principle has been tested to the 1×10^{-9} difference from relativity and the Cassini spacecraft has a measure of difference of 0.002 3% for gravitational redshift. What is significant is that from 10 properties with measurements, so many are at the 10^{-1} and 10^{-2} level.

CONCLUSION

This paper covers predominantly the period after 1928 to the present. From the three classical astronomical tests of general relativity (anomalous perihelion advance of the perihelion of Mercury, gravitational light bending and gravitational redshift), a plethora of other avenues has developed historically. Even the term relativistic astrophysics did not exist for the first 50 years following Einstein's publication of 1916. Topics covered are weak equivalence principle (equivalence of inertial and gravitational mass and gravitational redshift), orbital precession of a body in gravitational fields (the relativistic perihelion advance of the planets, the relativistic periastron advance of binary pulsars, geodetic precession and Lense-Thirring effect), light propagation in gravitational fields (gravitational optical light deflection, gravitational radio deflection due to the Sun, gravitational lensing, time dilation and atomic clocks) and strong gravity implications (Nordtved effect and potential gravitational waves). Each subject has been plumbed to determine the amount of measurement agreement with general relativity. Three questions were proposed as a guiding principle to this paper.

(i) How much evidence exists to support general relativity?

Einstein originally proposed that his concept could be tested by three astronomical tests. However, there was a significant hiatus between his 1916 publication and further experimentation. There was a need for technology to be developed and experimental techniques both invented and refined before more rigorous delving into the theory could ensue. Torsion balance data existed before 1916 but it continued to improve with better equipment. Lunar laser ranging and radar echoes from the inner planets improved the positioning of these solar system bodies. Allied with computer programs, scientists enhanced ephemerides and many of the perturbations were teased out to ascertain the contribution of each. By extending the reception of data from one station to several with a long base, scientists were able to use interferometry to tighten the uncertainty in their measurements. The introduction of spacecraft in Earth orbit and then venturing to the Moon and all the other planets opened up another methodology for experimentation. Precision was an essential requirement for the operation of these vehicles and so experimentation into relativity advanced. There promises to a burgeoning of data as planned spacecraft are put into service. However, with the myriad sets of results outlined in this article along with many tight constraints on the figures, general relativity has been tested well and not shown to be incorrect.

(ii) Is general relativity a reasonable way of thinking?

General relativity contains a number of simple ideas. From these, several predictions follow and these have been shown to be acceptable to usually better than a 1% level. It does not follow that general relativity is "correct" as other ideas may lead to the same forecasts. A model is judged by the fruitfulness of its operation. Against that criterion, general relativity has been shown to be superb.

A difficulty is that it does not square with notions people have, from their experience, of what reality is. However, experience tells us that the Earth neither spins nor orbits and that a body does not stay in constant motion. Yet, these ideas eventually won the day. People perceive space and time as absolute quantities and are more familiar with the geometry of Euclid than any other. Even though it is the province of scientists to understand the way the Universe operates, it is a task of all in the field to communicate these concepts to the public. Otherwise, the popularity of astrological signs in magazines and the reliance some people put on the ability of these to tell the future act as a signal of minds not thinking scientifically. General relativity is a successful concept and the public needs to have some appreciation of what it says.

(iii) What is the niche that general relativity should occupy?

Significant discussion abounds on the conflict between parts of general relativity and quantum mechanics. As a result, there is a search for a theory of everything. These models ought to be viewed as two of the greatest pieces of inspiration that have flowed from the mind of humans. It is imperative to celebrate such great thought. They are not reality but point to it. General relativity provides a worldview when masses are large and speeds approach that of the speed of light. Instead of seeing the disagreement between the two concepts, one may use whichever idea performs the role of explanation for each situation. This may involve a tension with some but the tension can be manageable. Light is light. On some occasions, its properties are better explained with a particle model and, at others, with a wave formulation. Neither holds a complete explanation; both are necessary to gain a perception of light. Perhaps, unification of general relativity and quantum mechanics may occur. In the meantime, Einstein's worldview may continually be applied to intriguing aspects of the Universe.

Formulated in 1916, general relativity was faced much later with a rapid succession of findings. In 1954 Cygnus A was a strong radio source associated with a distant galaxy that could not be detected optically. X ray sources entered the scene in 1962 followed by quasars in 1963, the 3 K background radiation in 1965, pulsars in 1967 and later further exotic objects of the cosmos. These features have been subsumed under the wing of general relativity and a scientific understanding of these phenomena would not currently exist without such a model.

Conflict of Interest

The author has not declared any conflict of interest.

REFERENCES

Anderson JD, Esposito PE, Martin W, Thornton CL, Muhleman DO (1975). "Experimental test of general relativity using time-delay data from Mariner 6 and 7". Astrophys. J. 200:221-233.

Bertotti B, less L, Tortora P (2003). "A test of general relativity using radio links with the Cassini spacecraft". Nature 425:374-376.

Biesbroeck G van (1953). "The relativity shift at the 1952 February 25 eclipse of the Sun". Astronol. J. 58(1207):87-88.

Bolton AS, Burles S, Koopmans LE, Treu T, Gavazzi R, Moustakas LA, Wayth R, Schlegel DJ (2008). "The Sloan lens ACS survey. V. the full ACS strong-lens sample". Astrophys. J. 682(2):964-984.

Brune RA Jr, Cobb CL, DeWitt BS, DeWitt-Morette C, Evans DS, Flloyd JE, Jones BF, Lazenby RV, Marin M, Matzner RA, Mikesell AH, Mitchell RI, Ryan MP, Smith HJ, Sy A, Thompson CD (1976). "Gravitational deflection of light: solar eclipse of 30 June 1973. I. description of procedures and final results". Astronol. J. 81(6):452-454.

Burgay M, D'Amico N, Possenti A, Manchester RN, Lyne AG, Joshi BC, McLaughlin MA, Kramer M, Sarkissian JM, Camilo F, Kalogera V, Kim C, Lorime DR (2003). "An increased estimate of the merger rate of double neutron stars from observations of a highly relativistic system". Nature 426:531-533.

Campbell L, Moffatt JW (1983). "Quadrupole moment of the Sun and the planetary orbits". Astrophys. J. 275:L77-79.

Campbell WW, Trumpler R (1928). "Observations made with a pair of five-foot cameras on the light-deflections in the Sun's gravitational field at the total solar eclipse of September 21, 1922". Lick Observatory Bulletin 397(13):130-160.

Ciufolini I, Pavlis EC (2004). "A confirmation of the general relativistic prediction of the Lense-Thirring effect". Nature 431:958-960.

Clemence GM (1943). "The motion of Mercury 1765-1937". Astronol. J. 1:126-127.

Damour T, Taylor JH (1991). "On the orbital period change of the binary pulsar PSR 1913 + 16". Astrophys. J. 366:501-511.

Dickey JO, Bender PL, Faller JE, Newhall XX, Ricklefs RL, Ries JG, Shelus PJ, Veillet C, Whipple AL, Wiant JR, Williams JG, Yoder CF (1994). "Lunar laser ranging: a continuing legacy of the Apollo program". Science 265(5171):482-490.

Duncombe RL (1956). "The motion of Venus 1750-1949". Astronol. J. 61:266.

Eddington AS (1919). "The total eclipse of 1919 May 29 and the influence of gravitation on light". Observatory 42:119-122.

Einstein A (1905). "On the electrodynamics of moving bodies". Annalen der Physik. 17:891-921.

EinsteinA (1907). "On the relativity principle and the conclusions drawn from it". Jahrbuch der Radioaktivität und Elektronik. 4:411-462.

Einstein A (1916). "The Foundations of the General Theory of Relativity". Annalen der Physik. 49:769-822.

Einstein A (1918)."Über Gravitationswellen" Sitzungsberichte der Königlich Preussischen Akademie der Wissenschaften. pp. 154–167.

Einstein A (1936). "Lens-like action of a star by the deviation of light in a gravitational field". Science 89(2188):506-507.

Einstein A, Grossmann M (1913). "Entwurf einer verallgemeinerten Relativitätstheorie und einer Theorie der Gravitation" (Outline of a generalised theory of relativity and of a theory of gravitation). Zeitschrift für Mathematik und Physik. 62:225–261.

Eötvös R v (1890). Mathematische und Naturwissenschaftliche Berichte aus Ungarn [Mathematical and Scientific Reports from Hungary] 8:65.

Eötvös R v (1909). l6th International Geodesic Conference in London.

Eötvös RV, Pekár D, Fekete E (1922). Beiträge zum Gesetz der Proportionalität von Trägheit und Gravität {Contributing to the law of proportionality of inertia and gravitation}. Ann. Phys. 68:11- 66.

Everitt CW, DeBra DB, Parkinson BW, Turneaure JP, Conklin JW, Heifetz MI, Keiser GM, Silbergleit AS, Holmes T, Kolodziejczak J, Al-Meshare M, Mester JC, Muhlfelder B, Solomonik V, Stahl K, Worden P, Bencze W, Buchman S, Clarke B, Al-Jadaan A, Al-Jibreen H, Li J, Lipa JA, Lockhart JM, Al-Suwaidan B, Taber M, Wang S (2011). "Gravity Probe B: final results of a space experiment to test general relativity". Phys. Rev. Lett. 106(22):1101-1105.

Fomalont EB, Sramek RA (1975). "A confirmation of Einstein's general theory of relativity by measuring the bending of microwave radiation in the gravitational field of the Sun". Astrophys. J. 199:749-755.

Fomalont EB, Sramek RA (1976). "Measurements of the solar gravitational deflection of radio waves in agreement with general relativity". Phys. Rev. Lett. 36(25):1475-1478.

Fomalont E, Kopeikin S, Lanyi G, Benson J (2009a). "Progress in measurements of the gravitational bending of radio waves using the VLBA". Astrophys. J. 699:1395-1402.

Fomalont E, Kopeikin S, Jones D, Honma M, Titov O (2009b). "Recent VLBA/VERA/IVS tests of general relativity" Proceedings IAU Symposium No. 261 Relativity in Fundamental Astronomy. 291-295.

Froeschlé M, Mignard F, Arenou F (1997). "Determination of the PPN parameter γ with the Hipparcos data". Proceedings of the ESA Symposium at Venice. 402:49.

Gamalath KW (2012). Einstein his life and works. P. 287.

Gefter A (July 2005). "Putting Einstein to the test" Sky and Telescope. pp. 33-40.

Hafele JC, Keating RE (1972). "Around-the-world atomic clocks: observed relativistic time gains". Science 177:168-170.

Heuvel EPJ van den (2003). "Testing time for gravity". Nature. 426:504-505.

Hewitt JN, Turner EL, Scneider DP, Burke BF, Langston GI, Lawrence CR (1988). "Unusual radio source MG1131 + 0456: a possible Einstein ring". Nature 333:537-540.

Hirate CM (2012). "Lecture ix: weak field tests of GR: the gravitational redshift, deflection of light, and Shapiro delay" Caltech.

Huchra J, Gorenstein M, Kent S, Shapiro I, Smith G (1985). "2237 + 0305: a new and unusual gravitational lens". Astrono. J. 90:691-696.

Hulse RA, Taylor JH (1975). "Discovery of a pulsar in a binary system". Astrophys. J. 195:L51-53.

Iorio L (2006). "A note on the evidence of the gravitomagnetic field of Mars". Classical Quantum Gravity 23(17):5451.

Janssen M (2002). "The Einstein-Besso manuscript: A glimpse behind the curtain of the wizard" Introduction to the Arts and Science.

Kramer M, Stairs IH, Manchester RN, McLaughlin MA, Lyne AG, Ferdman RD, Burgay M, Lorimer DR, Possenti A, D'Amico N, Sarkissian JM, Hobbs GB, Reynolds JE, Freire PCC, Camilo F (2006). "Tests of general relativity from timing the double pulsar". Science 314(5796):97-132.

Krisher TP, Anderson JD, Campbell JK (1990). "Test of the gravitational redshift effect at Saturn". Phys. Rev. Lett. 64(12):1322-1325.

Krisher TP, Morabito DD, Anderson JD (1993). "The Galileo redshift experiment". Phys. Rev. Lett. 70(15):2213-2216.

Lyne AG, Burgay M, Kramer M, Possenti A, Manchester RN, Camilo F, McLaughlin MA, Lorimer DR, D'Amico N, Joshi BC, Reynolds J, Freire PCC (2004). "A double-pulsar system: a rare laboratory for relativistic gravity and plasma physics". Science 303:1153-1157.

McNamara G (2008). Clocks in the Sky: the story of pulsars.112.

Michael, WH Jr, Tolson RH, Brenkle JP, Cain DL, Fjeldbo G, Stelzried CT, Grossi MD, Shapiro II, Tyler GL(1977). "The Viking radio science investigation". J. Geophys. Res. 82(28):4293-4295.

Morgan HR (1945). "The Earth's perihelion motion". Astronol. J. 51:127-129.

Morton DC (1956). "Relativistic advances of perihelions". J. Royal Astronol. Soc. Canada 1:223.

Muhleman DO, Ekers RD, Fomalont EB(1970). "Radio interferometric test of the general relativistic light bending near the Sun". Phys. Rev. Lett. 24:1377-1380.

Murphy TW Jr, Adelberger EG, Battat JBR, Carey LN, Hoyle CD, LeBlanc P, Michelsen EL, Nordtvedt K, Orin AE, Strasburg JD, Stubbs CW, Swanson HE, Williams E (2008). "APOLLO: the Apache Point Observatory Lunar Laser-ranging Operation:Instrument Description and First Detections". Publ. Astron. Soc. Pac. 120:20-37.

Pfister H (2007). "On the history of the so-called Lense-Thirring effect". Gen. Relativity Gravit. 39(11):1735-1748.

Pijpers FP (1998). "Helioseismic determination of the solar gravitational moment". Monthly Notices of the Royal Astronomical Society. 297:L76-80.

Pitjeva EV (2005). "Relativistic effects and solar oblateness from radar observations of planets and spacecraft". Astron. Lett. 31(5):340-349.

Pound RV, Rebka GA Jr (1959). "Gravitational redshift in nuclear resonance". Phys. Rev. Lett. 3(9):439-441.

Pound RV, Rebka GA Jr (1960a). "Apparent weight of photons". Phys. Rev. Lett. 4(7):337-341.

Pound RV, Rebka GA Jr (1960b). "Gravitational red-shift in nuclear resonance". Phys. Rev. Lett. 4(7):439-441.

Pound RV, Snider JL (1965). "Effect of gravity on gamma radiation". Phys. Rev. 140(3B):788-803.

Prince TA, Anderson SB, Kulkarni SR (1991). "Timing observations of the 8 hour binary pulsar 2127 + 11C in the globular cluster M15". Astrophys. J. 374:L41-44.

Reasenberg RD, Shapiro II, MacNeil PE, Goldstein RB, Breidenthal JC, Brenkle JP, Cain DL, Kaufman TM, Komarek TA, Zygielbaum AI (1979). "Viking relativity experiment: verification of signal retardation by solar gravity". Astrophys. J. 234:L219-221.

Renn J, Sauer T, Stachel J (1997). "The origin of gravitational lensing: A postscript to Einstein's 1936 science paper". Science. 275:184-186.

Renner J (1935). Matematikai és Természettudományi Értesítő. 13:542.

Robertson DS, Carter WE, Dillinger WH (1991). "New measurement of solar gravitational deflection of radio signals using VLBI" Nature. 349(6312):768-770.

Roll PG, Krotkov R, Dicke RH (1964). "The equivalence of inertial and passive gravitational mass". Ann. Phys. 26:442-517.

Schneider P, Ehlers J, Falco EE (1992). Gravitational Lenses. Springer, Astronomy and Astrophysics. Library XIV

Seidelmann PK ed. (2006). Explanatory Supplement to the Astronomical Almanac. 704.

Seielstad GA, Sramek RA, Weiler KW (1970). "Measurement of the deflection of 9.602-GHz radiation from 3C279 in the solar gravitational field". Phys. Rev. Lett. 24:1373-1376.

Shapiro II (1964). "Fourth test of general relativity". Phys. Rev. Lett. 13(26):789-791.

Shapiro II, Ash ME, Ingalls RP, Smith WB, Campbell DB, Dyce RB, Jurgens RF, Pettengill GH (1971). "Fourth test of general relativity: new radar result". Phys. Rev. Lett. 26(18):1132-1135.

Shapiro II, Reasenberg RD, Chandler JF, Babcock RW (1988). "Measurement of the de Sitter precession of the Moon: a relativistic three-body effect". Phys. Rev. Lett. 41(23):2643-2646.

Shapiro SS, Davis JL, Lebach DE, Gregory JS (2004). "Measurement of the solar-gravitational deflection of radio waves using geodetic very-long-basline interferometry data, 1979-1999". Phys. Rev. Lett. 92(12):1101-1104.

Sitter W de (1916). "On Einstein's theory of gravitation, and its astronomical consequences, second paper". Monthly Notices Royal Astron. Soc. 77:155-184.

Sitter W de (1917). "Planetary motion and the motion of the Moon according to Einstein's theory". Proceedings of the Royal Netherlands Academy of Arts Sci. 19(1):367-381.

Stairs IH, Arzoumanian Z, Camilo F, Lyne AG, Nice DJ, Taylor JH, Thorsett SE, Wolszczan A (1998). "Measurement of relativistic orbital decay in the PSR B1534 + 12 binary system". Astrophys. J. 505:352-357.

Staveley-Smith L, Wilson WE, Bird TS, Disney MJ, Ekers RD, Freeman KC, Haynes RF, Sinclair MW, Vaile RA, Webster RL, Wright AE(1966). "The Parkes 21 cm multibeam receiver". Publ. Astron. Soc. Australia.13:243-248.

Stockton A (1980). "The lens galaxy of the twin QSO 0957 + 561". Astrono. J. 242:L141-142.

Taylor JH (1992). "Pulsar timing and relativistic gravity". Philosophical Trans. Royal Soc. 341:116-134.

Taylor JH, Weisberg JM (1989). "Further experimental tests of relativistic gravity using the binary pulsar 1913 + 16". Astrophys. J. 345(1):434-450.

Treschman KJ (2014a). "Early astronomical tests of general relativity: the gravitational deflection of light". Asian J. Phys. 23(1-2):145-170.

Treschman KJ (2014b). "Early astronomical tests of general relativity: the anomalous advance in the perihelion of Mercury and gravitational redshift". Asian J. Phys. 23(1-2):171-188.

Valença J de (2008). "Gravitational Waves" bachelor project online.

Vankov AA (1915). "Einstein's Paper: Explanation of the perihelion motion of Mercury from General Relativity Theory" online, translated from German to English by Rydin RA.

Vessot RFC, Levine MW, Mattison EM, Blomberg EL, Hoffman TE, Nystrom GU, Farrel BF, Decher R, Eby PB, Baugher CR, Watts JW, Teuber DL, Wills FD (1980). "Test of relativistic gravitation with a space-borne hydrogen maser". Phys. Rev. Lett. 49(26):2081-2085.

Walsh D, Carswell RF, Weymann RJ (1979). "0957 + 561 A, B – twin quasi-stellar objects or gravitational lens?". Nature 279:381-384.

Weisberg JM, Taylor JH (2005). "The relativistic binary pulsar B1913 + 16: thirty years of observations and analysis". Binary Radio Pulsars

ASP Conference Series. 328:25-31.

Will CM (2006). "The confrontation between general relativity and experiment". Living Rev. Relativity 9(3):1-100.

Will M (1995). "Stable clocks and general relativity" in Dark Matter in Cosmology, Clocks and Tests of Fundamental Laws, Proceedings of the 30th Recontres de Moriond, Moriond Workshop ed. Guiderdoni B et al, pp. 417-427.

Williams JG, Turyshev SG, Boggs DH (2004). "Progress in lunar laser ranging tests of relativistic gravity". Phys. Rev. Lett. 93(26):1101-1104.

Williams JG, Turyshev SG, Boggs DH (2009). "Lunar laser ranging tests of the equivalence principle with the Earth and Moon". Int. J. Modern Phys. 518:1129-1175.

Wolszczan A (1991). "A nearby 37.9-ms radio pulsar in a relativistic binary system". Nature 350(6320):688-690.

Zhao SS, Xie Y (2013). "Parametrized post-Newtonian secular transit timing variations for exoplanets". Res. Astronol. Astrophys. 13(10):1231-1239.

Zwicky F (1937). "On the masses of nebulae and of clusters of nebulae". Astrophys. J. 86(3):217-246.

20

Geophysical investigation of the effects of sewage in the soil at university of Nigeria, Nsukka, Enugu State, Nigeria

Awalla, C. O. C.

Department of Geology and Mining, Faculty of Applied Natural Sciences, Enugu State University of Science and Technology, Enugu, Enugu State, Nigeria.

Geophysical investigation was carried out to detect the spread of sewage effluent and to locate the sources and delineate migration paths and the extent of leachate plume. The study was to find out the geological formations that are the most conductive layers in the sewage site for the free flow of the contaminants. Soil tests showed that hydraulic conductivity, bulk density and water retention were variable in sewage soil, but consistent in soil unaffected by sewage. In sewage soil, maize crops performance, organic matter, total nitrogen exchangeability, cations exchange capacity and sodium were significantly enhanced than in non-sewage soil. In sewage soil, electrical conductivity (EC), zinc (Zn), lead (Pb), copper (Cu), Cadmium (Cd), salt concentration and other saline properties, total faecal coliform and microbial activities were high. Twelve vertical electrical soundings (VES) points with Omega Terrameter were used in the Schlumberger Array configuration which the geoelectrical section from the resistivity data revealed seven subsurface layers. Three low resistivity zones were detected as correspondent zones to the plumes. The comparative evaluation of the 3-D stack model of the resistivity of layers with respect to depth suggested that these three low resistivity layers were contaminated, especially as the depth to water table is more than 100m. The flow direction of the contaminant plume is northwest to southwest as indicated by stack model of the low resistivity layers. Generally, the contaminant plume is a threat to ecosystem and a great health problem to people living around the sewage site.

Key words: Delineate, contamination, leachate, sewage effluent, saline, ecosystem.

INTRODUCTION

The sewage site at University of Nigeria, Nsukka is accessible through numerous routes within the campus, e.g. from Nnamdi Azikiwe new Library through Eni-Njoku street, Murtala Mohammed Way and Louis Mbanefo street. The UNN campus is situated on the hills of Obukpa town and Agu-ihe of Nsukka Local

Government Area in Enugu State, Nigeria.

Geophysical investigation was carried out with the aid of Omega Terrameter by the Department of Geology, UNN using Schlumberger Array configuration to detect the spread of sewage effluent; and also to locate and delineate the migration pathways and the extent of the leachate plume. The sewage site is envisaged as a potential source of groundwater contaminants in the campus and its environs. The soil tests were carried out at the Soil Science Unit and Department of Zoology, UNN.

The generation of the sewage effluent increased recently due to increase in the students population, and thus could be a possible source of contaminants to groundwater. The sewage effluent could be waste water generated mainly from toilets, bathrooms and laundry activities. The major sources of the waste water could be from students hostels, staff quarters, offices, medical centre, classrooms, banks, restaurants, primary and secondary schools in the campus.

Geologically, the sewage site is located on Ajali Sandstone which underlies Nsukka Formation. Ajali Sandstone within Nsukka and its environs is a sandstone unit with medium to coarse grained, moderately to poorly sorted, friable, whitish in colour with iron stains and clay lenses due to the overlying Nsukka Formation. The Nsukka Formation has sandstone, clayey shale and ironstone units (Oguamah, 1999). Ironstone is the prominent outcrop at the site.

The most outstanding threat to groundwater quality in the study area is the improper, unscientific, unacceptable and inadequate disposal, treatment and management of the sewage effluent generated. It is good to study the sewage disposal manipulations and handling from the oxidation ponds sewage systems. Due to the slow flow pattern of groundwater, the effluent plume spreads slowly as a contaminant (Montgomery, 2005), and thus the natural flow of groundwater disperses the sewage effluent as a plume of contamination. Thus, the study is to investigate the sewage effects in soil due to its indiscriminate wide spread on the soil surface for agricultural purposes.

GEOLOGY

The Benue Trough is a northeast-southwest trending sedimentary basin which consists of up to 5000 m of cretaceous sediments, contiguous with the rift basins of Niger, Chad and Sudan Republic (Akande, 2004) and extending to over 1000 kms from the Niger Delta to Lake Chad.

The sedimentary rocks of the Lower Benue Trough are the hosts of various igneous rocks (Obiora and Umeji, 2004). The main lithostratigraphic units that underlie the study area are the Ajali Sandstone and Nsukka Formation within the Lower Benue Trough. The Ajali Sandstone is of Maestrichtian Age and consists of white, medium to coarse grained sandstone and friable sands. It marks the height of regression that ended the Nkporo deposition cycle and the stratigraphic position of Ajali Sandstone between two paralic sequences of the underlying Mamu Formation and overlying

Nsukka Formation indicates a continental origin. This could be due to the evidence of tidal origin showed with the development of herringborne cross stratification, bimodal-bipolar, paleocurrent pattern, suspension deposit on forest laminae and mixed bedding. Ajali Sandstone is the main aquiferous unit within the campus. Ajali Sandstone is most often stained red, and thus overlain by thick red soil due to weathering.

Nsukka Formation is of the Upper Maestrichtian to Danian Age, known as Upper Coal Measure. It consists of less sands and less coal seams than Mamu Formation which has up to five coal seams. It is deposited under paralic conditions in a strand- plain marsh within shallow marine environment and occasional fluvial incursion. Nsukka Formation is divided into sandstone unit, especially at its boundary with Ajali Sandstone; clay and ironstone units at its upper contact with Imo Shale. The outcrop of the sandstone unit of Nsukka Formation exists at the opposite Green House, about 1.5 kms from University gate to Owerre-Ezu Orba, near UBA and Oceanic Bank. It is an old quarry site and also weathered with elevation of about 450 m and about 4 m thick. There is another outcrop at Hilltop Odenigwe with elevation of about 480 m and about 5.5 m thick.

RESULTS AND DISCUSSION OF THE ANALYSIS OF THE PHYSICAL CHARACTERISTICS OF THE SOIL ENVIRONMENT

The physical characteristics of the soil environment within the sewage site are classified into two, namely soil at the sewage disposal area and the soil within the non-sewage disposal area. The two categories of soils are derived from the Nsukka Formation (that is, False Bedded Sandstone). It is deep and excessively drained. Soil within the sewage disposal area is very dark reddish brown to reddish brown, while non-sewage disposal area has dark reddish brown to red in colour.

The soil colour variations within the sewage disposal site could be attributed to the existence of sludge and sewage effluent. This is because of the presence of organic matter in the soil, especially for a very long time without oxygenated environment, but with contaminated water as moisture occupying the soil- pore spaces it always changes the clayey-shale soil-colour from light grey to dark grey or dark brown with pungent smell, unless the accumulation of either salts or iron oxide in the soil modify and alter the decomposition processes that affect the colour and odour.

Table 1 helps to classify the soil with respect to some physical properties, especially after the disposal of sludge and sewage effluent. The texture or physical size analysis indicated mainly sand and loamy-sand on the top soil, but the subsoil has sandy loam. However, the coarse texture in all soils could be due to the existence of Nsukka Formation (that is, False Bedded Sandstone) that is dominance in the area. Clay content ranged from about 6 to 18%, but increased with depth in sewage soil and has no known trend in non-sewage soil. Thus, clay content is low as well as silt content which ranged from about 1 to 10%. The low contents of clay and silt indicate the degree of weathering and leaching which the soil has undergone. The low contents of clay and silt could also

Table 1. Comparative evaluation of some physical characteristics of sewage and non-sewage soil-types.

		Percentage						Percentage	
								(A) Sewage disposal area and soil-type				
Soil classification	Depth (cm)	Sand	Clay	Silt	Bulk density (g/cm^3)	Micro porosity	Macro porosity	Ratio of micro/macro porosity	Percent of total porosity
Sand	0-15	90	8	4	1.56	23	25	2:4	45
Loamy sand	15-36	87	7	7	0.82	39	8	5:1	47
Sandy loam	36-55	90	8	2	1.58	32	10	3:1	43
Sandy loam	55-105	80	18	2	1.55	35	12	5:4	40
Sandy loam	105-160	78	18	4	1.56	30	14	5:3	39
	Mean	85	12	4	1.41	32	14	-	43
					(B) Non-sewage Disposal area and soil-type				
Loamy soil	0-14	84	6	10	1.45	24	27	4:5	51
Sandy loam	14-37	82	14	4	1.48	24	26	4:5	50
Sandy loam	37-76	78	12	10	1.50	31	14	2:1	45
Sandy loam	76-90	76	18	6	1.53	24	16	3:2	40
Sandy loam	90-160	78	16	6	1.64	26	19	3:2	45
	Mean	80	13	7	1.52	26	20	-	46

be attributable to high detachability and transportability of these lighter materials (Obi and Ebo, 1995). Sand content was about 80% mean in non-sewage soil, and about 85% mean in sewage soil; but sewage soil may have sand content up to 90% due to easy dispersion of clay and silt fractions that were clearly washed away. The importance of clay dispersibility is a measure of soil structural integrity and implication of water infiltration and retention (Curtin et al., 1994).

For bulk density, the soil of sewage disposal area has a range from 0.82 to 1.58 g/cm^3, but for non- sewage disposal area, the soil has a range from 1.45 to 1.64 g/cm^3. The low bulk density in sewage soil could be due to the accumulation of humified sewage materials. Generally, organic matter has very low bulk density and hence the most likely cause of the low bulk density of the soil in the sewage disposal area. It was observed that there is an inconsistent pattern of bulk density with depth in sewage soil, but consistent pattern in non-sewage soil. This is because, sewage soil has discontinuity of soil profile due to the presence of humified sewage material layers. The density of the primary particles in sewage soil structures may be responsible for the low and high bulk density values variation. Generally, there was no positive improvement in bulk density and total porosity in both soil-types due to large application of sludge and sewage effluent over a long period of time, but short-time low application of sewage effluent to soil can yield positive improvement in the soil bulk density and porosity.

The high ratio of micro to macro porosity in sewage soil may make for aeration build-up and toxicity in plant roots

and microorganisms. Pagliai and DeNobili (1993) observed that adequate proportion of micro and macro porosity was necessary for the existence of continuous air diffusion pathways in the soil.

In Table 2, the sludge and sewage effluent soil has a range of 0.39-0.49 volumetric water content at saturation. The highest volumetric water content at saturation of 0.49 could be due to the water adsorption capacity of organic matter. About 70% water retained in sewage soil at 60cm tension was more than 50% of water held in non-sewage soil. The actual amount of water retained in the sludge and sewage effluent soil was contributed by the amount of organic materials and the dominance of micro-pores. The non-homogeneous continuity of the sewage soil could be contributed to high water retention capacity.

At the same depth of soil profile of both sewage soil and non-sewage soil, the saturated hydraulic conductivity is variable and consistent in sewage and non-sewage soils respectively. It is as low as 4.21 cm/h in sewage soil against 7.10 cm/h in non-sewage soil. The permeability class ranged from moderate to rapid for sewage soil and rapid to very rapid for non-sewage soil. The long-term application of sludge and sewage effluent could reduce soil hydraulic conductivity due to the formation of biological materials within the crust. It could also be due to the accumulation of solids filtered from the effluent and/or the collapse of soil structure because of dissolution of organic matter. The same observation was made by Lieffering and McLay (1996) because they observed that long-term application of organic waste such as sludge and sewage effluent significantly reduce

Table 2. Volumetric water content retention characteristics at saturation and 60cm tension, hydraulic conductivity and dispersion ratio of the sludge and sewage effluent and non-sludge and sewage effluent soils.

S/No	Depth (cm)	Volumetric water content at saturation	Moisture (gg/L) at 60cm tension	Hydraulic conductivity (cm/h)	Dispersion ratio (percentage)	Permeability class
			(A) Sludge and sewage effluent soil			
(i)	0 –15	0.47	0.26	20.52	96	Rapid
	15 – 35	0.47	0.40	4.21	98	Moderate
	35 – 55	0.40	0.23	7.89	98	Moderately rapid
	55 –105	0.42	0.28	13.15	91	Rapid
	105 -160	0.39	0.29	19.47	80	Rapid
	Mean	0.42	0.35	13.05	93	Rapid
(ii)	0-18	0.48	0.22	20.52	97	Rapid
	18-43	0.49	0.40	4.21	98	Rapid
	43-65	0.42	2.30	7.89	98	Moderately rapid
	56-80	0.46	0.27	19.15	88	Rapid
	Mean	0.47	0.32	13.05	92	Rapid
			(B) Non sludge and sewage effluent soil			
(i)	0-14	0.51	0.25	24.72	63	Very rapid
	14-37	0.53	0.25	7.10	75	Rapid
	37-76	0.45	0.30	16.78	42	Rapid
	76-90	0.40	0.32	19.99	67	Rapid
	90-160	0.44	0.33	19.47	80	Rapid
	Mean	0.47	0.30	20.52	62	Rapid

soil permeability and the low permeability was attributable to the accumulation of solids filtered from effluent and the collapsed soil structure due to the dissolution of organic matter.

Dispersion ratio is a measure of soil's structural integrity which is used mainly to identify the soils susceptibility to slaking, crusting, infiltration and erosion capabilities during rainfall. Thus, dispersion ratio is about 98% in sewage soil and about 80% in non-sewage soil. The high dispersion ratio in sewage soil causes aggregate breakdown and subsequent clay dispersion leading to pore blockage and surface crushing. This leads to low water infiltration, high retention and very high surface soil erosion (Table 3).

The effects of exchangeable sodium on the soil are based on sodium absorption ratio (SAR) and exchangeable sodium percentage (ESP). The saturated soil electrical conductivity (EC) is used to appraise the effect of soil salinity on plant growth. Salt concentration, total cations and osmotic pressure are used as indices of the wilting coefficient of soils. An index of the wilting coefficient of a soil is a measure of the quantity of water that the soil can supply to plants. A plant or flower wilts when it bends towards the ground because of either heat or lack of water.

The concentration values of SAR, ESP, EC, salt, total

cation and osmotic pressure of sludge and sewage effluent soil are higher than that of non-sludge and sewage effluent soil. For example, the SAR values of the sewage effluent soil ranged from 0.06 to 0.13, with 0.13 been the highest value which indicates that a high percentage of exchangeable sodium has been built in the soil, but in non-sewage effluent soil SAR ranged from 0.06 to 0.10 indicating no significant exchangeable sodium (Onah, 2012). Therefore, if the non-sewage effluent disposal soil is to be considered as the baseline for comparative evaluation, it is evident that long-term application of sludge and sewage effluent greatly increase the exchangeable sodium concentration in the top soil and subsoil of the sewage soils. High SAR causes an increase in soil dispersion.

The soil salinity and the high SAR in soils causes the yields of salt sensitive crops to be restricted. Furthermore, the salts may interfere with the absorption of water by plant hair roots through reduction in the soil osmotic water potential, and thus decrease the amount of water that would be readily available for the plant roots uptake and increase in the wilting coefficient of soils. Therefore, high salt concentration in soils through heavy application of municipal effluent most often interfere with the absorption of water by soyabeans through the reduction in the soil osmotic water potential.

Table 3. Results of SAR, ESP, EC, salt concentration, total cation concentration, osmotic pressure and salinity hazards of sewage soil.

S/No	Depth (cm)	SAR	ESP	EC (μ/cm)	Salt conc. (mg/L)	Total cations conc. (mg/L)	Osmotic pressure (atm)	Salinity hazards
					(A)	**Sludge and sewage effluent soil**		
(i)	0-15	0.13	2.0	1.04	665.60	10.40	0.37	Yields of many crops, especially sensitive crops may be restricted.
	15-35	0.12	2.25	3.15	2016.00	31.40	1.13	,,
	35-55	0.09	2.00	0.38	243.00	3.80	0.14	,,
	55 -105	0.09	2.00	0.21	134.00	2.10	0.08	Salinity effects negligible
	105-160	0.07	1.50	0.18	115.20	1.18	0.07	Salinity effects negligible
	Mean	0.10	2.00	0.99	634.88	9.92	0.36	Salinity effects negligible
(ii)	0-18	0.10	2.00	1.16	742.40	11.60	0.42	Salinity effects negligible
	18-43	0.13	2.22	3.05	1925.00	30.60	1.10	Salinity effects negligible
	43-65	0.09	2.25	0.40	256.00	4.00	0.14	Salinity effects negligible
	65-80	0.10	2.00	0.21	134.00	2.10	0.08	Salinity effects negligible
	80-150	0.06	1.50	0.20	128.00	2.00	0.07	Salinity effects negligible
	Mean	0.01	2.00	1.00	642.56	10.06	0.36	Salinity effects negligible
					(B) Non – sludge and sewage effluent soil			
	0 – 14	0.08	1.50	0.09	57.60	0.90	0.03	Salinity effects negligible
	14 – 37	0.10	1.80	0.06	38.40	0.60	0.02	Salinity effects negligible
	37 – 76	0.08	1.76	0.03	19.20	0.30	0.01	Salinity effects negligible
	76 – 90	0.07	1.50	0.02	12.80	0.20	0.01	Salinity effects negligible
	90 – 160	0.06	1.50	0.02	12.80	0.20	0.01	Salinity effects negligible
	Mean	0.08	1.62	0.04	28.16	0.44	0.20	Salinity effects negligible
	Waste water	1.89	1209.60	18.90	0.68	Yields of very sensitive crops may be restricted	Yields of very sensitive crops may be restricted	Salinity effects negligible

CONCLUSION

The spread and migration pathways of sludge and sewage effluent in soil physical environment has variable and consistent hydraulic conductivity, bulk density and water retention in sewage soil zone and in soil unaffected by sewage respectively. From the geophysical study, the low resistivity zones are correspondent zones to the plumes. The 3-D stack model of the resistivity of layers with respect to depth suggest that the three low resistivity layers were contaminated. The study of flow direction of the sludge and sewage effluent is NW to SW as indicated by stack model of the low resistivity layers.

The permeability of the sludge and sewage effluent into the soil is moderate to rapid for sewage soil, but rapid to very rapid for non-sewage soil. Therefore, long application of sludge and sewage effluent reduce soil hydraulic conductivity. This is because of the dissolution and accumulation of solid particles filtered from the sewage effluent and organic matter.

Dispersion ratio which measures soil structural integrity helps to identify the susceptibility of the soil to slaking, infiltration and erosion. It is about 98 percent in sewage soil and about 80 percent in non-sewage soil. The high value of dispersion ratio in sewage soil causes clay dispersion that blocks the soil pores leading to surface crushing, low water infiltration and high surface erosion. Salt concentration, total cations and osmotic pressure in soil are the qualities applied as measures of the wilting coefficient that determines the quantity of water the soil can supply to plants. Both SAR and ESP are higher (0.06-0.13 and 1.50-2.25) in sludge and sewage effluent soil than (0.06-0.10 and 1.50-1.80) in non-sludge and sewage effluent soil. The implication is that there is no significant exchangeable sodium in non-sewage soil, but long-term application of sludge and sewage effluent can increase the exchangeable sodium concentration in topsoil and subsoil. Therefore, to improve and enhance the natural condition of the soil within the sewage disposal site, there must be scientific and sanitary approaches in handling the wastes to reduce the accumulation of both SAR and ESP.

Conflict of Interests

The author(s) have not declared any conflict of interests.

REFERENCES

Akande SO (2004). Cretaceous source and thermal maturation in the Nigeria sedimentary basins. Implication for offshore petroleum prospects. NAPE News. 3(5):20.

Curtin D, Campbell CA, Zentner RP, Lafond GP (1994). Long term management and clay dispersibility in the Haploborolls in Saskatchewan. Soil Sci. Amer. J. 58: 962-967. http://dx.doi.org/10.2136/sssaj1994.03615995005800030046x

Lieffering RE, McLay CDA (1996). The effect of high pH liquid wastes on soil properties: Aggregate stability and hydraulic conductivity. In first Int. Conf. on contaminant and the soil environment. Extended abstract, Adelsaide, pp. 253-254.

Montgomery CW (2005). Environmental Geology. McGraw-Hill Publ., USA. P. 970.

Obi ME, Ebo PO (1995). The effect of Organic and inorganic amendments on soil physical properties and maize production in severely degraded sandy soil in Southeastern Nigeria. Bioresource Tech. J. 51:117-123. http://dx.doi.org/10.1016/0960-8524(94)00103-8

Obiora SC, Umeji AC (2004). Petrographic evidence for regional burial metmamorphism of the sedimentary rocks in the Lower Benue Rift. J. Africa. Earth Sci. 38:269-277. http://dx.doi.org/10.1016/j.jafrearsci.2004.01.001

Oguamah VH (1999). The Geology of Nsukka and its environs in Nsukka L.G.A. of Enugu State. Unpublc B.Sc. Thesis, U.N.N. P. 83.

Onah DI (2012). Environmental and Geophysical Investigations of Sewage disposal and management at UNN. Unpubl. PGD Thesis, GLM Dept., ESUT. P. 134.

Pagliai M, DeNobili M (1993). Relationships between soil porosity, root development and soil enzyme activity in cultivated soils. Geoderma. 50:243-248. http://dx.doi.org/10.1016/0016-7061(93)90114-Z

Permissions

All chapters in this book were first published in IJPS, by Academic Journals; hereby published with permission under the Creative Commons Attribution License or equivalent. Every chapter published in this book has been scrutinized by our experts. Their significance has been extensively debated. The topics covered herein carry significant findings which will fuel the growth of the discipline. They may even be implemented as practical applications or may be referred to as a beginning point for another development.

The contributors of this book come from diverse backgrounds, making this book a truly international effort. This book will bring forth new frontiers with its revolutionizing research information and detailed analysis of the nascent developments around the world.

We would like to thank all the contributing authors for lending their expertise to make the book truly unique. They have played a crucial role in the development of this book. Without their invaluable contributions this book wouldn't have been possible. They have made vital efforts to compile up to date information on the varied aspects of this subject to make this book a valuable addition to the collection of many professionals and students.

This book was conceptualized with the vision of imparting up-to-date information and advanced data in this field. To ensure the same, a matchless editorial board was set up. Every individual on the board went through rigorous rounds of assessment to prove their worth. After which they invested a large part of their time researching and compiling the most relevant data for our readers.

The editorial board has been involved in producing this book since its inception. They have spent rigorous hours researching and exploring the diverse topics which have resulted in the successful publishing of this book. They have passed on their knowledge of decades through this book. To expedite this challenging task, the publisher supported the team at every step. A small team of assistant editors was also appointed to further simplify the editing procedure and attain best results for the readers.

Apart from the editorial board, the designing team has also invested a significant amount of their time in understanding the subject and creating the most relevant covers. They scrutinized every image to scout for the most suitable representation of the subject and create an appropriate cover for the book.

The publishing team has been an ardent support to the editorial, designing and production team. Their endless efforts to recruit the best for this project, has resulted in the accomplishment of this book. They are a veteran in the field of academics and their pool of knowledge is as vast as their experience in printing. Their expertise and guidance has proved useful at every step. Their uncompromising quality standards have made this book an exceptional effort. Their encouragement from time to time has been an inspiration for everyone.

The publisher and the editorial board hope that this book will prove to be a valuable piece of knowledge for researchers, students, practitioners and scholars across the globe.

List of Contributors

Lim Eng Aik
Institut Matematik Kejuruteraan, Universiti Malaysia Perlis, Perlis, Malaysia

Doğan KALAFAT
National Earthquake Monitoring Center, Kandilli Observatory and Earthquake Research Institute, Boğaziçi University, 34680 Çengelköy, Istanbul, Turkey

Gündüz HORASAN
Department of Geophysical Engineering, Engineering Faculty, Sakarya University, 54040, Sakarya, Turkey

Ch'ng Han Siong
School of Applied Physics, Faculty of Science and Technology, Universiti Kebangsaan Malaysia 43600 UKM Bangi, Selangor D. E. Malaysia

Shahidan Radiman
School of Applied Physics, Faculty of Science and Technology, Universiti Kebangsaan Malaysia 43600 UKM Bangi, Selangor D. E. Malaysia

Mohamed Metwaly
Department of Geology and Geophysics, Faculty of Science, King Saud University, Saudi Arabia
National Research Institute of Astronmy and Geophysics (NRIAG), Cairo, Egypt

Eslam Elawadi
Department of Geology and Geophysics, Faculty of Science, King Saud University, Saudi Arabia
Nuclear Materials Authority (NMA), Cairo, Egypt

Sayed S. R. Moustafal
Department of Geology and Geophysics, Faculty of Science, King Saud University, Saudi Arabia
National Research Institute of Astronmy and Geophysics (NRIAG), Cairo, Egypt

F. Al Fouzan
King Abd Alaziz City for Science and Technology (KACST), Saudi Arabia

S. Mogren
Department of Geology and Geophysics, Faculty of Science, King Saud University, Saudi Arabia

N. Al Arifi
Department of Geology and Geophysics, Faculty of Science, King Saud University, Saudi Arabia

C. OKONKWO Austin
Department of Geology and Mining, Enugu State University of Science and Technology, Enugu, Nigeria

I. UJAM Isaac
Department of Geology and Mining, Enugu State University of Science and Technology, Enugu, Nigeria

Kadiri Umar Afegbua
Department of Earthquake Seismology, Centre for Geodesy and Geodynamics, P. M. B. 11, Toro, Bauchi State, Nigeria

F. O. Ezomo
Department of Physics, Faculty of Physical Sciences, University of Benin Ugbowo, Benin City, Edo State, Nigeria

P. L. Verma
Department of Physics, Government Vivekan and P. G. College Maihar Satna M. P. India

Blake T. Dotta
Biophysics Section, Biomolecular Sciences Program, Laurentian University Sudbury, Ontario, Canada

Nirosha J. Murugan
Biophysics Section, Biomolecular Sciences Program, Laurentian University Sudbury, Ontario, Canada

Lukasz M. Karbowski
Biophysics Section, Biomolecular Sciences Program, Laurentian University Sudbury, Ontario, Canada

Michael A. Persinger
Biophysics Section, Biomolecular Sciences Program, Laurentian University Sudbury, Ontario, Canada

M. Siejka
Department of Land Surveying, University of Agriculture in Krakow, Poland

M. Ślusarski
Department of Land Surveying, University of Agriculture in Krakow, Poland

M. Zygmunt
Department of Land Surveying, University of Agriculture in Krakow, Poland

A. O. Adelus
Department of Applied Geophysics, Federal University of Technology, Akure (FUTA), Ondo State, Nigeria

A. A. Akinlalu
Department of Applied Geophysics, Federal University of Technology, Akure (FUTA), Ondo State, Nigeria

A. I. Nwachukwu
Department of Applied Geophysics, Federal University of Technology, Akure (FUTA), Ondo State, Nigeria

Shinsuke Hamaji
Hyama Natural Science Research Institute, 403, Daiichi-Kiriya Building, 5-2, Chuo 2-chome, Nakano-ku, Tokyo 164-0011, Japan

A. M. Abd-Alla
Mathematics Department, Faculty of Science, Taif University, Saudi Arabia

T. A. Nofal
Mathematics Department, Faculty of Science, Taif University, Saudi Arabia

S. M. Abo-Dahab
Mathematics Department, Faculty of Science, Taif University, Saudi Arabia
Mathematics Department, Faculty of Science, Qena, Egypt, Qena 83523, Egypt

A. Al-Mullise
Mathematics Department, Faculty of Science, Taif University, Saudi Arabia

Abraham Bairu
Tigray Water Resources Bureau, Mekelle, Tigray, Ethiopia

Yirgale G/her
Tigray Water Resources Bureau, Mekelle, Tigray, Ethiopia

Gebrehiwot G/her
Tigray Water Resources Bureau, Mekelle, Tigray, Ethiopia

Francisca N. Okeke
Department of Physics and Astronomy, University of Nigeria, Nsukka, Enugu State, Nigeria

Esther A. Hanson
Centre for Basic Space Science, University of Nigeria, Nsukka, Enugu State, Nigeria

Eucharia C. Okoro
Department of Physics and Astronomy, University of Nigeria, Nsukka, Enugu State, Nigeria

B. C. Isikwue
Department of Physics, University of Agriculture, Makurdi, Benue State, Nigeria

Oby J. Ugonabo
Department of Physics and Astronomy, University of Nigeria, Nsukka, Enugu State, Nigeria

H. Goodarzian
Islamic Azad University, Mahdishahr Branch, Mahdishahr, Iran

T. Armaghani
Islamic Azad University, Mahdishahr Branch, Mahdishahr, Iran

M. Okazi
Islamic Azad University, Mahdishahr Branch, Mahdishahr, Iran

C. Kurtulus
Department of Geophysics, Kocaeli University, Engineering Faculty, Izmit-Kocaeli, Turkey
A. Bozkurt ABM Engineering Co., Izmit-Kocaeli, Turkey

H. Endes
Department of Geophysics, Kocaeli University, Engineering Faculty, Izmit-Kocaeli, Turkey

Sebastian Toplak
Faculty of Civil Engineering, University of Maribor, Smetanova 17, 2000 Maribor, Slovenia

Andrej Ivanic
Faculty of Civil Engineering, University of Maribor, Smetanova 17, 2000 Maribor, Slovenia

Primoz Jelusic
Faculty of Civil Engineering, University of Maribor, Smetanova 17, 2000 Maribor, Slovenia

Samo Lubej
Faculty of Civil Engineering, University of Maribor, Smetanova 17, 2000 Maribor, Slovenia

Keith John Treschman
51 Granville Street Wilston 4051 Australia

C. O. C. Awalla
Department of Geology and Mining, Faculty of Applied Natural Sciences, Enugu State University of Science and Technology, Enugu, Enugu State, Nigeria